Perinatal and Infant Brain Imaging:
Role of Ultrasound and Computed Tomography

Perinatal and Infant Brain Imaging: Role of Ultrasound and Computed Tomography

Carol M. Rumack, M.D.
Associate Professor of Radiology and Pediatrics
Director of Pediatric Radiology
Codirector of Pediatric CT and Ultrasound
University of Colorado Health Sciences Center
Denver, Colorado

Michael L. Johnson, M.D.
Associate Professor of Radiology and Medicine
Director of Division of Ultrasound and Section of Body CT
University of Colorado Health Sciences Center
Denver, Colorado

ybmlp ®

YEAR BOOK MEDICAL PUBLISHERS, INC.
CHICAGO

Reprinted, September 1984

Library of Congress Cataloging in Publication Data

Rumack, Carol M.
    Perinatal and infant brain imaging.

    Includes index.
    1. Brain—Diseases—Diagnosis.   2. Brain—Abnormalities—Diagnosis.   3. Fetus—Diseases—Diagnosis.
4. Infants (Newborn)—Diseases—Diagnosis.   5. Infants—Diseases—Diagnosis.   6. Ultrasonics in obstetrics.
7. Diagnosis, Ultrasonic.   8. Tomography.   I. Johnson, Michael L. (Michael Lord), 1945-   .   II. Title.
[DNLM: 1. Brain—Radiography.   2. Brain diseases—In infancy and childhood.   3. Infant, Newborn, Diseases—Diagnosis.   4. Tomography, X-ray computed—In infancy and childhood.   5. Ultrasonics—Diagnostic use.

WS 340 R936p]
RG629.B73R86   1984   618.92′8047543   83-21656
ISBN 0-8151-7458-6

Sponsoring editor: Daniel J. Doody
Editing supervisor: Frances M. Perveiler
Production project manager: Sharon W. Pepping
Proofroom supervisor: Shirley E. Taylor

To my parents, whose love and lifestyle convinced
me that I could do it.
To Barry, Becky and Marc, whose wonderful love and
support made it possible.
C.M.R.

To my wonderful children, Bryan and Katie,
who have shown me the beauty of the world.
M.L.J.

Foreword

"By the delicate, invisible webb you wove -
The inexplicable mystery of sound."
T.S. ELIOT
to WALTER DE LA MARE

THE INITIAL ACHIEVEMENTS in computed tomographic imaging of the neonatal brain were soon followed in a similar manner by ultrasound. The ease and convenience of ultrasound, the increasing sophistication and sensitivity of its machinery, and the experience of its operators soon needed melding with the anatomical spacial resolution, tissue differentiation and physician experience within CT. Early practitioners of CT were not well versed with the then relatively crude or now precise ultrasound images; whereas present practitioners of sophisticated ultrasound are not usually experienced in advanced CT. I believe there is a pressing need for a unified diagnostic approach that is clinical, rather than technically oriented.

An up-to-date text such as this has, in great part, accommodated this need and is therefore most useful and educational for someone in either of the above groups or indeed in any other. This is a clinical text in which techniques are merely means to a diagnostic end and not each a panacea for the diagnostic ills of the neonatal brain. Carol Rumack, a talented pediatric radiologist doing CT, and Michael Johnson, both expert and experienced ultrasonologists, are part of a new group of enviable and complete physicians of diagnostic imaging of the neonatal brain, a practice that is truly an art and a science.

The gift of expression, the degree of energy and the detail and quality of images evident in this book are impressive. A book of this sort is much needed not so much for each part, but for the whole perspective and diagnostic protocol towards this the most complex of organs, and in these the most difficult of patients, the newborn. Its elucidation of technique with comparative CT and ultrasound anatomy smoothly leads into comprehensive sections on disturbances of development and disorders of structure. Those concerned with cerebral development, congenital anomalies and hydrocephalus are especially noteworthy in their CT and ultrasound correlation. The comparison between the clinical status and subsequent ultrasound and CT in the distressed postpartum infant, especially the premature, is most valuable and timely.

The fruits of this opus, over and above its purpose to detail and compare and contrast the respective images, provide both answers and foundations for further understanding of many complex disorders of formation, feature and function that afflict this member royal of the developing body organs. It satisfactorily complements previous publications separately concerned with neonatal CT or ultrasound.

The advances in technique and knowledge of CT and ultrasound relative to the neonatal brain in particular, an organ previously so inviolate from early neuroradiological techniques, have been overwhelming. We need the opportunity to pause, reflect and then further pursue our evaluation of the neonatal brain. The comprehensive and comparative contents of this book provide just this. It also shows that CT is less likely to be capricious tomography, and for ultrasound it establishes that T.S. Eliot's mystery of sound and delicate woven webb are indeed neither inexplicable nor invisible.

DEREK C. HARWOOD-NASH, M.B., CH.B., F.R.C.P.(C)
PROFESSOR OF RADIOLOGY, UNIVERSITY OF TORONTO
RADIOLOGIST-IN-CHIEF, HOSPITAL FOR SICK CHILDREN

Preface

THE IMPETUS for this book came from work on the frontiers of radiology—at the bedside of the critically ill infant—and at the controls of new radiologic equipment allowing us to study the newborn brain in the detail that was never previously available. We began our investigations as computerized tomography (CT) became available to visualize events in the living neonatal brain that were previously only studied at postmortem examination. An unbelievably high incidence of intracranial hemorrhage in premature newborn infants led us to focus on this major cause of neurologic damage. As we followed these infants with CT and ultrasound, we began to realize how much of this frontier could be defined by ultrasound. The development of small, portable, high-resolution, real-time ultrasound which could sector 90° through the fontanelle opened up an acoustic window. It allowed the study of a critical infant at the time of the insult. Serial studies could be done easily to identify a possible cause and to follow the natural course of a disease process.

By continued study of the neonatal brain, it is hoped that preventive measures may be developed to decrease the incidence of hemorrhage, to know when to intervene in hydrocephalus, to know exactly where to put the shunt or remove masses intraoperatively, and to define anomalies for genetic counseling and prognosis. Intrauterine diagnosis will lead to better care for the fetus.

We have come a long way with a multispecialty group of investigators and have seen with it an explosion of information across the radiologic community. Radiology is now stretching the frontiers, first with CT, now with ultrasound, and next with nuclear magnetic resonance imaging, positron emission tomography scanning, and Doppler ultrasound.

With most diseases, development of potential therapeutic measures follow the definition of a disease entity. As we begin to understand the natural history of neonatal brain diseases and can diagnose them during life, they can be studied in detail for pathogenesis. Now we can plunge into new frontiers and define new problems. It is an exciting era in neonatal neurologic understanding.

CAROL M. RUMACK, M.D.

MICHAEL L. JOHNSON, M.D.

Acknowledgments

MANY PEOPLE DESERVE acknowledgment for their help in this endeavor.

William R. Hendee, Ph.D., our chairman, was a major source of encouragement to our writing this book and to our careers.

Daniel Doody, our editor, has been both enthusiastic and patient through this long process. His guidance and calm encouragement through all of our difficulties are greatly appreciated.

Several contributors have added greatly to the quality of this book. The chapter on normal anatomy was written by Larry Mack, M.D. and his colleagues at the University of Washington in Seattle. Dr. Marilyn McDonald, pediatric hematologist at the University of Colorado Health Sciences Center, offered special expertise on the clinical aspects of intracranial hemorrhage. The chapter on fetal intracranial diseases was a combined effort spearheaded by Delores Pretorious, M.D., who summarized a great deal of information from the literature and from our experience at the University of Colorado.

Our fellows in pediatric radiology and ultrasound over several years have been involved in the pursuit of our understanding of these phenomena. In the past year we have had the pleasure to work with Steven L. Christensen, Joy A. Johnson, Margaret A. Montana, M'Lou F. Wallis, and Gail R. Weingast.

Through many drafts and major changes, our secretarial assistants have been an encouraging part of this process. Dorothy Mueller and Mary Ann Isenhart have researched many references and carefully prepared the majority of the manuscript. Judy Banjevic and Jacqueline Gilliam have typed several chapters.

Our thanks extend to everyone who had a part in these investigations, including our clinical colleagues Mary Ann Guggenheim, M.D., Marilyn McDonald, M.D., William Hathaway, M.D., and Beverly Koops, M.D.

Our special thanks to our superb diagnostic medical sonographers: Kevin Appareti and Terese Kaske, the principal technological investigators, and Susan Kulp, Carol Sundgren, Julia Drose and the excellent student sonographers.

Grant support from McNeil Consumer Products Division gave us the ability to study intracranial hemorrhage in detail and allowed us to pursue this field in great depth.

Prototype real-time ultrasound equipment was made available to us by the generosity of Advanced Technology Laboratories (ATL) and Technicare Corporation.

Our medical illustrator, Jennifer Cole, deserves special mention for her excellent original art work.

CAROL M. RUMACK, M.D.

MICHAEL L. JOHNSON, M.D.

Contents

CHAPTER 1

Introduction

Carol M. Rumack, M.D.
Michael L. Johnson, M.D.

WITH THE DEVELOPMENT of angiography, ventriculography, and pneumoencephalography, pediatric neuroradiology was defined by many investigators as a specialty. Noninvasive imaging with computerized tomography (CT) and nuclear medicine then made neonatal brain investigation a possibility in even the tiniest infants. The state of the art was defined by the classic text written by Derek C. Harwood-Nash and Charles R. Fitz, *Neuroradiology in Infants and Children*, published in 1976.[1]

Ultrasound had been used to determine midline shifts since the 1960s, but A-mode ultrasound had no cross-sectional detail. It was technically difficult and thus operator dependent. In 1974[2] and 1975[3] George Kosoff and his colleagues demonstrated the ability of high-resolution gray scale ultrasound to evaluate the newborn brain with research equipment. It took commercially available gray scale ultrasound equipment before other investigators would catch up with the Australian pioneers. Diane S. Babcock and Bokyung K. Han then defined the state of the art of pediatric neurosonology in *Cranial Ultrasonography of Infants*, published in 1981.[4]

Embarking on our own prospective studies of intracranial hemorrhage, we discovered that both CT and high-resolution ultrasound could be used to study many facets of the neonatal brain. Our goal in this book is to define the role of CT and ultrasound in neonatal brain imaging and to delineate certain questions that may yet be answered by Doppler ultrasound, NMR, and positron emission tomography scanning.

In some conditions such as intracranial hemorrhage, the anatomy and pathology are well understood, although the etiology and prognosis are not yet explicitly defined. Infarction, edema, and vascular disorders are only beginning to be understood and should become the basis of a great deal of research with pulsed Doppler and cross-sectional ultrasound. Myelination defects will be understood more thoroughly as NMR develops. We have reached a peak in our knowledge of neonatal brain imaging, but there are many more levels of investigation. New modalities are ushering in a new era. Where do we stand now with respect to neonatal brain imaging? What factors are well understood and can be used as a basis for further investigation?

We have moved from outlining the brain surfaces and sulci by angiography and the internal surface by pneumoencephalography. Now brain parenchymal patterns can be well evaluated with CT and ultrasound, and ventricular anatomy can be visualized in its entirety. We are able to recognize major malformations that previously could be diagnosed only at autopsy. However, we are still not able to define these malformations histologically or functionally. Neonates in coma can have a normal CT scan and yet major areas of hemorrhage may be clinically silent because they are not yet functionally significant. Malformation syndromes and anomalies previously thought to be extremely rare and only diagnosable at autopsy are now recognizable in life, and we are beginning to understand the whole spectrum of many diseases.

This book presents the current understanding of the anatomy and pathology in the perinatal and infant brain as imaged with ultrasound and CT. The focus is on the developing brain in utero, the neonatal brain, and the infant brain until the fontanelle closes at about age 18 months. In hydrocephalus, the skull may be thin enough for several years so that one can study the ventricles longer but parenchymal detail is poor after the fontanelle closes. Intraoperative neurosonology will eventually allow the brain to be studied at all ages with ultrasound.

The neonatal period is critical. Once past this turning point, children rarely undergo a stress so severe. Ninety percent of brain growth occurs in the first 2 years of life which is the period of focus of the studies described in the following chapters. In the words of a pathologist who has studied these problems in great depth,[5]

No other organ changes as much as the brain does from the 24th week of gestation to 1 year of age. . . . Birth is only an incident which occurs while the brain is growing and developing.

Margaret G. Norman
Perinatal Brain Damage

REFERENCES

1. Harwood-Nash D.C., Fitz C.R.: *Neuroradiology in Infants and Children.* St. Louis, C. V. Mosby Co., 1976.
2. Kosoff G., Gerrett W.L., Radavanovick G.: Ultrasonic atlas of the normal brain of infants. *Ultrasound Med. Biol.* 1:259–266, 1974.
3. Garrett W.J., Kosoff G., Jones R.F.C.: Ultrasonic cross-sectional visualization of hydrocephalus in infants. *Neuroradiology* 8:279–288, 1975.
4. Babcock D.S., Han B.K.: *Cranial Ultrasonography of Infants.* Baltimore, Williams & Wilkins Co., 1981.
5. Norman M.G.: Perinatal brain damage, In *Perspectives in Pediatric Pathology,* Chicago, Year Book Medical Publishers, Inc., 1978.

CHAPTER 2

Neonatal Scanning Techniques

Carol M. Rumack, M.D.
Kevin Appareti, B.S., R.D.M.S.*
Michael L. Johnson, M.D.

NEONATAL BRAIN SCANNING has given us a tremendous amount of insight into neo-natal brain pathology. However, one must remember that a neonate is a relatively fragile person. Neonates are extremely dependent on a controlled environment to maintain their stability. Premature infants in particular are unable to maintain a normal body temperature and are very sensitive to hypothermia. They may require high levels of oxygen because of respiratory disease, but the oxygen must be mixed appropriately with air (Fig 2–1).

Accommodation must be made for maintaining these life support systems when neonates undergo investigative procedures. Careful handling and monitoring of the neonate are done preferably by a neonatal nurse or physician and with pediatric resuscitation equipment immediately available.

CARE OF THE INFANT DURING SCANNING

Sedation

Sedation is rarely required in neonatal scanning.[1] Good restraint can be achieved by taping the head and enclosing the arms with a warm blanket; these measures are usually sufficient for the weak neonate. Feeding the infant as little as 1 hour before the examination should be routine so that the infant is not agitated from hunger.

Sedation of neonates with any drug is associated with increased risks for adverse reaction and enhanced effects.[2] If sedation is absolutely required it is prudent to select an agent that can be readily reversed. The only group of agents with readily available pharmacologic antagonists are the opiates. The safest opiate with the few-est side effects is morphine sulfate. Since extreme lethargy, respiratory depression, and prolonged effect may occur in neonates, morphine should be used only if ap-

*Division of Diagnostic Ultrasound, Department of Radiology, University of Colorado Health Science Center.

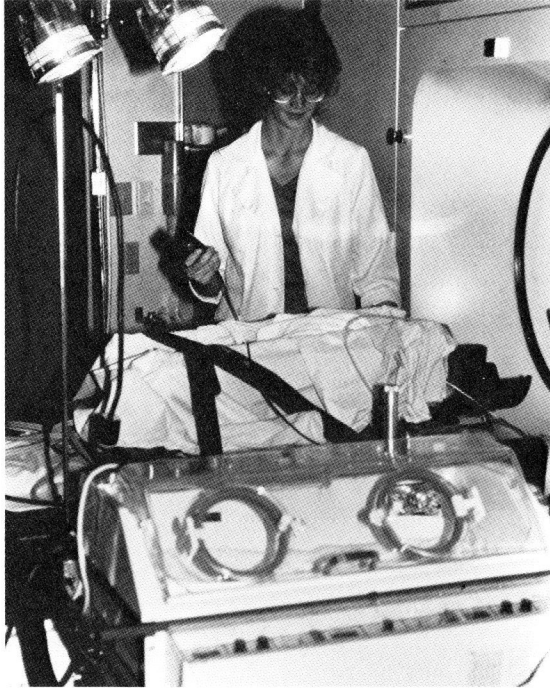

Fig 2–1.—Premature infant has been brought to the scan room in a transport isolette *(foreground)*. Overhead heating lamps and sidemounted radiant warmers are used to maintain normal body temperature. The infant is wrapped in a warm blanket and carefully monitored during the procedure.

propriate precautions are taken. These precautions include the capacity to monitor the heart and respirations, readily available intubation equipment, a secure intravenous (IV) access route, naloxone (Narcan®), and the presence of one of the infant's attending physicians. Once these conditions have been met, morphine may be administered at dosage of 0.1 mg/kg IV. Naloxone at a dosage of .01 mg/kg may be used IV to reverse adverse side effects. If one dose of naloxone is required, it must be assumed that additional doses may be required, since the agonist effects of morphine can outlast the antagonist effects of naloxone. The infant should be symptom free for 4–6 hours after the last naloxone dose is administered before it is safe to discontinue monitoring. If adverse effects occur, but not to the degree that naloxone is required, the infant should be monitored for several hours, until it is asymptomatic. Once naloxone has been administered the infant must be observed for return of symptoms. Naloxone should not be used routinely after morphine, since once the infant becomes alert spontaneously, monitoring can be discontinued without concern for recurrent problems.

Morphine should not be given intramuscularly. Extreme prolongation of effect and erratic absorption are frequent with this mode of administration making it difficult to achieve the desired effect and hazardous to declare the infant safe from adverse effects. The use of phenobarbital, pentobarbital, chloral hydrate, or other drugs or mixtures of drugs is not recommended in neonates. Respiratory disease and apnea are so common in neonates that using sedatives that cannot easily be reversed may be too risky.

Real-time ultrasound does not require sedation because the transducer can be held on the fontanelle so that it moves with the head and motion of the infant does not interfere with the formation of the image.

Temperature Control

The neonate is very sensitive to hypothermia, and therefore the scan room should not be air-conditioned to the usual low degree. The ambient room temperature should be 75–78°. Special warming devices may be required whenever the neonate is scanned in the department. Radiant heaters are frequently the best choice because of the multiple life support monitors, tubes, and catheters attached to the neonate. A servomonitor should be attached to maintain the body temperature during the procedure (see Fig 1).

Scanning in the isolette with real-time ultrasound will prevent hypothermia, and the life support systems need not be disturbed.

EQUIPMENT SELECTION FOR NEONATAL SCANNING

Portability and Cost

Computerized tomography (CT) or static ultrasound imaging cannot be done with portable equipment. Real-time ultrasound equipment has been designed to be portable, and this is one of the main advantages of real-time ultrasound. Because it is portable, it saves transport time for support personnel, doesn't require moving a critically ill infant, and allows the infant to stay within the life support systems and intensive care monitoring. Procedural time is saved because four infants can be scanned per hour in the nursery, whereas a maximum of two can be scanned per hour in the ultrasound department because two transport teams are required. If infants are well enough to be moved to the department without a transport team and life support equipment, it would be more efficient to bring them to one central area.

The cost of a real-time ultrasound examination is about one-fourth the cost of CT. The cost of the examination is related both to the differential cost of the equipment and to the amount of time required for the examination.

When serial scans are necessary, they can be done easily and safely because of the low cost of the examination, portability of the equipment, and lack of any known hazards from ultrasound.

Vascular Detail: Contrast or Real-Time

Contrast enhancement for vascular detail is not necessary for most neonatal lesions. If contrast material is required, 3 cc/kg of 60% iodinated contrast material may be given. The maximum safe dose is 4 cc/kg. The infant should be well hydrated prior to contrast administration. Contrast reactions are very rare in children; a survey by the Society for Pediatric Radiology reported only five major reactions in 12,419 cases and no deaths.[3-8]

Real-time ultrasound scans can show vascular pulsations which, when correlated with pulsed Doppler studies or visually compared with pulsations from the opposite side, may reflect important vascular abnormalities. Cystic structures may be evaluated with Doppler imaging techniques to establish whether they are truly cystic, and if they are vascular, whether arterial or venous.

ULTRASOUND SCAN TECHNIQUES AND INSTRUMENTATION

Basic Principles of Ultrasound Imaging

Ultrasound imaging is based on mapping regional variations in backscattered echoes.[9-11] A brief ultrasound pulse is sent through the tissue from the transducer, which then receives the backscattered echoes. The ultrasound beam is typically pulsed out into the tissue for only 1/100 of the time that the transducer receives. The tissue reflectors are mapped as a scan by assuming a known velocity. Only tissues that are reflective enough to backscatter echoes are recorded. Each returning echo deforms the transducer slightly, and due to the transducer's piezoelectric properties, an electric pulse is displayed on the scanner.

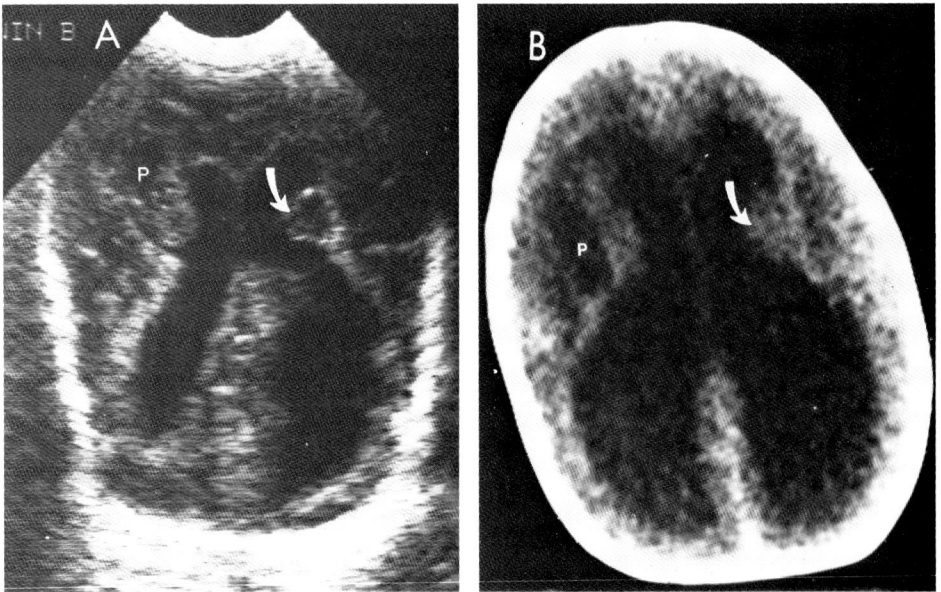

Fig 2–2.—A, Ultrasound scan obtained in 2-week-old infant shows left subependymal hematoma *(white arrow),* which has a different echogenic pattern than the surrounding parenchyma because it reflects brain architecture and density. **B,** CT scan in same infant shows left subependymal hematoma *(white arrow),* which is isodense with brain but does not reflect the difference in architecture. Both scans demonstrate the resolving right intraparenchymal hematoma, which is becoming porencephaly *(P).*

The reflective quality of a tissue depends on tissue density and structural configuration.[12] Interfaces are displayed well when the acoustic impedance between two tissues varies. Acoustic impedance is the product of the velocity of sound in a medium and the density of that medium. The difference in echo strength of the returning signals is displayed on the scan converter as various shades of gray. The stronger reflector will be brighter and more "echogenic."

Tissues with different atomic composition but similar structural composition may have a similar appearance on ultrasound but two different densities on CT. White and gray matter differentiation in neonates is possible on CT due to differences in density—because white matter regions have more water. There is little structural difference between neonatal gray or white matter, so this interface is not well displayed on ultrasound. Two structurally different tissues with the same atomic composition may look different on ultrasound but may have a similar density on CT. The most common example is a hematoma of greater than 7–10 days' duration, which will be isodense with brain on CT but will look markedly different on ultrasound due to changes in the brain architecture from the hematoma (Fig 2–2).

Ultrasound evaluation of the brain is much more dependent on the ability of the examiner and the type of equipment being used than is CT. Therefore, more time must be devoted to the technique of scanning and to selecting the best equipment for this study.

The technique used to visualize the neonatal brain depends on the type of instrument used. There are three general types of ultrasonic scanners available that can be used to visualize the neonatal brain—real-time scanners, contact scanners, and automated water-delay scanners. We will concentrate on the real-time scanner

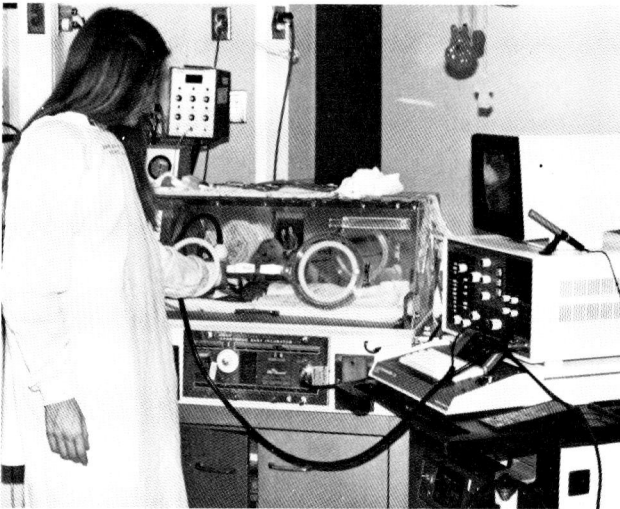

Fig 2–3.—Real-time ultrasound scan being performed in the nursery through holes in the isolette without disturbing the life support systems of the newborn infant. The key requirements are a portable real-time scanner with a 5-MHZ or higher frequency, a small transducer face, maximum sector angle, a frame rate of 15–30 frames per second, a character generator, a portable camera, and a videotape player.

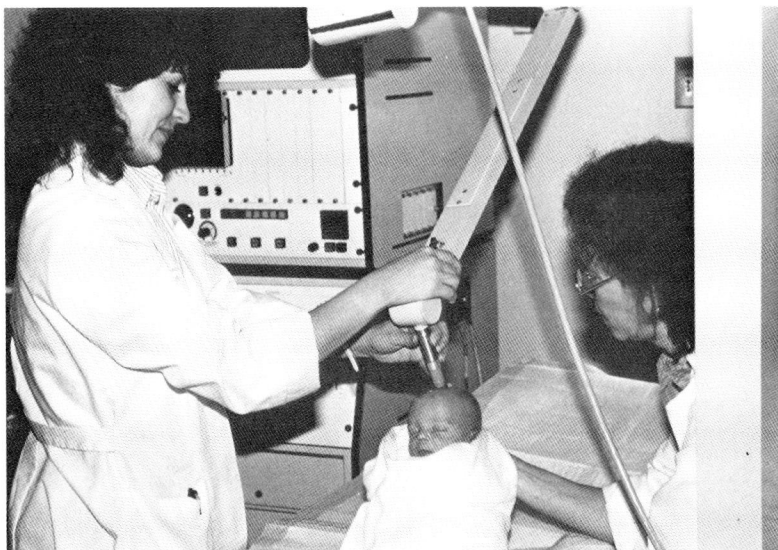

Fig 2–4.—Static ultrasound scan being performed in the department. An assistant holds the infant in position so that the fontanelle is easily accessible and the infant does not move. The key requirements are a 5-MHZ or higher frequency transducer with a medium focus, a 6-mm-diameter or smaller transducer face, a character generator, and a multiformat camera.

(Fig 2–3) and the contact scanner (Fig 2–4), as they are the more commonly used instruments. The real-time scans illustrated throughout this book were obtained with an ATL real-time scanner model 300IC or MARK III and a Technicare MCU. Contact scans were obtained with a Technicare model EDP1200, a Technicare EP or the Picker DI.

REAL-TIME ULTRASOUND SCANNERS

Real-time scanning generates images with updated information at a rate of 15 to 60 frames per second. This enables the examiner to see how structures move as a function of time and to see continuity of structures by moving the transducer back and forth. By following vessels and normal landmarks, the examiner can see how they are spatially oriented with respect to surrounding structures. This ability becomes very important for following structures such as the choroid plexus in the lateral ventricle. There are two types of real-time scanners: sector scanners and linear sequential array scanners.

Basic Equipment Requirements

The basic requirements for a real-time scanner include a small transducer face that will fit easily into the fontanelle. The sector image obtained should be as wide as possible to maximize the information from this small acoustic window. A rapid frame rate to eliminate flicker and loss of information is necessary and should be at least 30 frames per second. A character generator is essential to identify the scans. An internal measuring device is valuable, but many such devices are not accurate and must be calibrated with each use. Hard copy capability is essential and must be available on the scanner, since the quality of reproduction is degraded by imaging from videotape. Videotape is essential, however, to store the nearly maximum sampling of brain that occurs with each sweep.

Intraoperative scanning requires the smallest possible transducer to fit into a standard burr hole, which may range from 1.0 to 1.9 cm.

Cleansing

Between uses, the transducer should be cleaned with iodine, allowed to dry, and wiped with alcohol. Intraoperative technique requires a sterile cover, which can be a manufactured cover, a sterile glove with stockinette, or a sterile trash bag with sterile adhesive tape. Gel must be applied between the bag and the transducer as well as between the bag and the patient. Saline should be used between the bag and the brain.

Linear Array Scanners

The linear sequential array scanners are composed of 32–64 crystals arranged in a linear fashion. Groups of 3–5 crystals are fired in sequence to generate one frame at a rate of up to 60 frames per second. Most real-time units work at a rate of 30 frames per second. While this type of transducer produces fine images and has been used extensively in imaging other areas of the body, its use in visualizing the neonatal brain is very limited. The highest-quality images are produced when sound is sent through the open fontanelle and sutures. Since these areas are very small, the size and shape of the linear array transducers are just not adequate. Information gathering occurs only directly below the transducer, so only a thin slice can be obtained. Therefore, the size of the actual image depends on the size of the fontanelle. The bigger the fontanelle, the bigger will be the rectangular field of view. Small linear array scanners may prove to be more useful than the typical long transducer.

Sector Scanners

Sector scanners are well suited to brain imaging. These transducers produce pie-shaped images with the apex at the transducer face. The transducer face is usually small, and good contact can be maintained when it is positioned on the small fontanelle, so the size of the image is less dependent on the size of the fontanelle (Fig 2–5). There is a limit, however, to how small a fontanelle can be before it affects the quality and size of the image. The smaller the fontanelle, the more difficult it becomes to image the brain without getting dropout of echoes from bone interfaces (Fig 2–6). Since there is no bone to penetrate when the transducer is positioned in the fontanelle, 5- or 7.5MHz, medium focus (4–7 cm focus) transducers are usually employed. The relatively long transducer placed against the small curved head produces good contact only for a short distance, and therefore a limited view is obtained away from the fontanelle over bone.

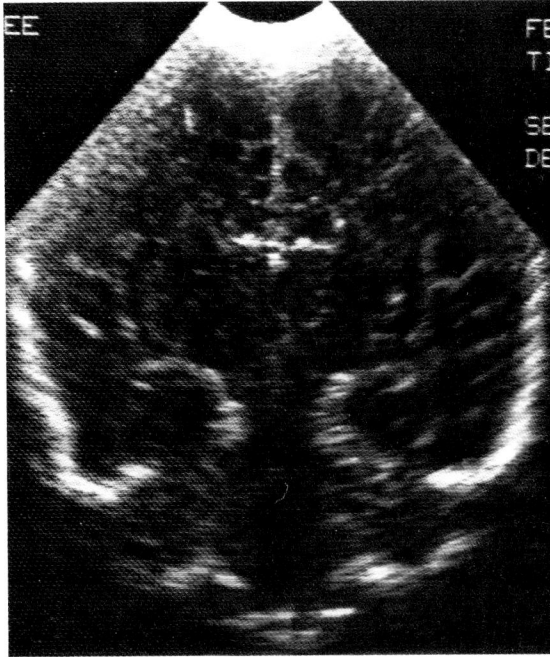

Fig 2–5.—Normal coronal ultrasound scan of the brain. Note the sector configuration and how it is ideal for visualizing through the fontanelle so the majority of the brain can be seen. This is plane C in Figure 2–11. Note the initial 1-cm transducer artifact at the top of the scan.

Sector scanners are of two types, mechanical and electronic. Mechanical sector scanners use one to three crystals which vibrate or rotate to form sector angles of 15°, 30°, 45°, 60°, or 90°. The advantage of mechanical sectors are that they are less expensive and have better signal-to-noise ratios than electronic sectors. Electronic sectors use 32–64 crystals lined up on a very small transducer face. These crystals are fired in a phase and sequence such that the sound waves emitted are steered to form a sector. These transducers also have the ability to change the angle of sector from 15° to 90°. The advantage of the electronic sector is there are no moving parts. However, the price is very high, and there is a poor signal-to-noise ratio.

The resolution of the newer real-time scanners is very near the resolution of the contact scanner. This eliminates the fear of missing important information if a good real-time scanner is used.

TECHNIQUES FOR USING A REAL-TIME SECTOR SCANNER

Ultrasound Monitoring - Standard Projections and Artifacts

Videotaping availability is essential to record several sweeps of the brain with accurate narration for the interpreter. Freeze-frame images should be photographed on a multiformat camera for permanent records. Recording both still im-

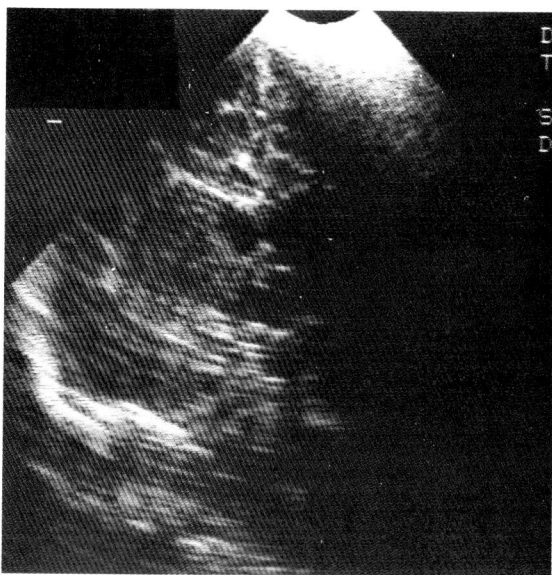

Fig 2–6.—Coronal ultrasound scan with bone interference on the left resulting in a shower of echos (reverberation) and loss of all brain detail below the bone (acoustic shadowing).

ages of selected areas and complete sweeps of the brain in both coronal and sagittal projections eliminate many false positive or false negative examinations. The permanent hard copy images may be the only information that must be stored for a permanent record. However, the nearly total sampling of the brain acquired by videotaping the sweeps should be reviewed before the final diagnosis is made. The standard projection may be slightly off a lesion that could be better demonstrated. The standard projection may also incorrectly suggest a lesion if it is made in slight obliquity to a normal structure.

Artifacts from bone attenuating the sound are encountered with realtime scanners as well as contact scanners. Liberal amounts of gel and obtaining proper contact within the fontanelle will eliminate many of these artifacts. However, some infants have such small fontanelles that bone artifact cannot be avoided. The smallest transducer face should be used, but with real-time transducers there is a limit as to how small the transducer face can be (Fig 2–7). Angulation or rocking of the transducer is often required rather than sliding the transducer across the fontanelle.

Poor contact due to lack of gel or improper positioning can cause loss of information. Maximum gain is usually required to see the deep brain structures. Noise of 60 cycles per second is frequently present in the nursery and may be seen on scans (Fig 2–8). This can be reduced by having the ultrasound machine properly grounded. Beam width averaging may occasionally create a problem. An example is the appearance of brain within the occipital horn in the presence of very thin, normal ventricles (Fig 2–9). Beam width averaging is independent of scan technique as there is a finite lateral beam width in all transducers that is not affected by gain controls.

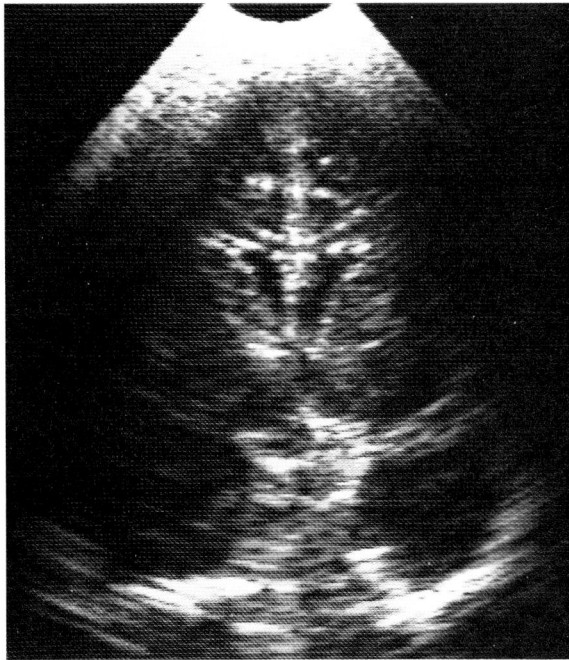

Fig 2–7.—Shadowing caused from bone interference on both the right and left sides. This often occurs with very small fontanelles.

All three views obtained using the real-time scanner show the same anatomy seen on static scans, but there are the added benefits of portability, maneuverability, detection of motion, and continuity. Real-time ultrasound imaging has become the modality of choice in evaluating the neonatal head.

Although initial studies[13-15] were done in an axial plane to compare ultrasound results with CT results, it was the recognition of the anterior fontanelle as a fortuitous acoustic window that led to a marked improvement in neonatal brain imaging.[16-19]

Coronal Views (Real-Time)

The real-time examination takes about 10 minutes and can be done at the bedside or in the laboratory in conjunction with static scanning. Acoustic gel is applied to the anterior fontanelle. The transducer is placed on the fontanelle in a coronal position (Fig 2–10). A 5- to 7.5-MHz (medium focus) transducer is usually employed. The time gain compensation (TGC) controls are set to obtain adequate penetration with even near and far field parenchymal echoes. Slight attenuation of the near field for about 1 cm and then maximum gain through the rest of the brain is required. The field of view (sector depth) is chosen to display the brain to the maximum size without losing information. The image should be oriented to display the infant's left side on the right of the screen. Different coronal planes are gener-

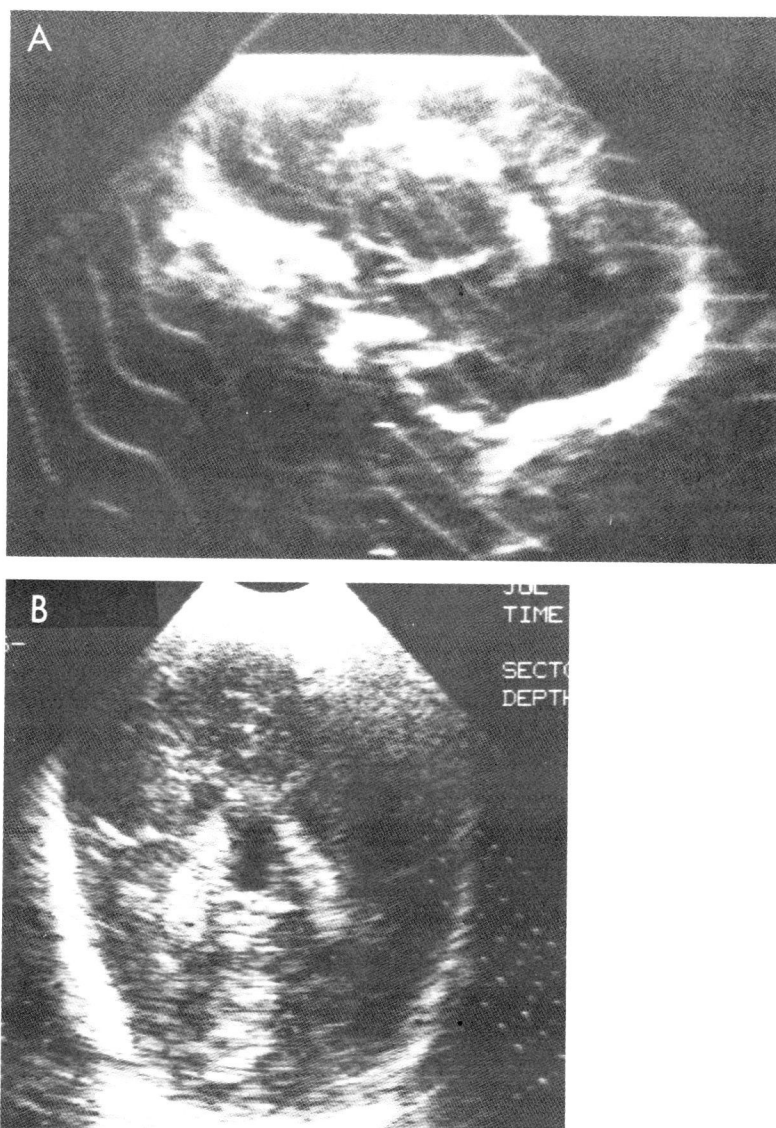

Fig 2–8.—Noise of 60 cps appears as streaking lines **(A)** or as a discrete pattern of dots **(B)**. This noise often interferes with image quality. It is caused by the very sensitive receivers in the ultrasound machine picking up line noise from other equipment.

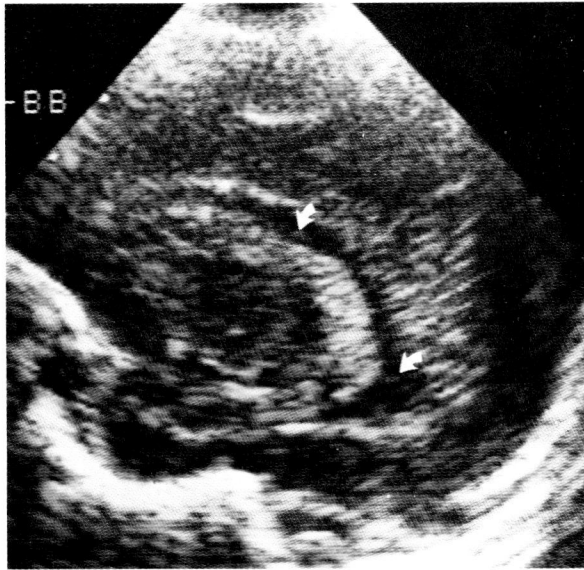

Fig 2–9.—Beam width averaging causes the lateral ventricle to be filled with artifactual echoes *(arrows)* arising from brain parenchyma that falls within the beam diameter when the ventricle is small.

ated by the relative angle of the transducer to the ventricular system (Fig 2–11).

The transducer is first angled posteriorly toward the occipital horns (Fig 2–12), then it is slowly angled anteriorly toward the frontal horns (Fig 2–13). As the transducer sweeps from posterior to anterior, care must be taken to ensure that the image remains symmetric. It is very easy to rotate and twist the transducer out of the correct plane during the transducer sweep. Figure 2–14 illustrates the problem of asymmetry.

Once the frontal horns are visualized, the transducer should be swept back to the occipital horns. This should be done at least twice to ensure adequate visualization of the entire brain. Next, freeze-frame images should be taken at specific coronal sections from posterior to anterior. These images should include the occipital horns, trigone, midbody of the lateral ventricles with the temporal horns, and the frontal horns.

Sagittal Scans

By rotating the transducer 90° to the coronal plane, a sagittal plane is produced (Fig 2–15). A sweep from one lateral ventricle through the midline to the opposite ventricle should be done first. The examiner must remember the oblique planes the ventricles lie in to determine the rotation of the transducer as the sweep is performed (Fig 2–16). Next, freeze-frame images of each lateral ventricle and midline structures should be recorded (Fig 2–17); asymmetry should be avoided (Fig 2–18).

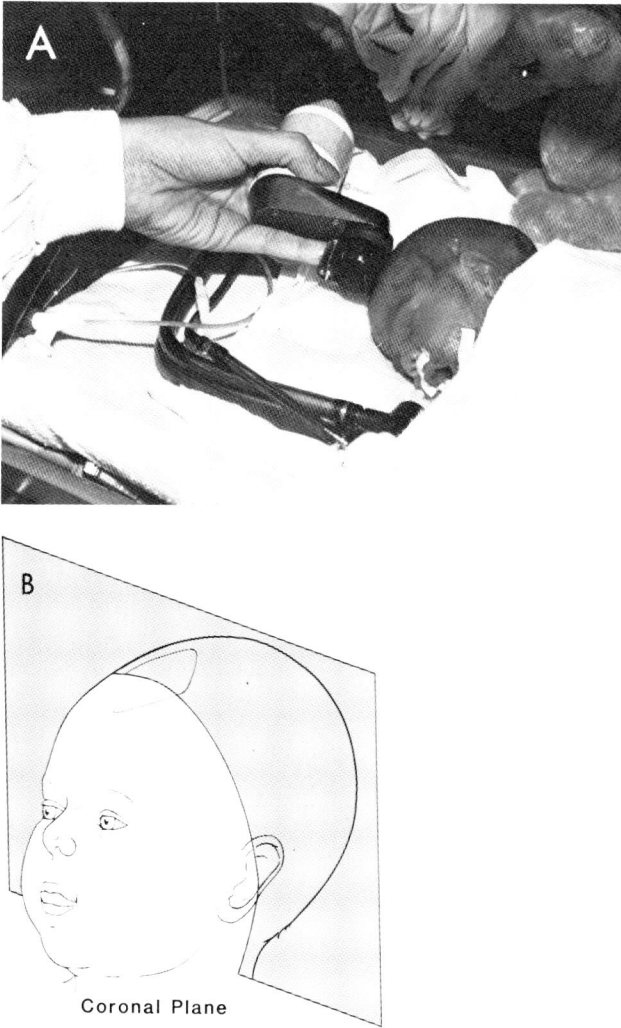

Fig 2–10.—Coronal planes are generated by placing the transducer on the fontanelle **(A)** such that the plane extends towards the feet **(B)**, as when looking directly at the infant.

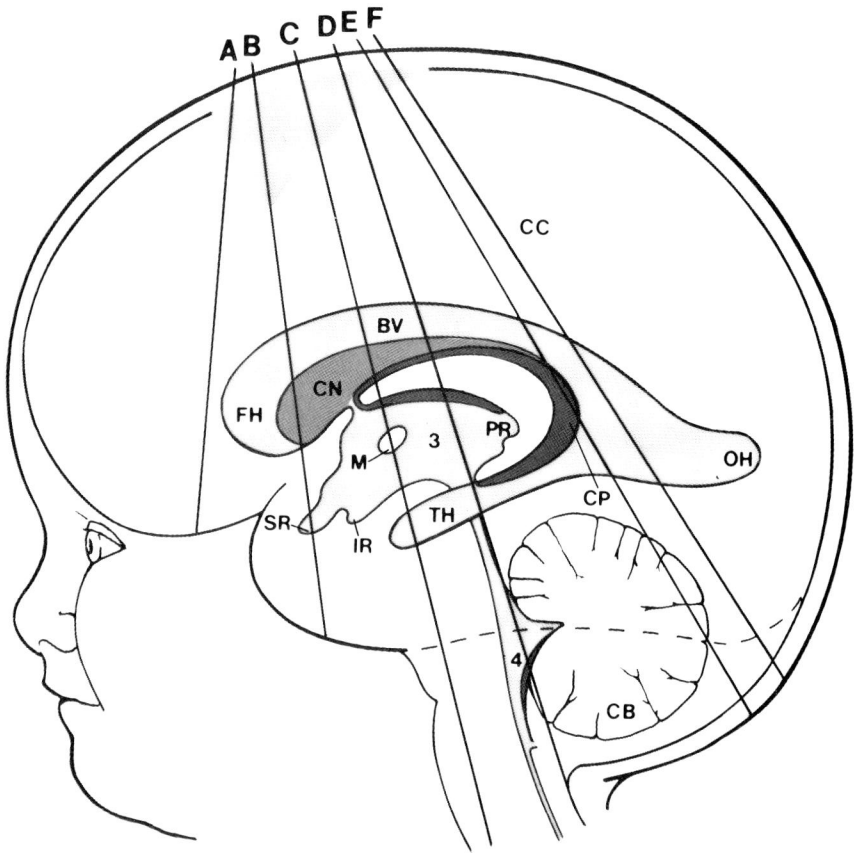

Fig 2–11.—Schematic representation of angled planes used in coronal scans. *BV,* body of ventricle; *CB,* cerebellum; *CC,* cerebral cortex; *CN,* caudate nucleus; *CP,* choroid plexus; *FH,* frontal horn; *IR,* infundibular recess; *M,* massa intermedia; *OH,* occipital horn; *PR,* pineal recesses; *SR,* supraoptic recess; *TH,* temporal horn; *3,* third ventricle; *4,* fourth ventricle.

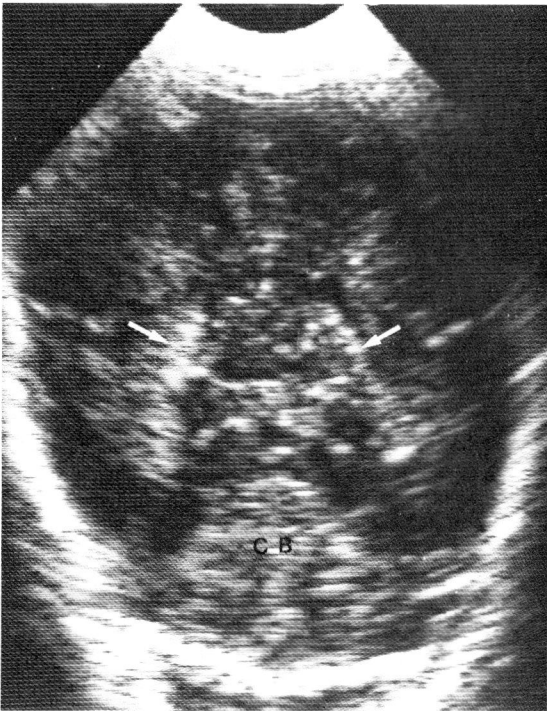

Fig 2–12.—Coronal scan obtained by angling the transducer posteriorly in the region of the trigone just anterior to the occipital horn. This is plane E in Figure 2–11 and shows the choroid plexus *(arrows)* and cerebellum *(CB)*.

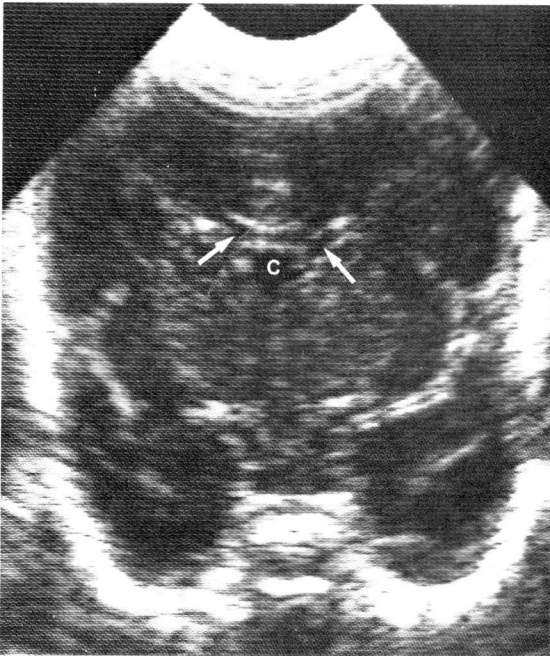

Fig 2–13.—Coronal scan obtained by angling the transducer anteriorly in the region of the frontal horns *(arrows)*. The cavum septi pellucidi *(C)* lies between frontal horns. This is plane B in Figure 2–11.

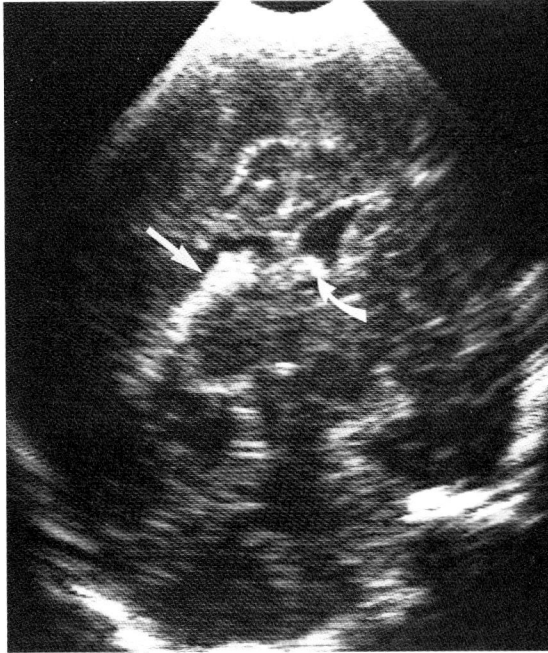

Fig 2–14.—Asymmetry in the coronal plane. Note the difference in bone contour from left to right side. Also note how the right choroid at the trigone *(straight arrow)* is much longer than the choroid at the midbody of the left lateral ventricle *(curved arrow)*.

Axial Scans

Finally, scanning should be done in axial planes by positioning the transducer 15° to 20° above the canthomeatal line (Fig 2–19). Starting just above the external auditory meatus (EAM), the transducer should be moved rostrally. When the lateral ventricles are in view just above the choroid plexus, the examiner photographs the ventricles to measure the lateral ventricular ratio (Fig 2–20). The side away from the transducer can always be measured, but the near side may be obscured by transducer artifact. Therefore, to obtain both lateral ventricular measurements, the neonate must be scanned from both the right and left sides.

COMPOUND SCANNERS

When extremely high-resolution images are required the contact scanner should be used. Sending sound through bone to visualize soft tissues has its problems since bone has strong attenuation and produces many artifacts. Consequently, sectoring the transducer in the anterior fontanelle and other open sutures results in very high-resolution images without bony artifacts.

When the contact scanner is used to penetrate bone from an axial plane, a 3.5-MHz transducer should be used. The focal depth should be picked to correspond to the neonate's head size. When the fontanelle is used to visualize the brain a 5-

Sagittal Plane

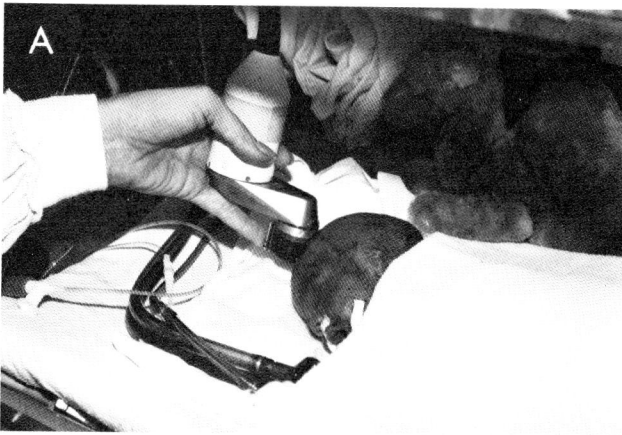

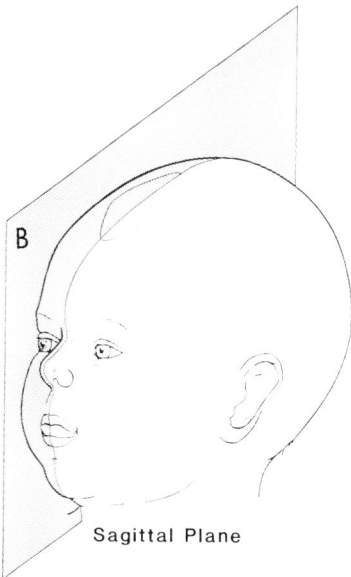

Fig 2–15.—**A,** Close-up of transducer position in a sagittal plane; **B,** Drawing demonstrating sagittal plane.

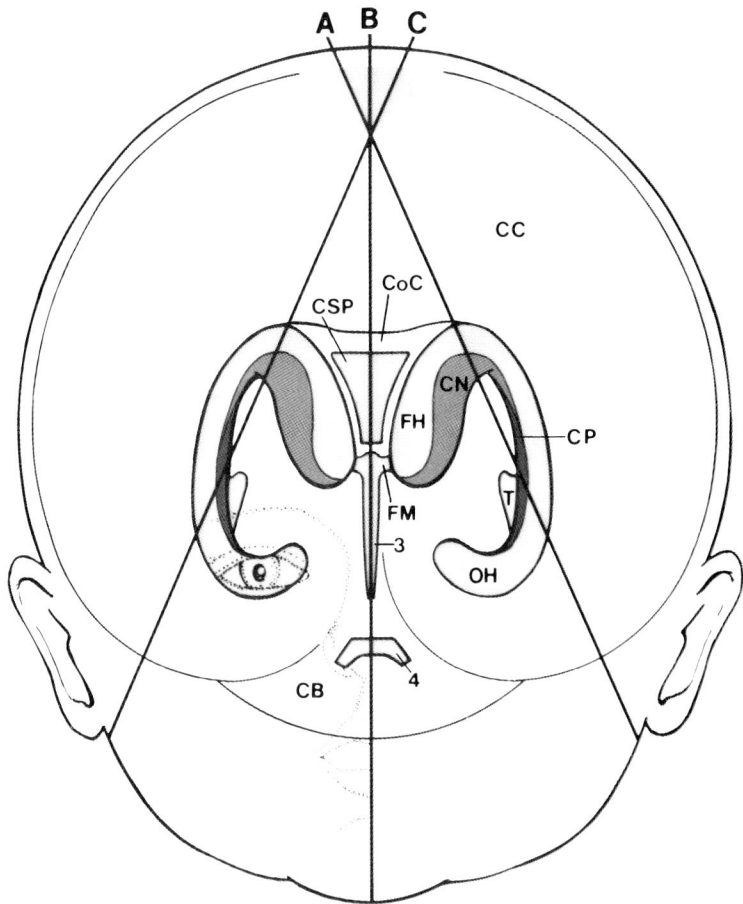

Fig 2–16.—Schematic or sagittal planes. *CB,* cerebellum; *CC,* cerebral cortex; *COC,* corpus callosum; *CN,* caudate nucleus; *CP,* choroid plexus; *CSP,* cavum septi pellucidi; *FH,* frontal horn; *FM,* foramen of Monro; *OH,* occipital horn; *T,* temporal horn; *3,* third ventricle; *4,* fourth ventricle.

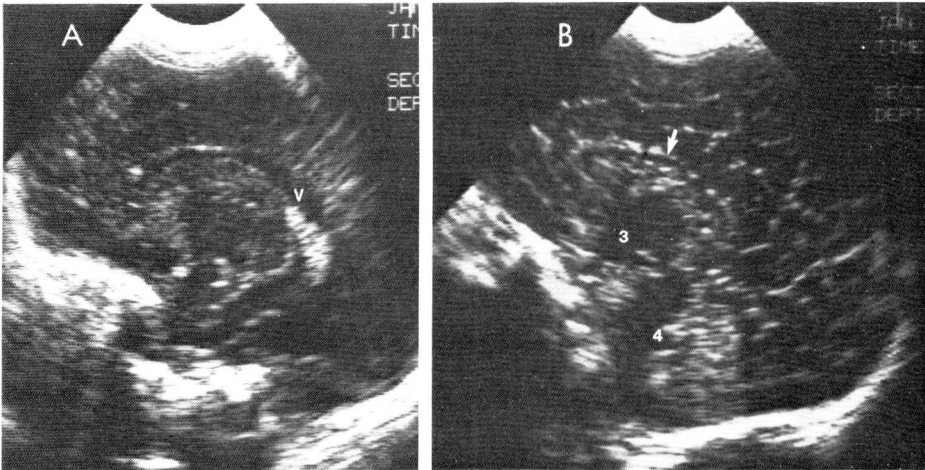

Fig 2–17.—A, Real-time image of lateral ventricle *(V)* in the sagittal plane with face of infant to the left. **B,** Real-time image of midline structures in the sagittal plane. Third *(3)* and fourth *(4)* ventricles are in front of the cerebellum and corpus callosum *(arrow)*.

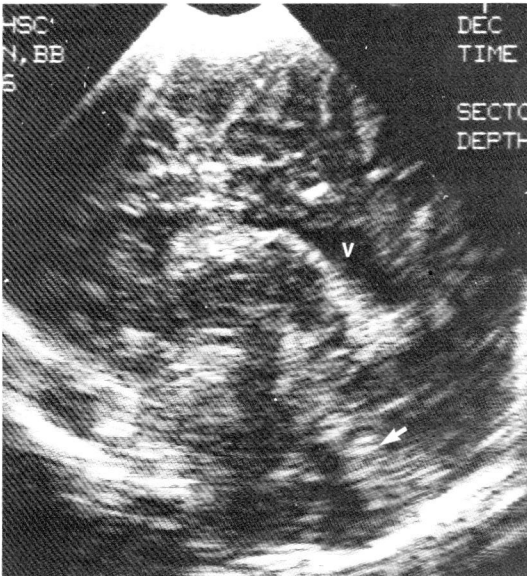

Fig 2–18.—Asymmetry in a sagittal plane. The lateral ventricle *(V)* and the cerebellum *(arrow)* should not be visualized in the same plane.

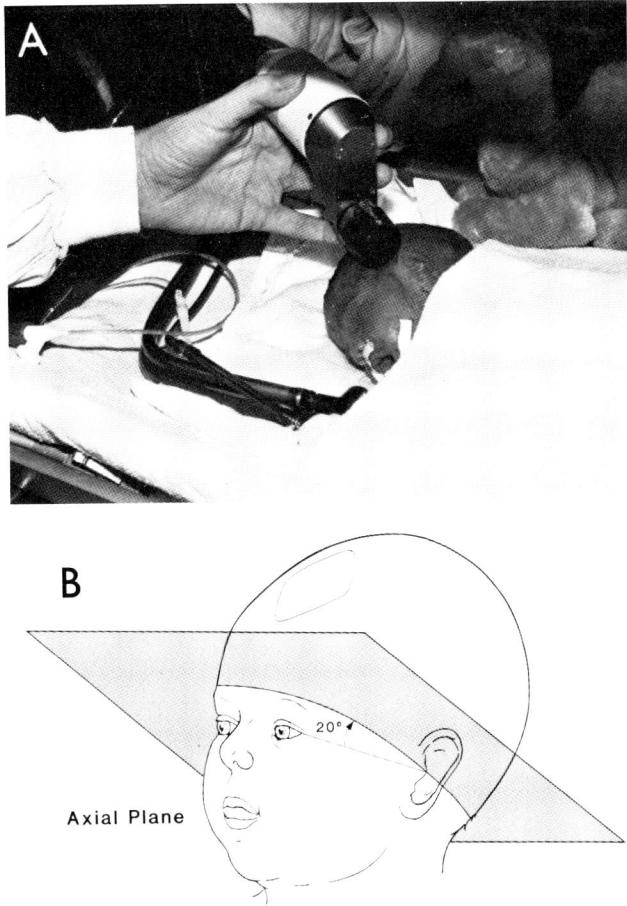

Fig 2–19.—The transducer is placed 15–20° above the canthomeatal line **(A)** to generate an axial plane **(B)**.

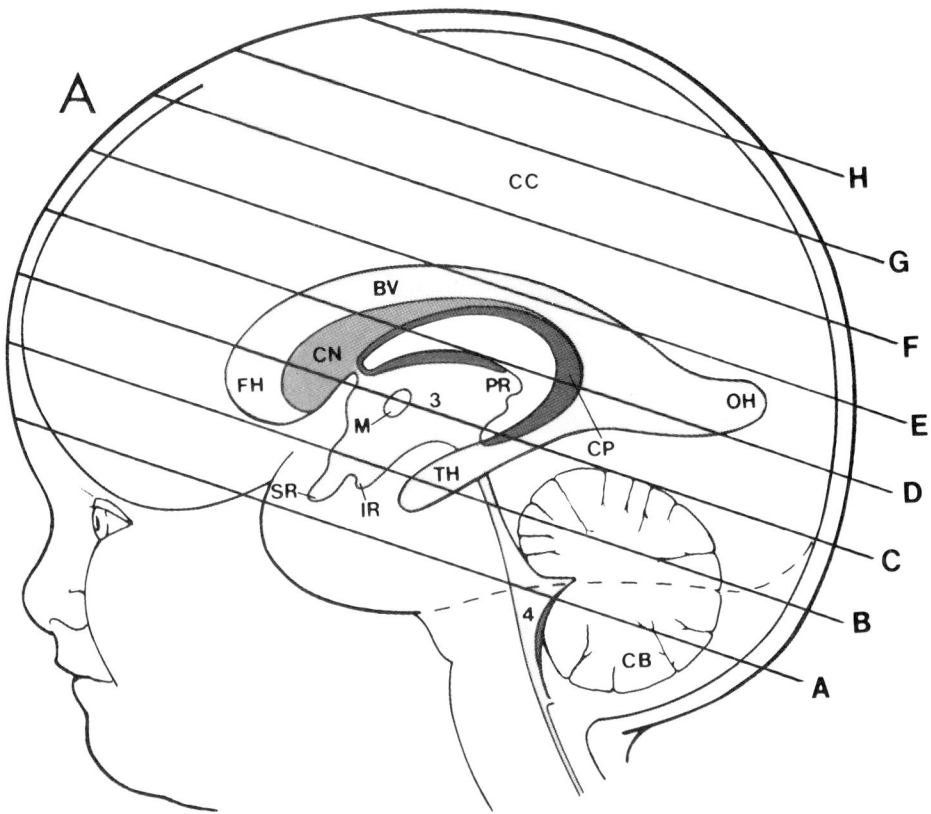

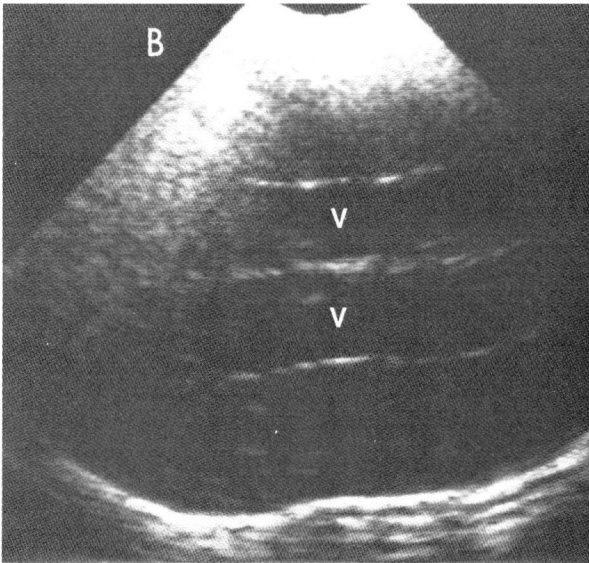

Fig 2–20.—A, Schematic representation of axial planes. **B,** Real-time image of lateral ventricles *(V)* seen in an axial plane.

to 7.5-MHz transducer should be used. Because the fontanelle can be very small, a narrow-diameter transducer should be used. We routinely use a 6-mm-diameter transducer. On the fontanelle, acoustic gel is used instead of mineral oil to provide a thicker medium that will stay in place. The TGC settings should be positioned so that echoes of an even level are received from near and far fields. This usually is a flat slope with relatively high initial gain compared to real-time scanning, because there is less near field noise with compound scanners.

A disadvantage to using the contact scanners is that it sometimes takes as long as 30 minutes to complete the examination. It also takes more technical skill to perform contact scanning than real-time scanning.

Technique For Using A Static Compound Scanner

Coronal Scans

The newborn infant can be held in a sitting position or the head may be elevated on a very thick sponge for the coronal scans (Fig 2–21). Depending on the particular machine and the laboratory setup, either method will work. Because the anterior fontanelle may be very small or very large, the scanning arm cannot be held in one position for the whole study. Depending on the size and shape of the fon-

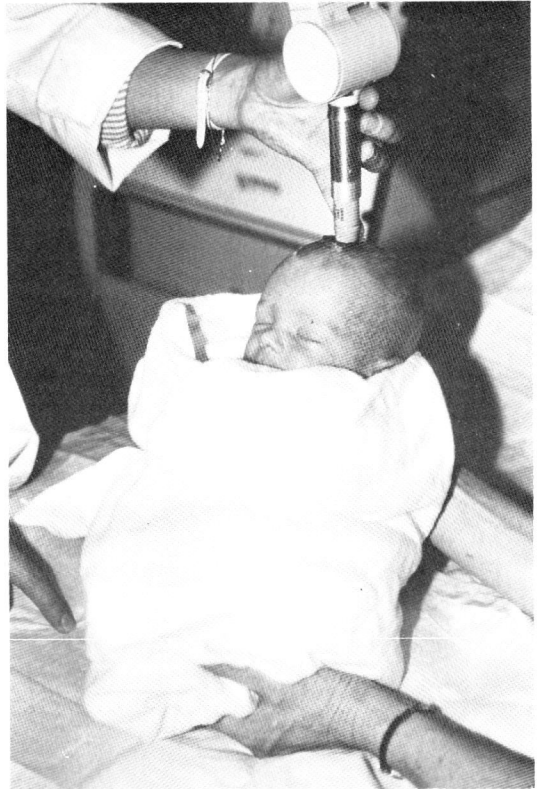

Fig 2–21.—Normal setup for performing coronal sections with a static imager. The infant is held in a sitting position and the transducer is sectored in the fontanelle.

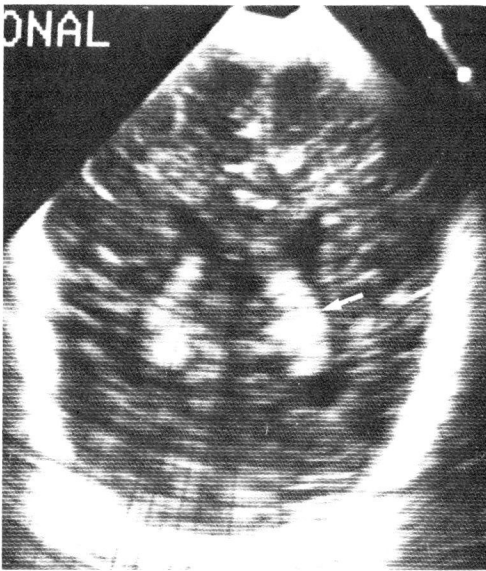

Fig 2–22.—Coronal scan demonstrating the use of a single sector sweep used in static scanning. *Arrow* points to choroid plexus.

tanelle, a number of modified coronal angles must be used. Starting anteriorly, the transducer is angled and moved posteriorly, while staying within the fontanelle. Anterior sections are scanned with the transducer plane angled toward the patient's face while posterior sections require the transducer to be angled toward the back of the head. Figure 2–22 illustrates single sector sweeps employed in all coronal scans. Because of the number of sources of unwanted movement, overwriting and compounding should be avoided.

Coronal sections of the head usually take approximately 10–15 minutes.

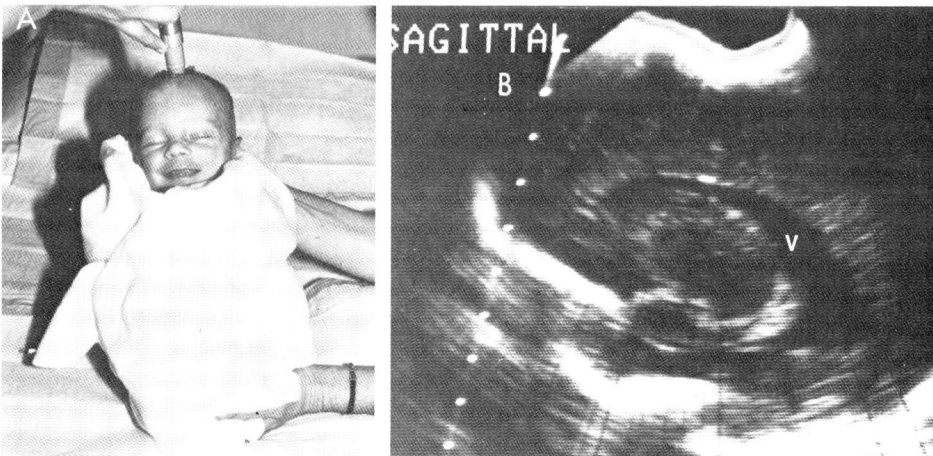

Fig 2–23.—**A,** Normal setup for performing sagittal sections with a static imager. The infant is held in a sitting position and the transducer is sectored from an anteromedial position to a posterolateral position. Refer to Figure 2–16. **B,** Sagittal scan of lateral ventricle made with static scanner.

Sagittal Views

By moving the transducer 90° to the coronal plane, a sagittal section can be obtained (Fig 2–23). With the anterior fontanelle as a window, the lateral ventricles lie in an anteromedial to posterolateral plane. Therefore, when setting the scanning arm the examiner should follow this same plane for each of the ventricles. Best results are obtained when the transducer is angled from the opposite side of the midline from the ventricle that is being visualized.

Axial Scans

Axial scanning with a contact scanner offers important information about the relative sizes of the ventricles but parenchymal detail is hindered by the bony calvarium. The infant is placed on the examining table with the head in a true lateral position. An assistant is essential to stabilize the infant's head during the examination. Mineral oil is spread on the head and the canthomeatal line is noted. The transducer is aligned to 25° above this line to obtain sections that correspond to the normal CT sections. The highest frequency transducer should be used that will permit adequate penetration, usually a 3.5-MHz, medium focus transducer. The TGC curve should be set so the knee of the slope is positioned at the echo arising from the far side of the calvarium. The initial gain is then set to produce an even gray level of echoes from the brain parenchyma in the near field to the far field. Gain settings must be set high to produce low to medium level echoes from the brain parenchyma while leaving fluid-filled structures anechoic. Maximum gain levels are necessary on presently available equipment.

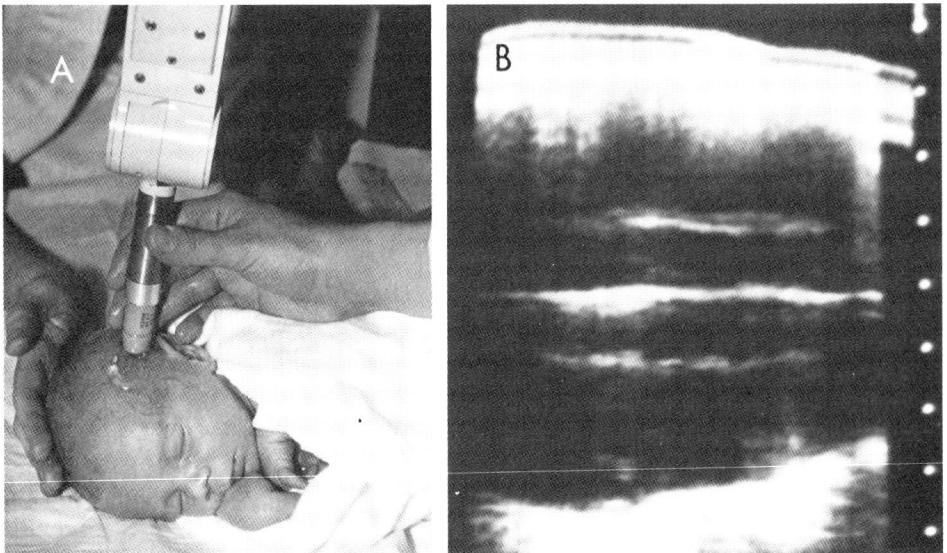

Fig 2–24.—Normal setup for performing axial sections with a static scanner. **A,** The infant's head is placed in a lateral position and held stable by an assistant. **B,** The transducer is moved in linear and sector sweeps to generate the axial scan of the ventricles.

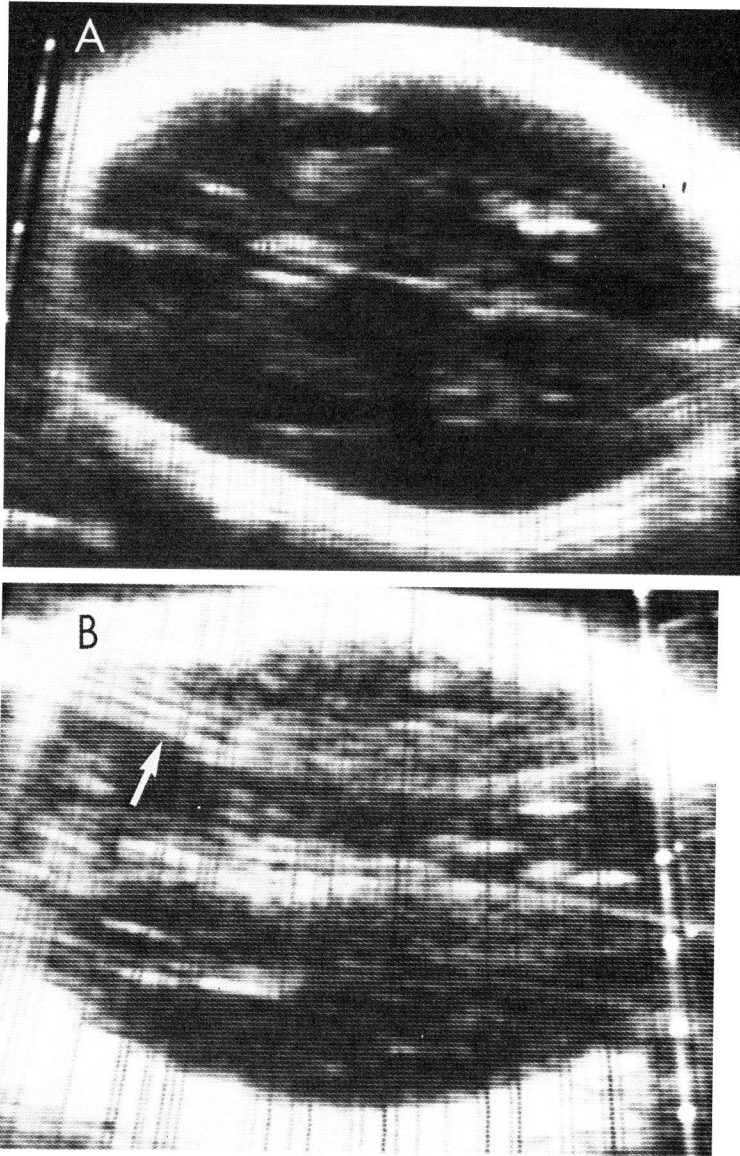

Fig 2–25.—Axial scan of the brain using a water path system. **A,** This system enables better visualization of the frontal and occipital areas. **B,** Artifacts do arise, however, and this is the result *(arrow)* if the depth of the water is less than the diameter of the head.

Axial scans are taken at 5-mm intervals, starting approximately 5 mm above the external auditory meatus and moving rostrally. The transducer must move back and forth in linear and sector sweeps to get the needed information (Fig 2–24). The flatter the infant's head, the more linear the technique, and a single sweep will be adequate. When overwriting (sweeping back across part of the same area) cannot be avoided, it will be very important for the assistant to maintain the infant's head position. The inner table of the skull and the first centimeter of brain near the face of the transducer are not well visualized as a result of the initial transducer artifact, often referred to as the "main bang" artifact. Therefore, information on the near or up side hemisphere cannot be adequately obtained. Consequently the infant is turned 180° and the procedure is repeated. If there are intraventricular catheters or other tubing connected to the baby, care must be taken not to disturb them.

Axial Water Path Scans

Axial scans can also be done using a water path system. Water is warmed to body temperature and allowed to stand to eliminate bubbles. A small amount of liquid soap can be used as a surfactant to help eliminate bubbles. It is important to have the depth of the water slightly greater than the biparietal diameter so there will be no artifacts within the brain image. With the increased distance from brain to transducer face, a longer focal length is required. Mineral oil is applied to the infant's head and to the plastic membrane. The water path assembly is positioned over the head, care being taken not to put too much pressure on the skull. One advantage of using a water path system is that it enables better visualization of the frontal and occipital areas and may result in better image quality (Fig 2–25,A).

However, the setup is cumbersome and eliminating artifacts can be tricky. Two additional types of artifacts are associated with the water path system. One is a result of an improper depth of water (Fig 2–25,B). The other artifact arises from reverberations off the sides of the plastic water path membrane.

Artifacts with Compound Scanners

Artifacts may occur with movement of the infant and bone shadowing. A second pair of hands often is the solution to movement artifacts. If the infant has a very small fontanelle, it can be difficult to angle the transducer without hitting a bony surface. Using the smallest-diameter transducer and liberal amounts of gel is the best solution. Other artifacts encountered are similar to those encountered in real-time scanning.

Automated Water-Delay Scanners

The first neonatal brain scanning was performed by Kossoff et al. with an automated water-delay scanner.[13] With this technique the infant lies on a plastic membrane. Oil is used as a coupling medium, and there is water between the membrane and transducers. Multiple automated transducers scan the infant, creating a contact scan, but with better information along the curves of the skull than the usual static scanner, resulting in a more complete scan along the periphery (Fig 2–26). The real-time scanners viewing through the anterior fontanelle have the advan-

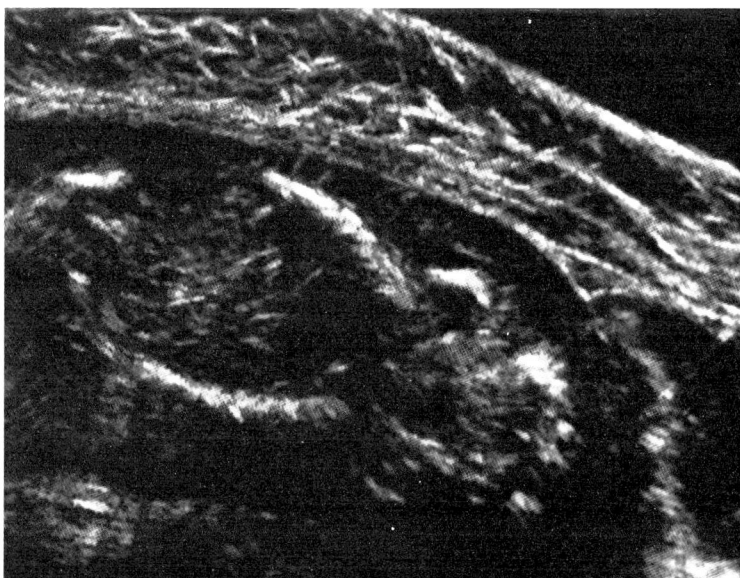

Fig 2–26.—*Octason* scan of infant with anencephaly. See chapter on fetal intracranial diagnosis for details.

tage of less interference from bone and do not have the overwriting problems of any type of contact scanner. Automated ultrasound scanners do not have the portability of real-time scanners. However, heat loss is not a problem because the water bath is heated to body temperature and the infant can be covered with a blanket during the procedure. Additional details are described in other publications on this technique.[20, 21]

CT SCAN TECHNIQUES

Basic Principles of CT Imaging

CT scans are images based on mapping regional variations in x-ray attenuation. The CT image can distinguish structures that differ in atomic number and electron density. A highly collimated x-ray beam is sent through the patient. As the beam rotates around the patient, crystal detectors arrayed about the patient record the amount of radiation left after absorption and scattering of the photons. By using a complex algorithm developed by Hounsfield, CT numbers related to absorption densities for individual pixels (.5 mm × .5 mm × slice thickness) can be calculated. CT images are typically based on a 512 × 512 matrix of pixels and will display the relative tissue characteristics in multiple shades of gray, depending on the center (window level) and range (window width). Hounsfield units (HU) are calibrated at zero for water and extend from −1,000 for air to +1,000 for metal. Most organs are of soft tissue density and vary from 40 to 60 HU. Brain tissue is typically slightly lower, particularly in neonates, due to the presence of relatively large amounts of water. Fat density averages approximately −15 to 0 HU. Freshly clotted blood averages 60–80 HU. Calcium usually measures at least 80 and often over 100 HU. Bone, which contains large amounts of calcium, is nearly always 120 HU or more. Small flecks of calcium may measure only 60–80 HU and thus may be confused with blood. All of these measurements are approximate and vary with the machines used, but the equipment should be calibrated daily so that CT numbers can be as accurate as possible.

Certain artifacts result from the techniques of CT scanning. The basic problems include beam hardening and slice thickness averaging. Beam hardening is caused by absorption of the lower energy photons by the skull. This raises the effective energy of the beam and decreases the gray/white matter contrast of the brain. Partial volume effect due to slice thickness and pixel size has also improved with newer equipment, but there is a tendency for CT to underestimate the edges of structures that are cystic due to averaging of brain at the interface.

More detailed understanding of the physics of CT scanning can be obtained from several excellent publications dedicated specifically to this topic.[9, 22, 23]

Head Position

The head rest commonly used in CT scanners usually must be modified for infants. The entire neonate may fit in the adult head-holding device. Additional sponges or a totally separate head-holder may be required to immobilize the tiny infant head and support the body. Radiant warmers mounted beside the patient on nonopaque material can be used as the basic support for the neonate, with its head placed on the head-holder (Fig 2–27).

Circle Size

We commonly use a radius of approximately 150 mm in newborns, although it may have to be larger in some patients. It is valuable to use the same diameter on follow-up examinations to facilitate the diagnosis of a change in size of a lesion.

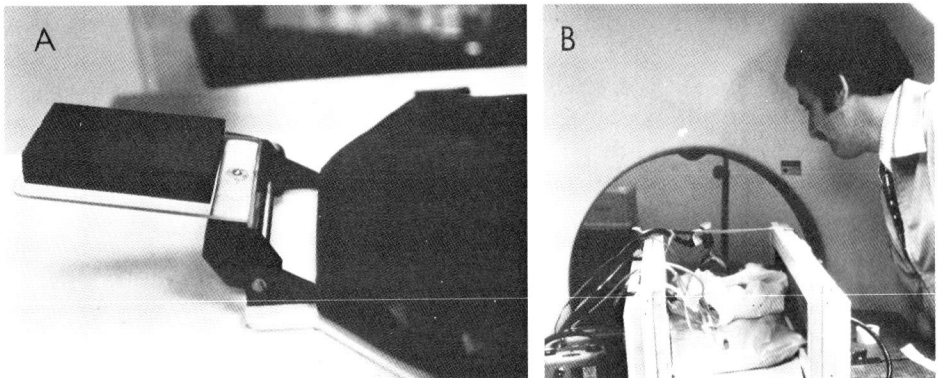

Fig 2–27.—Head-holding device **(A)** is too large for neonate. Side-mounted radiant warmers **(B)** are used to maintain body temperature and support the infant.

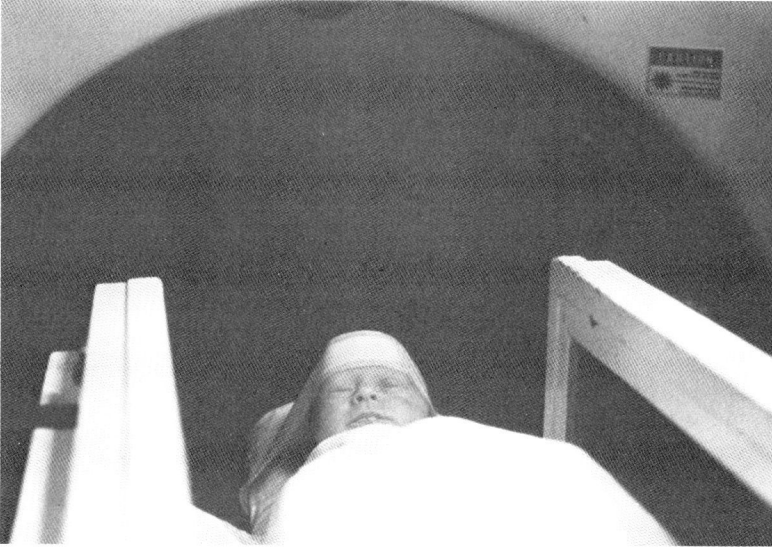

Fig 2–28.—Neonate with head taped in axial position.

CT Scan Planes

Axial scans are angled approximately 25° to the canthomeatal line (see Fig 19). These are usually the best views obtained on CT because they are associated with the least artifact. In complicated cases two views should be obtained, usually an axial (Fig 2–28) and a coronal view (Fig 2–29).[24] Two angled views may be necessary if the lesion is not in the axial plane. One should be perpendicular to the lesion. The other should be chosen in a parallel plane, such as parallel to the clivus for the brain stem and fourth ventricle. Scans should begin above the orbits unless it is necessary to visualize the base of the skull.

Fig 2–29.—Neonate with head taped and foam pads to stabilize the coronal position.

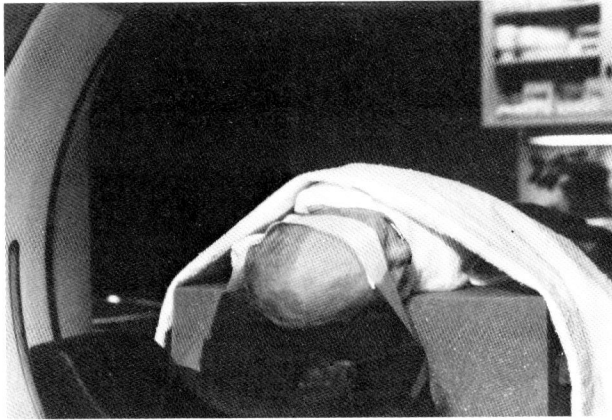

Fig 2–30.—Sagittal positioning in CT scanner.

Sagittal views, whether direct (Fig 2–30) or reconstructed, may be useful for unusual lesions, but artifacts are prominent in this view.[25] Unusual scan planes may be confusing but good anatomical atlases are available, particularly for axial and coronal planes. The atlas by Matsui and Hirano[26] is recommended. Follow-up scans should be done in the same scan plane so as not to distort the anatomy.

Slice Thickness and Time

For neonates, 8-mm slices at 9-mm intervals are best. Thinner slices may be necessary (e.g., 4-mm slices at 5-mm intervals) if there is a specified area of interest such as the posterior fossa. Very thin slices may be useful, but they become increasingly coarse or grainy.

Reconstructions require multiple thin slices, which increases the radiation dose but will allow multiple angled projections to be reconstructed at the exposure for just one set of scans. Direct sagittal and coronal views are almost always attainable in neonates.

Short scan times of 5 seconds or less are necessary to acquire sufficient speed to decrease motion artifacts. Slow scan times may be better for maximum information.

Monitoring CT: Standard Settings and Artifacts

The scan should be directly monitored during the examination because changes in window level and width will bring out specific lesions that might otherwise be missed. This is particularly true for peripheral hematomas, which tend to blend into bone. The window level will generally be lower than for adult head scans. A standard setting of the window level (approximately 70 HU) is obtained initially for the basic contrast. Then the window width should be changed until the brain appears relatively high in contrast (approximately 25–30 HU). If this is not done, too much smoothing of the gray and white matter may occur and subtle low-density areas may be obscured. In the normal premature neonate there should be low-

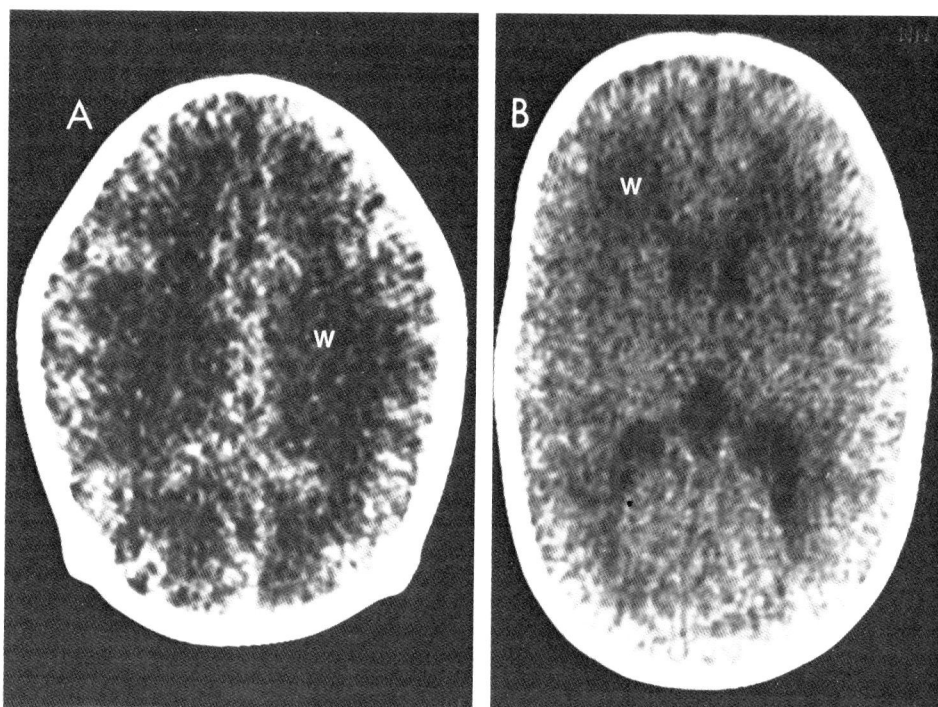

Fig 2–31.—A, Normal low-density white matter *(W)* in 28-week premature infant is very prominent. By 40 weeks' gestation **(B),** only small areas of low-density white matter are present adjacent to the frontal horns.

density white matter bifrontally anterior to the frontal horns, biparietally just above the lateral ventricles, and adjacent to the trigone bilaterally. In the term infant, these low-density areas will become progressively smaller, with the frontal low-density areas the last to disappear at several months after birth (Fig 2–31).

One must take care to measure any possible areas of hemorrhage because if the white matter is of very low density, the gray matter may appear hemorrhagic. This results from setting the window level to make white matter appear of normal brain density, and thus the normal gray matter appears white. One should check the Hounsfield numbers even though they are not exact and develop a standard window width for each machine. Actual numbers will keep one from making the diagnosis of hemorrhage when the real problem is edema or confusing normal low-density white matter in premature infants (see Fig 31).

Brain stem hemorrhage is very difficult to diagnose at the level of the foramen magnum because the large amount of bone causes an apparently dense brain stem even in normal patients (Fig 2–32). In fact, any area where there are large amounts of bone, and particularly where there is any motion, may cause high-density artifacts. These areas include the petrous pyramids. The edges of the fontanelle and sutures may also cause dense, streaking artifacts.

The most frequent artifacts in newborns result from poor restraints. Paper tape

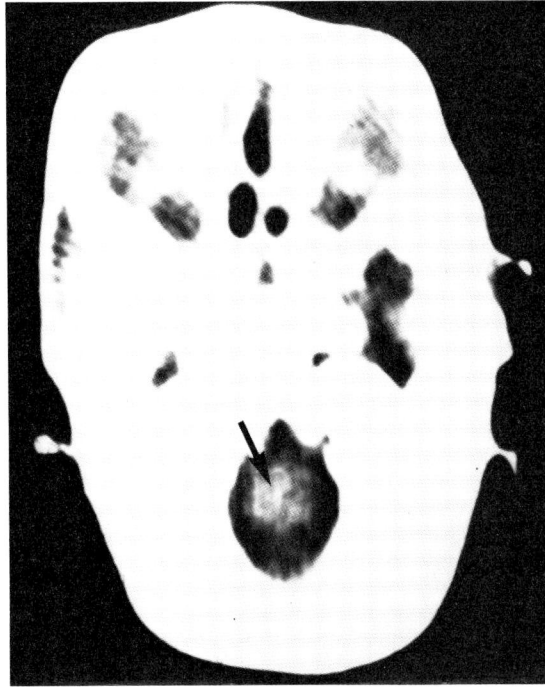

Fig 2–32.—Artifact from surrounding bone causes the brain stem to have a high-density appearance *(arrow),* falsely suggesting hemorrhage.

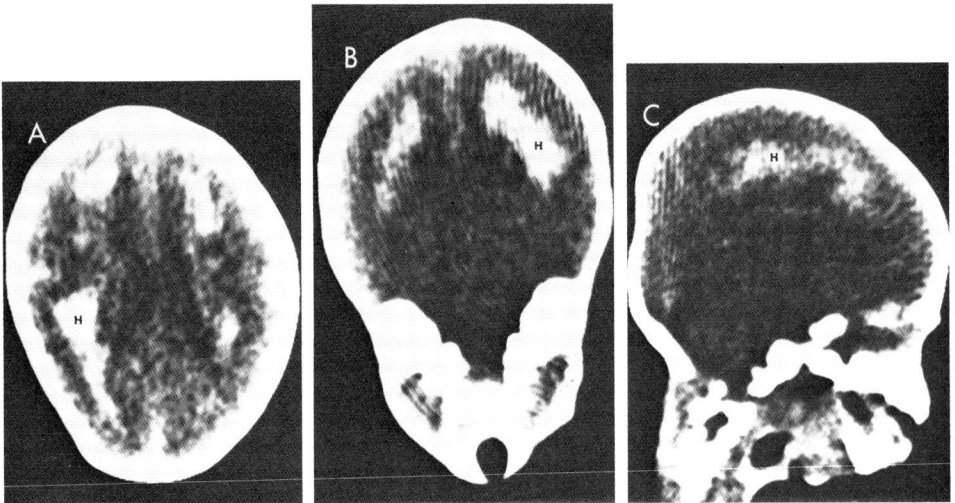

Fig 2–33.—Neonate with periventricular hemorrhage *(H)* best demonstrated on axial view **(A).** Coronal **(B)** and sagittal **(C)** views have linear artifacts from bone but much less than would occur in adult scans.

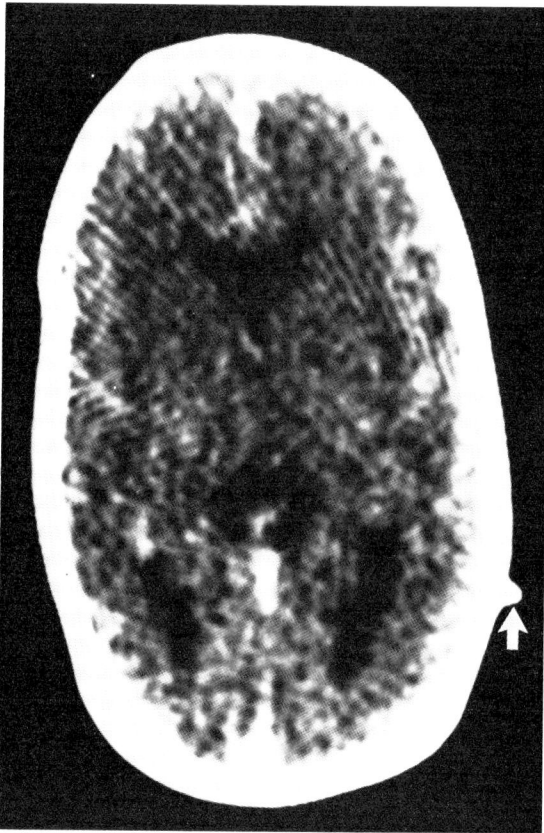

Fig 2–34.—Metal artifact caused by a needle *(arrow)* on the scalp results in diverging lines across the scan.

to protect the skin and then adhesive tape for firm positioning may eliminate motion artifacts, but occasionally sedation is necessary. Adjusting restraints and repositioning during the procedure may be necessary to ensure a good scan. If motion can be stopped, there are generally fewer artifacts from bones due to the thinner calvarium than one might see in an adult scan (Fig 2–33). Sagittal and coronal CT scans have more bone artifact than axial scans but can be very valuable.

Artifacts can result from the monitoring devices and particularly the endotracheal tube attachments, which are frequently metal. IV catheters should be avoided on the scalp, although this is a common site for catheter placement in neonates. There will not always be an artifact, but a spray of lines will result if heavy plastic or metal is scanned on present equipment, although there has been marked improvement since the initial scanners (Fig 2–34).

Asymmetry is best evaluated early in scan monitoring by using bony landmarks at the base of the skull (Fig 2–35).

Ventricular size may appear slightly smaller on CT than on ultrasound due to averaging of brain into the ventricular wall. This is not a technical artifact that can be changed but a basic problem in scanning.

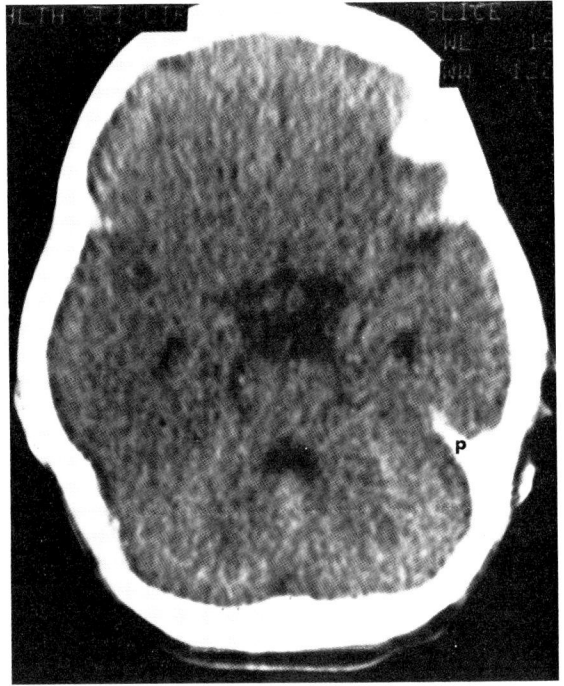

Fig 2–35.—The petrous pyramid and roof of the orbit are seen on the left and not on the patient's right, indicating asymmetry.

Radiation Dose

Comparison of radiation dose for CT scanners has been reported by Brasch and Cann.[27] With fast scanners and good reconstructions, the radiation dose can be minimized and is generally equivalent in dose to a skull series. The doses vary from .47 rad to 3.64 rad in the central plane of section for a standard series of 10 axial CT scans, depending on the actual scanner. The CT images in this book were obtained on AS&E 0450 and GE 8800 scanners.

REFERENCES

1. Anderson R.E., Radmehr A., Osborn A.G., et al., Impact of a "fast" scanner on image quality in pediatric computed tomography. *Radiology* 134:251-252, 1980.
2. Mitchel A.A., Louik C., Lacouture P., et al.: Risks to children from computed tomographic scan premedication. *JAMA* 247:2385-2388, 1982.
3. Gooding C.A., Berdon W.E., Brodeur A.E., et al.: Adverse reactions to intravenous pyelography in children. *AJR* 123:802-804, 1975.
4. Baker D.H., Berdon W.E.: The use of safety of "high" dosage in pediatric urography: A survey of the Society for Pediatric Radiology. *Radiology* 103:371-373, 1972.
5. Wood B.P., Smith W.L.: Pulmonary edema in infants following injection of contrast media for urography. *Radiology* 139:377-379, 1981.
6. Scott W.R.: Seizures: A reaction to contrast media for computed tomography of the brain. *Radiology* 137:359-361, 1980.
7. Shehadi W.H., Toniolo G.: Adverse reactions to contrast media: A report from the Committee on Safety of Contrast Media to the International Society of Radiology. *Radiology* 137:299-302, 1980.

8. Fischer H.W.: Occurrence of seizure during cranial computed tomography. *Radiology* 137:563-64, 1980.
9. McDicken W.N.: *Diagnostic Ultrasonics: Principles and Use of Instruments.*, ed. 2. New York, John Wiley & Sons, 1981.
10. Wells P.N.T.: *Biomedical Ultrasonics.* New York, Academic Press, 1977.
11. Fullerton G.D., Zagzebski J.A. (eds): *Medical Physics of CT and Ultrasound: Tissue Imaging and Characterization.* AAPM Monograph No. 6. New York, American Institute of Physics, 1980.
12. Birhnolz J.: Physical foundations of ultrasound imaging of intraventricular hemorrhage. Presented at the Perinatal Intracranial Hemorrhage Conference, Washington, D.C., Dec., 1980.
13. Kossoff G., Garrett W.J., Radananovich G.: Ultrasonic atlas of normal brain of the infant. *Ultrasound Med. Biol.* 1:259, 1974.
14. Johnson M.L., Mack L.A., Rumack C.M., et al.: B-mode echoencephalography in the normal and high risk infant. *AJR* 133:375-381, 1979.
15. Johnson M.L., Rumack C.M.: Ultrasonic Evaluation of the Neonatal Brain. *Radiol. Clin. North Am.* 18: 117:131, 1980.
16. Babcock D.S., Han B.K., Le Quesne G.W.: B-mode gray-scale ultrasound of the head in the newborn and young infant. *AJR* 134:457, 1980.
17. Ben-Ora A., Eddy L., Hatch G., et al.: The anterior fontanelle as an acoustic window to the neonatal ventricular system. *JCU* 8:65, 1980.
18. Dewbury K.C., Aluwihare A.P.R.: The anterior fontanelle as an ultrasound window for the study of the brain: A preliminary report. *Br. J. Radiol.* 53:81, 1980.
19. Lipscomb A.P., Blackwell R.J., Reynolds E.O.R., et al.: Ultrasound scanning of brain through the anterior fontanelle of newborn infants. *Lancet* 2:39, 1979.
20. Haber K., Watcher R.D., Christenson R.C., et al.: Ultrasonic evaluation of intracranial pathology in infants: A new technique. *Radiology* 134:173, 1980.
21. Dubbins P.A., Goldberg B.B.: Automated ultrasound, in Haller J.O., Shkolnih A. (eds.): *Clinics in Diagnostic Ultrasound.* New York, Churchill-Livingstone, 1981, pp. 231-246.
22. Hendee W.R., *Physical Principles of Computed Tomography.* Boston, Little, Brown & Co., 1983.
23. Newton T.H., Potts D.G., (eds.): *Radiology of the Skull and Brain.* Vol. 5: *Technical Aspects of Computed Tomography.* St. Louis, C.V. Mosby Co., 1981.
24. Byrd S.E., Harwood-Nash D.H., Barry J.F., et al.: Coronal computed tomography of the skull and brain in infants and children: Parts I and II. *Radiology* 124:705-714, 1977.
25. Anderson R.E., Roehler P.R.: An accessory patient table for multidirectional scanning. *Radiology* 130:802-803, 1979.
26. Matsui T., Hirano A.: *An Atlas of the Human Brain for Computed Tomography.* New York, Igaku-Shoin, 1978.
27. Brasch R.C., Cann C.E.: Computed tomographic scanning in children: II. An updated comparison of radiation dose and resolving power of commercial scanners. *AJR* 138:127-133, 1982.

CHAPTER 3

The Neonatal Brain: Normal Appearances

Laurence A. Mack, M.D.*
Ellsworth C. Alvord, Jr., M.D.†
Dale R. Cyr, R.D.M.S.*
A. George F. Aitken, M.D.*
Thomas E. Richardson, M.D.*

SONOGRAPHIC IMAGING of the neonatal brain has evolved in conjunction with innovations in sonographic instrumentation. Initial work with articulated arm and water path scanners produced images in parallel planes of section approximating the anatomical view provided by computerized tomographic (CT) scans.[1,2] The introduction of mechanical sector scanners allowed use of the anterior fontanelle acoustic window for production of a new neuroanatomical view of the brain with radially oriented coronal and angled sagittal images.[3-5] Interpretation of these images, unique in radiology, requires a thorough understanding of the three-dimensional relationships of the ventricular system and adjacent structures. This chapter considers these structures in detail and their relationships in coronal, angled sagittal, and axial projections. CT appearances in the axial projection are also described.

VENTRICULAR SYSTEM

With its easily identifiable, reliable landmarks, the ventricular system assists in identification of adjacent brain structures similar to the way in which arterial and venous structures in the upper abdomen assist in identification of adjacent parenchymal organs (Fig 3–1).

The lateral ventricles are the largest of the CSF-filled cavities. Each lateral ventricle has been divided arbitrarily into four segments: frontal horn, body, occipital horn and temporal horn.

The foramen of Monro divides the frontal horn anteriorly from the body of the

*Section of Ultrasound, Department of Radiology, University of Washington School of Medicine.
†Section of Neuropathology, Department of Pathology, University of Washington School of Medicine.

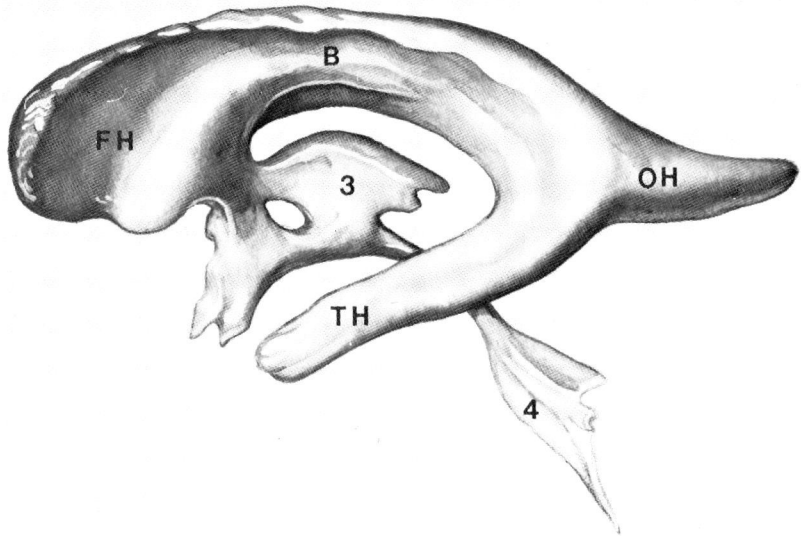

Fig 3–1.—Lateral view of the ventricular system demonstrating the frontal horn *(FH)*, body *(B)*, occipital horn *(OH)*, and temporal horn *(TH)* of the lateral ventricle. Also seen are the third ventricle *(3)*, aqueduct of Sylvius *(AS)*, and fourth ventricle *(4)*. (From Mack and Alvord.[5] Reproduced with permission.)

ventricle posteriorly. When viewed in cross section, the frontal horn is triangular with a concave lateral wall. The roof of the ventricle is formed by the corpus callosum, the medial wall is formed by the septum pellucidum, and the lateral wall is formed by the head of the caudate nucleus.

The body of the lateral ventricle extends from the foramen of Monro back to the trigone, formed by the junction with the temporal and occipital horns posteriorly. Viewed in cross section, this portion of the lateral ventricle is crescentic with two concavities in the outer margin. The roof continues to be formed by the corpus callosum and the medial wall by the septum pellucidum. The thalamus abuts the inferior concavity of the lateral ventricular surface and the body of the caudate nucleus borders the superior concavity. The notch between these structures is occupied by the echogenic choroid plexus posteriorly and the thalamostriate vein anteriorly.

The temporal horn extends anteriorly from the trigone through the temporal lobe as an arcuate structure with its concavity directed medially and inferiorly. The roof of this horn of the ventricle is formed by the temporal lobe white matter and by the tail of the caudate nucleus. The medial wall is formed by the hippocampus.

The occipital horn extends posteriorly from the trigone as a pyramid-shaped structure. The medial wall is formed by the occipital cortex and white matter. The proximal roof and lateral wall are formed by the corpus callosum.

The third ventricle is connected to the lateral ventricles via the foramen of Monro. The lateral walls of this slit-like structure are formed by the thalami superiorly and the hypothalami inferiorly. The anterior margin is bordered by the lamina terminalis and anterior commissure. The floor of the third ventricle is formed by the hypothalmus and optic chiasm. Extending from the anteroinferior aspect of the ventricle are two recesses: the supraoptic and infundibular. The pineal recess can be visualized at the posterosuperior margin of the ventricle, especially if the ventricle is dilated. The third ventricle is usually bridged by a soft tissue structure, the massa intermedia, best visualized in the presence of ventricular dilatation.

The aqueduct of Sylvius connects the third and fourth ventricles. It is rarely imaged except in the presence of massive ventricular enlargement. The fourth ventricle, a thin but broad structure, has as its floor the medulla oblongata. The cerebellar vermis and the posterior medullary vellum form the roof of the ventricle.

CAUDATE NUCLEUS

The caudate nucleus is divided into three sections (Fig 3–2). The largest and most anterior section is the head of the caudate nucleus, which lies in the concavity of the lateral surface of the frontal horn of the lateral ventricle. The caudate nucleus tapers as it extends posteriorly to become the body, with the division between the head and body occurring at the level of the foramen of Monro. The body of the caudate nucleus occupies the superior concavity of the lateral wall of the body of the lateral ventricle as it sweeps posteriorly. The body of the caudate nucleus blends into the tail as it extends inferiorly and then anteriorly in the roof of the temporal horn of the ventricle. This portion of the caudate nucleus is rarely imaged in ultrasound or CT examinations.

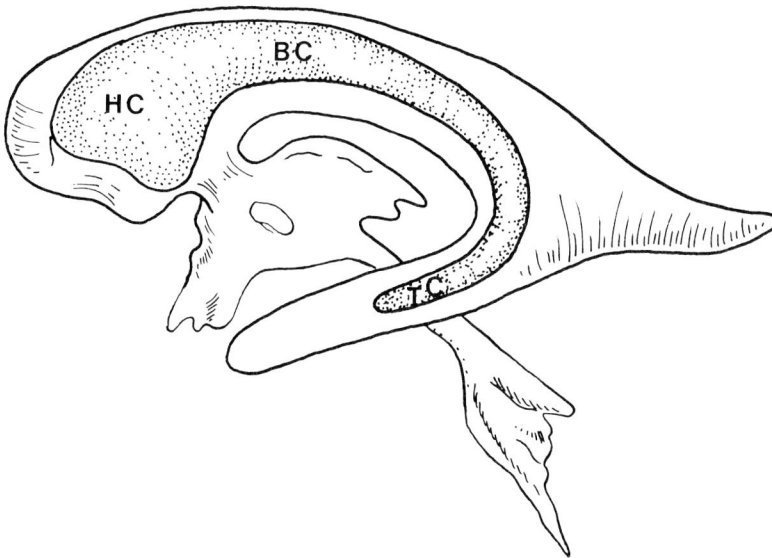

Fig 3–2.—The caudate nucleus fitting into the lateral concavity of the lateral ventricle. *HC,* head of caudate nucleus; *BC,* body of caudate; *TC,* tail. (From Mack and Alvord.[5] Reproduced with permission.)

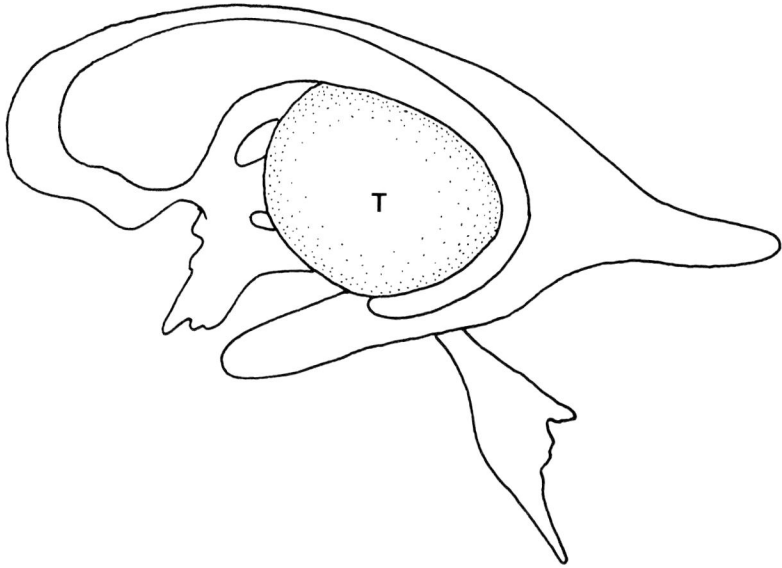

Fig 3–3.—Thalamus *(T)* in relationship to the lateral ventricle. (From Mack and Alvord.[5] Reproduced with permission.)

THALAMUS

The thalami are paired, egg-shaped structures that lie on either side of the third ventricle (Fig 3–3). They occupy the inferior concavity of the lateral surface of the ventricle. Anteriorly they are bounded by the foramen of Monro and columns of the fornices, as well as by the head of the caudate; posteriorly they are bounded by the trigone of the ventricle, and inferiorly they are bounded by the hypothalamus.

CAVA SEPTI PELLUCIDI ET VERGAE

The cava septi pellucidi and Vergae lie between the frontal horns and bodies of the two lateral ventricles (Fig 3–4). The fornix is a landmark arbitrarily dividing this single structure into the cavum septi pellucidi anteriorly and the cavum Vergae posteriorly. During the 6th month of gestation, the cavum Vergae begins to close from posterior to anterior. This closure progresses to complete obliteration of the cavum septi pellucidi in 85% of infants by age 2 months.[6] The cava septi pellucidi and Vergae normally do not connect with either subarachnoid or ventricular fluid. Care must be taken not to confuse either the cavum septi pellucidi with the interhemispheric fissure or the cavum Vergae with the third ventricle or with the cavum velli interpositi (CVI). The CVI, occasionally seen in conjunction with hydrocephalus, is triangular and lies between the cavum Vergae (or the fornices if the cavum Vergae is obliterated) and the roof of the third ventricle.

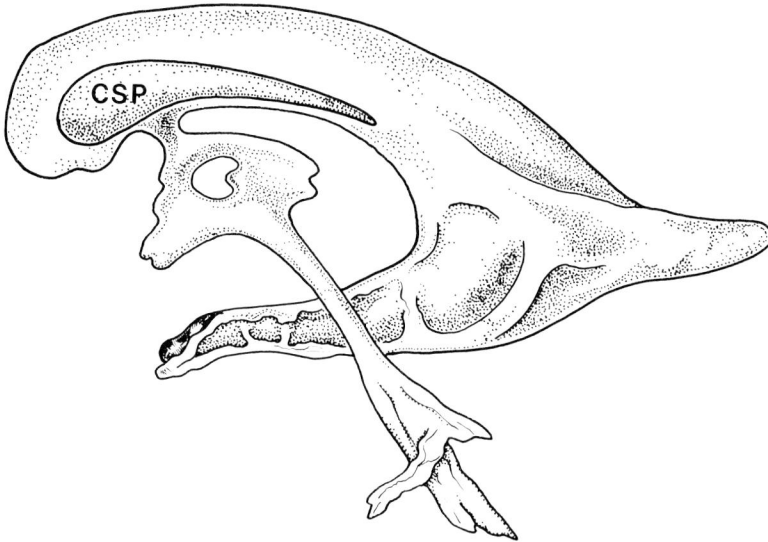

Fig 3–4.—The cava septi pellucidi and Vergae *(CSP)* as it projects on the medial surface of the lateral ventricles.

FORNIX

The fornix has been divided into three sections: the crura, the bodies, and the anterior columns (Fig 3–5). The crura starts posteriorly as a continuation of the fimbria of the hippocampus in the medial part of the temporal lobe, where they form part of the floor and medial wall of the temporal horn of the lateral ventricle.

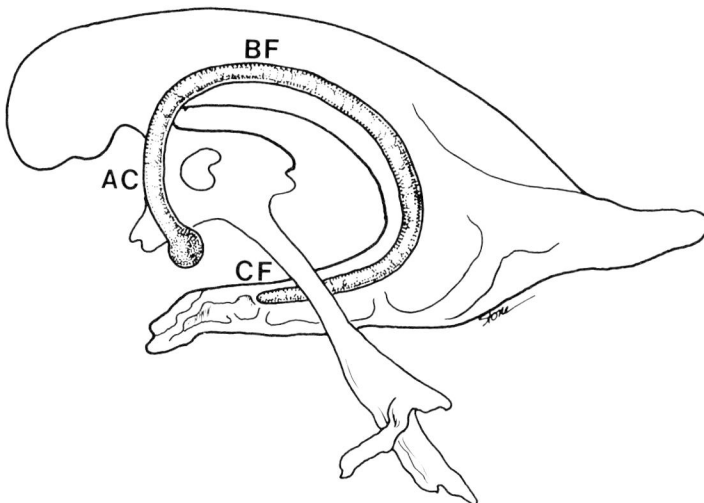

Fig 3–5.—The fornix as it relates to the medial surface of the lateral ventricle. *AC,* anterior column of fornix; *BF,* body of fornix; *CF,* crura of fornix.

The crura arch up over the thalami to form the hippocampal commissure (psathrium) and then the bodies of the fornices. The bodies touch each other in the midline, forming the floor of the cavum Vergae and the medial aspect of the floor of the body of the lateral ventricle. Anteriorly the bodies curve sharply inferiorly as columns which diverge slightly to enter the hypothalamus, ending in the mammillary bodies. The columns form the anterior boundary of the interventricular foramen of Monro and lie just behind the anterior wall of the third ventricle.

CHOROID PLEXUS

The choroid plexus of the lateral ventricle runs posteriorly from the tip of the temporal horn to arch around the thalami, then anteriorly in the floor of the body of the lateral ventricle, and to the foramen of Monro (Fig 3–6). It then passes through the foramen and with its counterpart runs posteriorly in the roof of the third ventricle to the suprapineal recess. In the temporal horn the plexus adheres to its roof medially and is thin, but as it passes around the thalami, in the region of the trigone opposite the occipital horn, it thickens to form the glomus. From the glomus, as it passes into the body of the lateral ventricle, it becomes thin again. At no time does the choroid plexus pass into the frontal or occipital horn.

GERMINAL MATRIX

The germinal matrix cannot be imaged by either ultrasound or CT (Fig 3–7). However, because the germinal matrix is the site of the vast majority of subependymal hemorrhages, understanding the location and evolution of this structure is important.

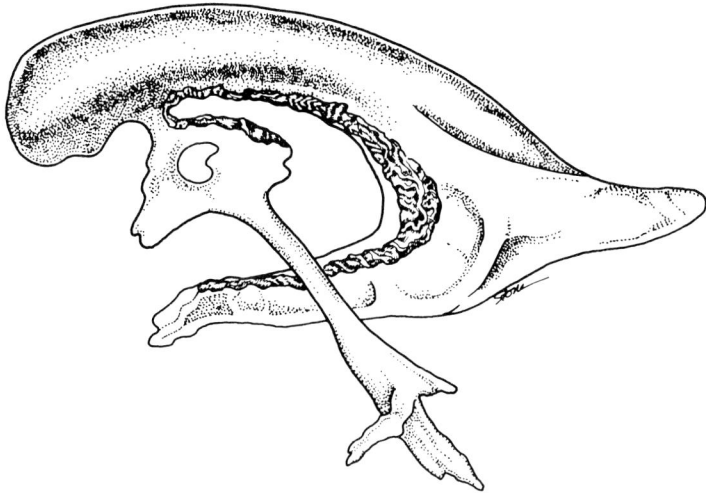

Fig 3–6.—The choroid plexus as it courses through the lateral and third ventricle.

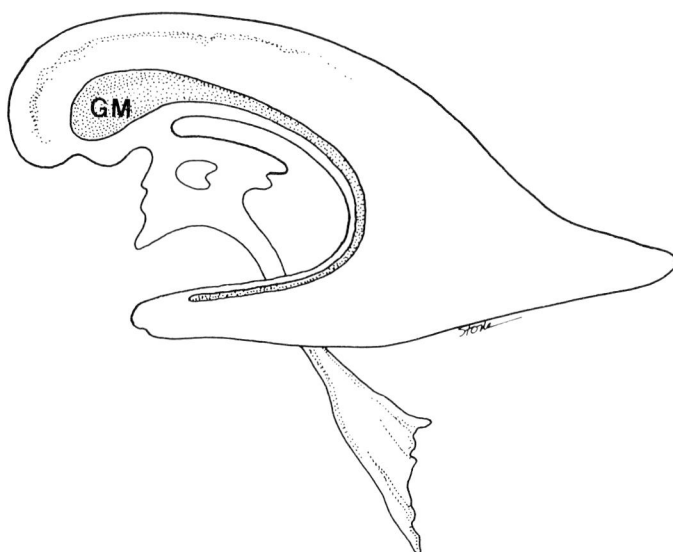

Fig 3–7.—The remnant of the germinal matrix *(GM)* projected over the medial surface of the lateral ventricle at 30–32 weeks' gestation. The largest area is in the region of the head of the caudate nucleus, with smaller remnants adjacent to the body and tail.

The germinal matrix develops deep to the ependyma and consists of loosely organized, proliferating cells that give rise to the neurons and glia of the cerebral cortex and basal ganglia. Its vascular bed is the most richly perfused region of the developing brain. Vessels in this region form an immature vascular rete of fine capillaries, extremely thin-walled veins, and larger irregular vessels which cannot be classified by light microscopy.[7] The capillary network is best developed on the periphery of the germinal matrix and becomes less well developed toward the central glioblastic mass.

Early in gestation, the germinal matrix forms the entire wall of the ventricular system. After the third month of gestation the germinal matrix regresses, first around the third ventricle, then around the temporal and occipital horns and trigone. By 6 months' gestation, the germinal matrix persists only over the head of the caudate nucleus and to a lesser extent over the body of the caudate. This regression continues until by birth it ceases to exist as a discrete structure and the immature vascular rete has been remodeled to form adult vascular patterns.[8, 9]

CROSS-SECTIONAL IMAGES

The coronal sections described in the following section are those which would be obtained by imaging through the anterior fontanelle. Each ultrasound section is correlated with a gross anatomical section and a labeled anatomical drawing.

Scan Through the Anterior Frontal Horn

At this level, the frontal horns are seen as arcuate fluid-filled spaces separated, in part, by the cavum septi pellucidi (Fig 3–8). The echogenic head of the caudate nucleus nestles in the lateral ventricular concavity. Lateral and inferior is a second echogenic structure, the putamen, which is separated from the caudate nucleus by a linear, less echogenic area, the anterior limb of the internal capsule. Sufficient sound is transmitted through the thin orbital roof to outline the margins of the orbits.

Scan Through the Posterior Frontal Horn

The frontal horns of the lateral ventricles continue to be seen as crescentic fluid-filled spaces, with the caudate nucleus filling the lateral concavity (Fig 3–9). The

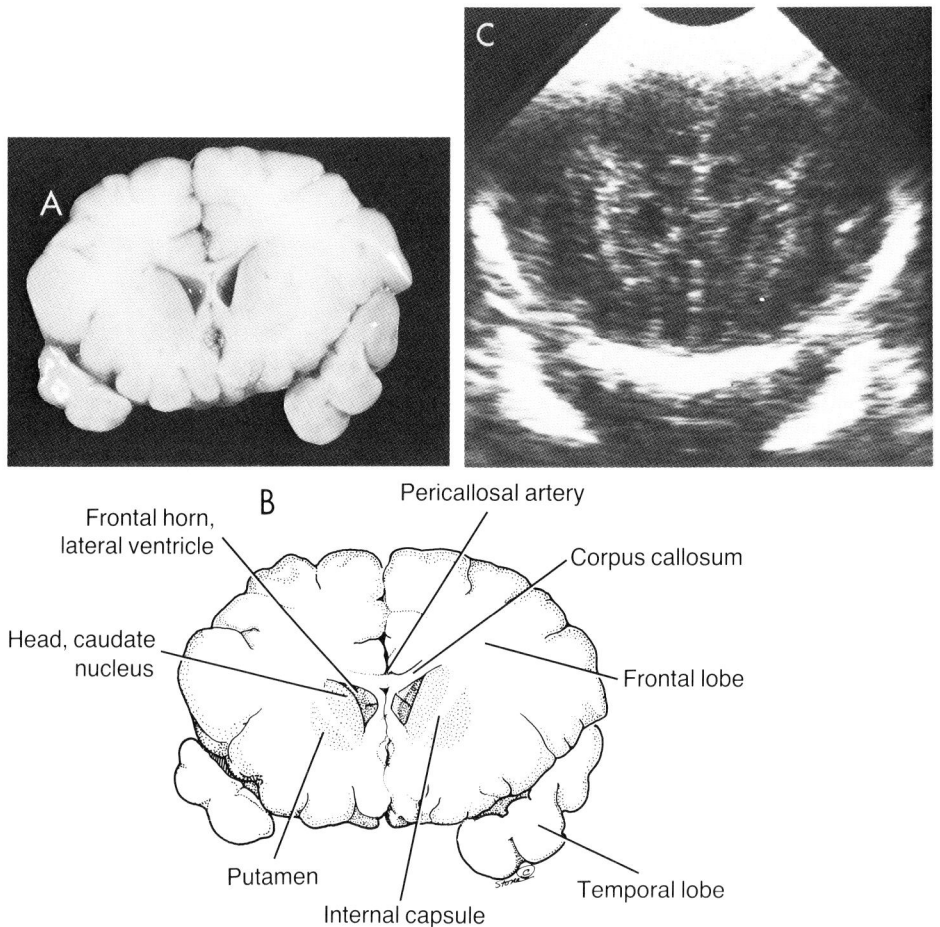

Fig 3–8.—Coronal scan through the anterior frontal horn of the lateral ventricle. **A,** Anatomical specimen; **B,** Anatomical drawing; **C,** Sonogram.

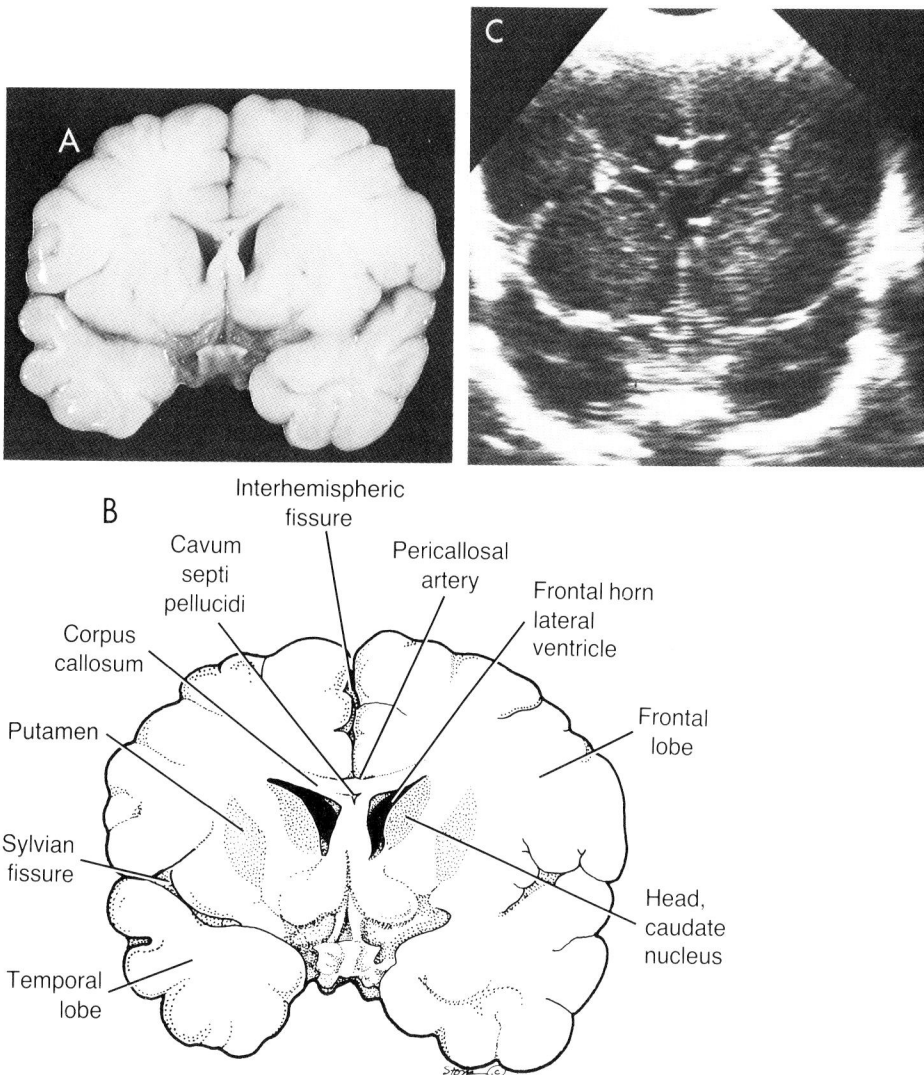

Fig 3–9.—Coronal scan through the posterior frontal horn. **A,** Anatomical specimen; **B,** Anatomical drawing; **C,** Sonogram.

corpus callosum forms the roof of the lateral ventricle as a horizontal echopenic (very few echoes) structure with its rostral margin defined by the pulsations of the pericallosal artery. The putamen is seen laterally and inferiorly to the caudate nucleus as an echogenic structure. The parenchyma of the frontal and temporal lobes is visualized with relatively low echogenicity. The anterior clinoids and planum sphenoidale are seen as highly echogenic structures. Centrally, pulsations from the internal carotid artery may be seen on real-time examination, and at this level pulsations of the anterior and middle cerebral arteries may also be detected in the interhemispheric and sylvian fissures, respectively.

Scan Through the Foramen of Monro

The lateral ventricles are seen connecting to the slit-like third ventricle via the foramen of Monro (Fig 3–10). When normal in size, the third ventricle may be difficult to image in this projection. The cavum septi pellucidi remains as a triangular space between the lateral ventricles. Just inferior to the cavum septi pellucidi the columns of the fornix may be seen as oval, soft tissue–density structures. The lateral recess of the body of the lateral ventricle contains the echogenic body of the caudate nucleus. At the lateral aspect of the lateral ventricle, the periventricular veins and arteries are seen in the periventricular white matter. The middle cerebral

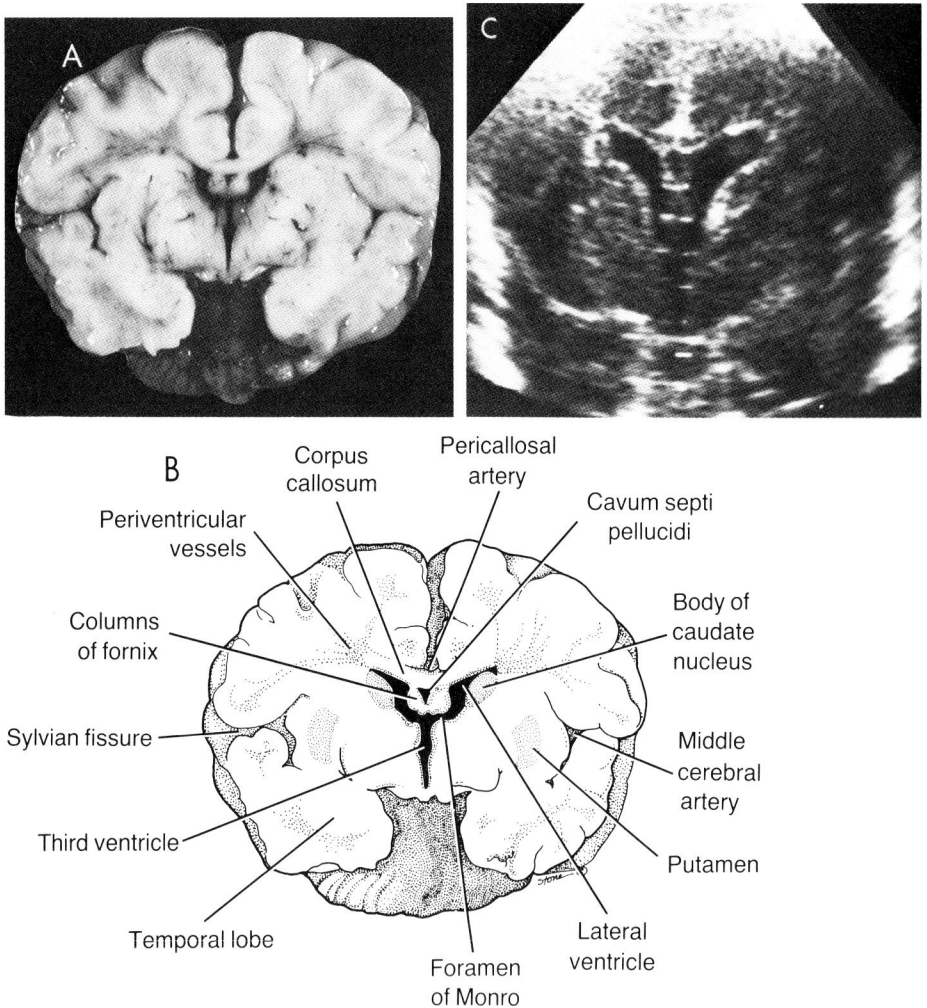

Fig 3–10.—Coronal scan through the foramen of Monro. **A,** Anatomical specimen; **B,** Anatomical drawing; **C,** Sonogram. (Part **A,** from Shuman et al.[3]; reproduced by permission. Part **C,** from Mack and Alvord[5]; reproduced with permission.)

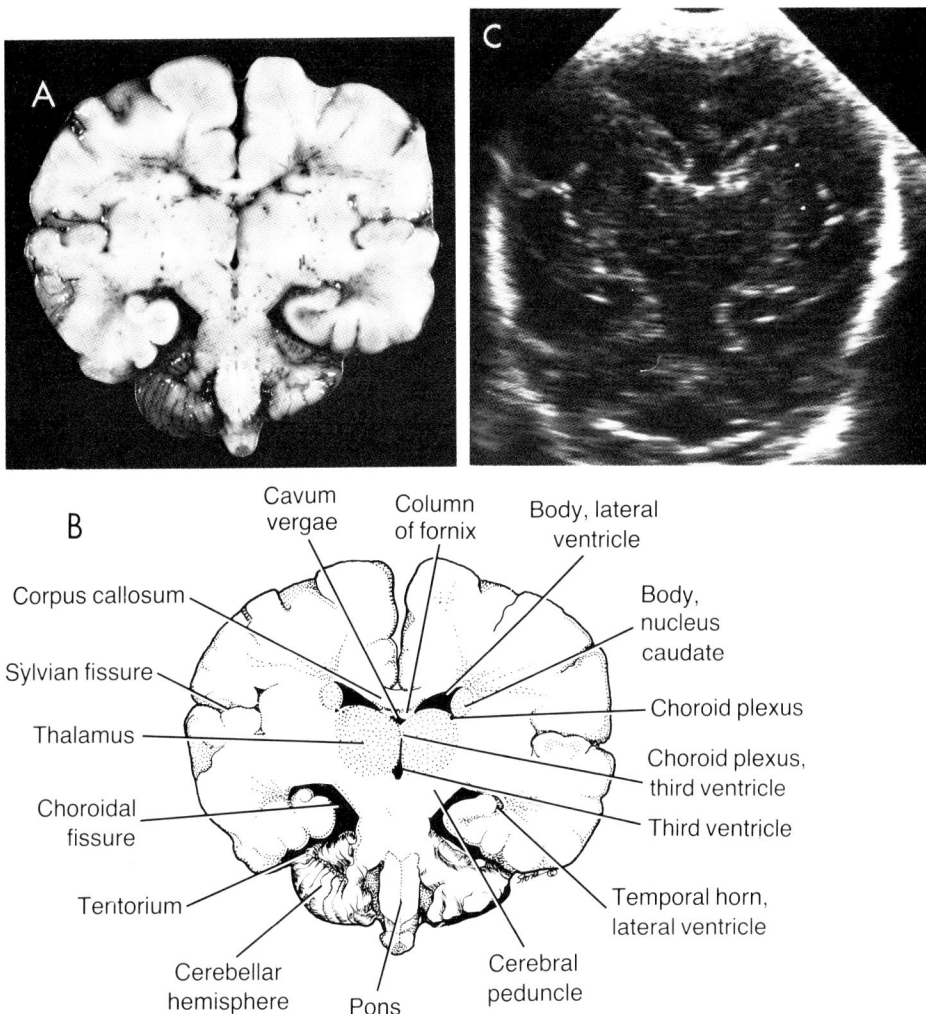

Fig 3–11.—Coronal scan through thalami. **A,** Anatomical specimen; **B,** Anatomical drawing; **C,** Sonogram. (Part **A,** from Mack and Alvord.[5] Reproduced with permission.)

artery is seen pulsating in the sylvian fissure, as is the pericallosal artery in the interhemispheric fissure at this level.

Scan Through the Thalami

In this plane, the lateral ventricles are seen as arcuate fluid-filled structures, with the body of the caudate nucleus lying laterally in the superior concavity and the thalami lying inferiorly in the lower concavity (Fig 3–11). In the groove between these lies the anterior extent of the choroid plexus. The columns of the fornix may be seen inferior to the corpus callosum and at the medial aspect of the lateral ventricle. The thalami are separated from each other by the slit-like third ventricle.

This portion of the ventricle may be difficult to image, but the echogenic choroid plexus along its roof is often seen as a landmark.

The cerebral peduncles are seen inferiorly as a Y-shaped echopenic structure which fuses at the level of the pons. Laterally, the tentorium cerebelli and the choroidal fissure are present as highly echogenic structures. Inferior to the tentorium is the echogenic cerebellum.

Scan Through the Quadrigeminal Cistern

The quadrigeminal cistern and cerebellum appear together as an echogenic tree-shaped structure (Fig 3–12). The superior extent of the quadrigeminal cistern ana-

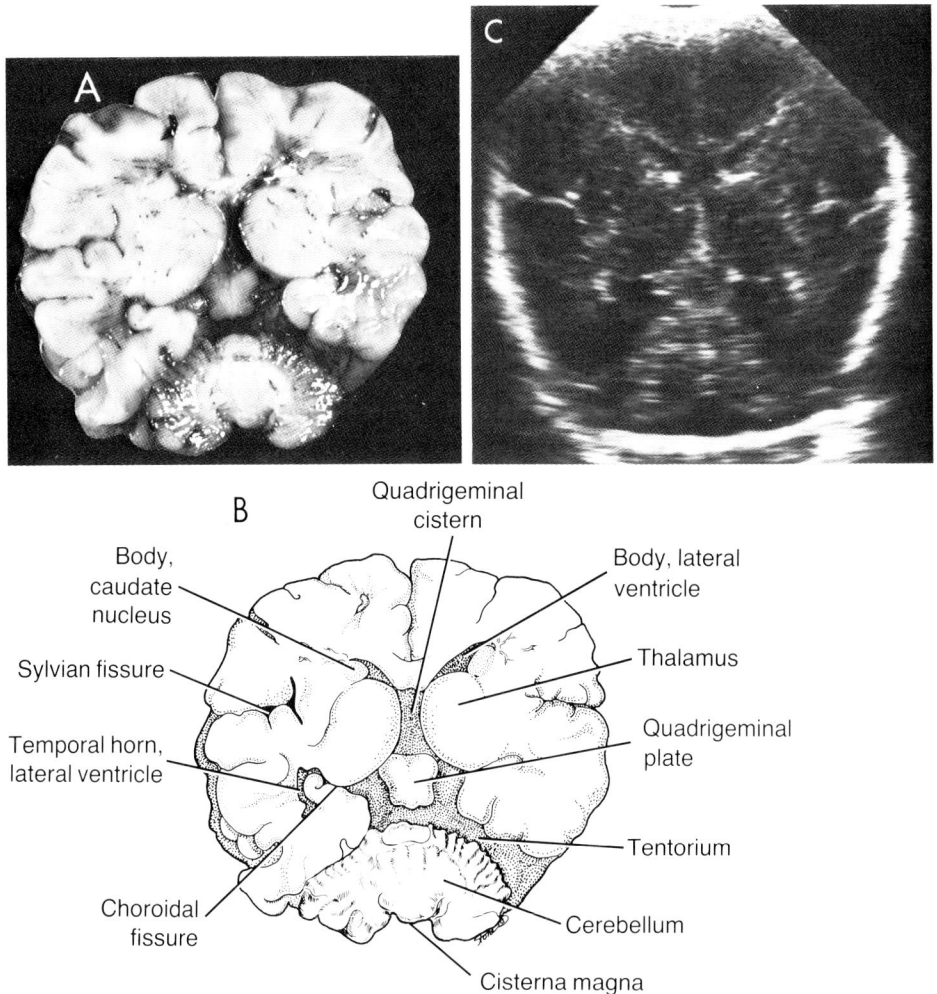

Fig 3–12.—Coronal scan through quadrigeminal cistern. **A,** Anatomical specimen; **B,** Anatomical drawing; **C,** Sonogram. (Part **A,** from Shuman et al.[3] Reproduced with permission.)

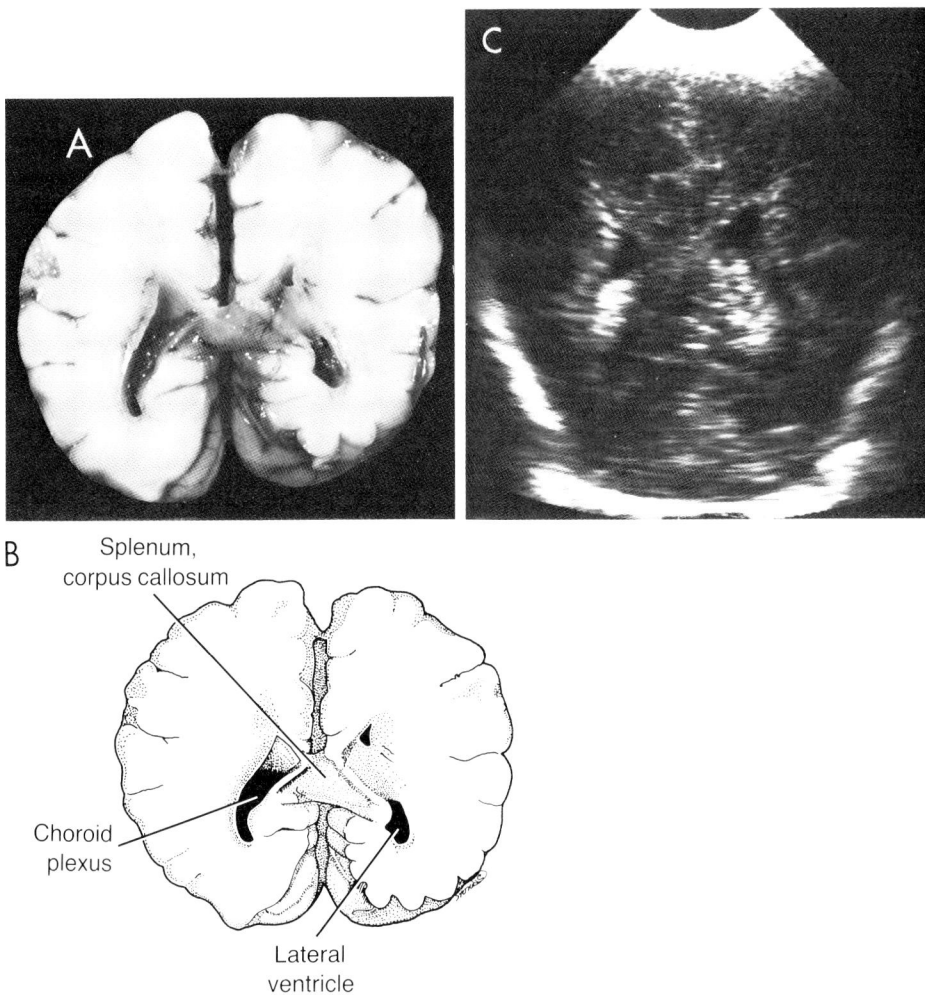

Fig 3–13.—Coronal scan through the trigone and lateral ventricle. **A,** Anatomical specimen; **B,** Anatomical drawing; **C,** Sonogram. (Part **A,** from Shuman et al.[3] Reproduced with permission.)

tomically is the cavum velli interpositi, but the highly echogenic choroid plexus of the third ventricle undoubtedly contributes to this structure on sonograms. Laterally there are extensions into the choroidal fissures and more inferiorly the tentorium. The quadrigeminal cistern contains multiple arachnoid adhesions which are thought to account for its high level of echogenicity.[10] The base of the structure is the echogenic cerebellum, whose level of echogenicity is the same as that of the quadrigeminal cistern. The body of the lateral ventricle is seen superiorly and the temporal horn of the lateral ventricle is seen inferiorly. The vascular structures on this section include the anterior choroidal artery, the middle cerebral artery, and the pericallosal artery in the choroidal, sylvian, and interhemispheric fissures, respectively.

Plane Through the Trigone of the Ventricle

The lateral ventricles are seen as laterally diverging, fluid-filled spaces filled with the highly echogenic choroid plexus (Fig 3–13). The splenium of the corpus callosum is seen between the superior portions of the ventricle. The tentorium forms a V-shaped structure, beneath which is seen the echogenic cerebellum.

SAGITTAL SECTIONS

Midline sagittal images are produced by turning the transducer 90 degrees to the position used to make coronal images. Because the posterior portion of the lateral ventricle is more laterally placed than the anterior portion, the plane of section must be angled to include a greater portion of ventricle as more lateral images are made.

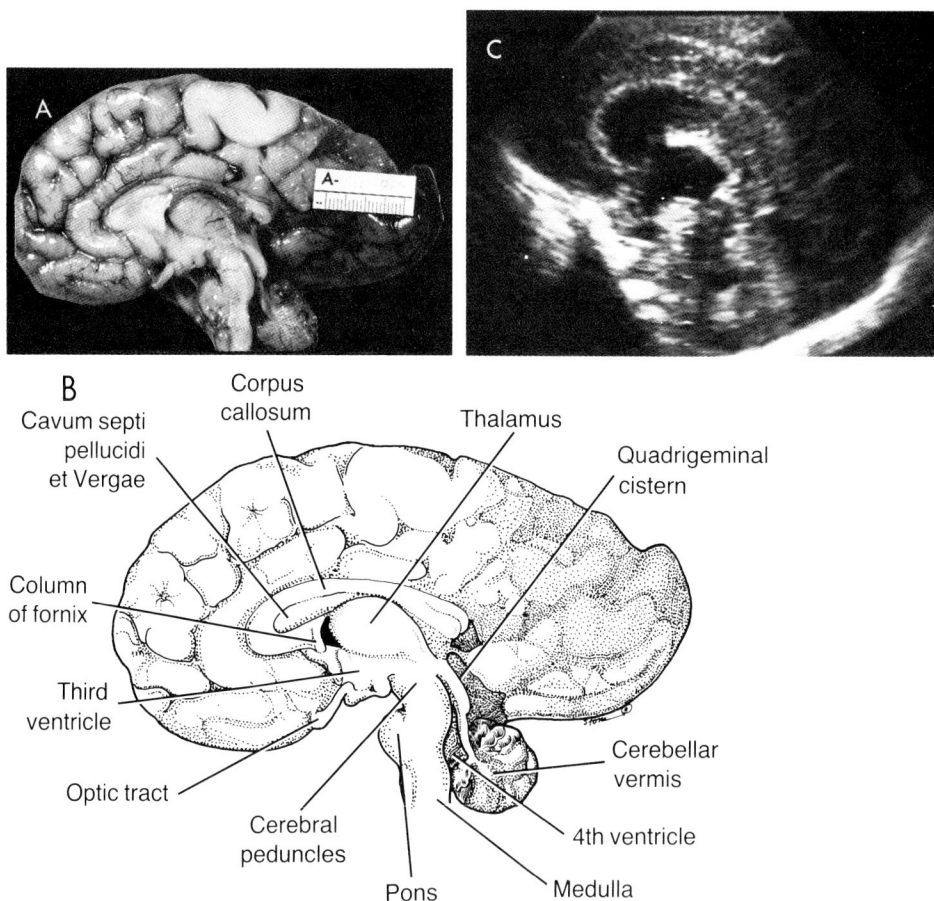

Fig 3–14.—Sagittal midline section. **A,** Anatomical specimen; **B,** Anatomical drawing; **C,** Sonogram. (Part **A,** from Shuman et al.[3]; reproduced with permission. Part **C,** from Mack and Alvord[5]; reproduced with permission.)

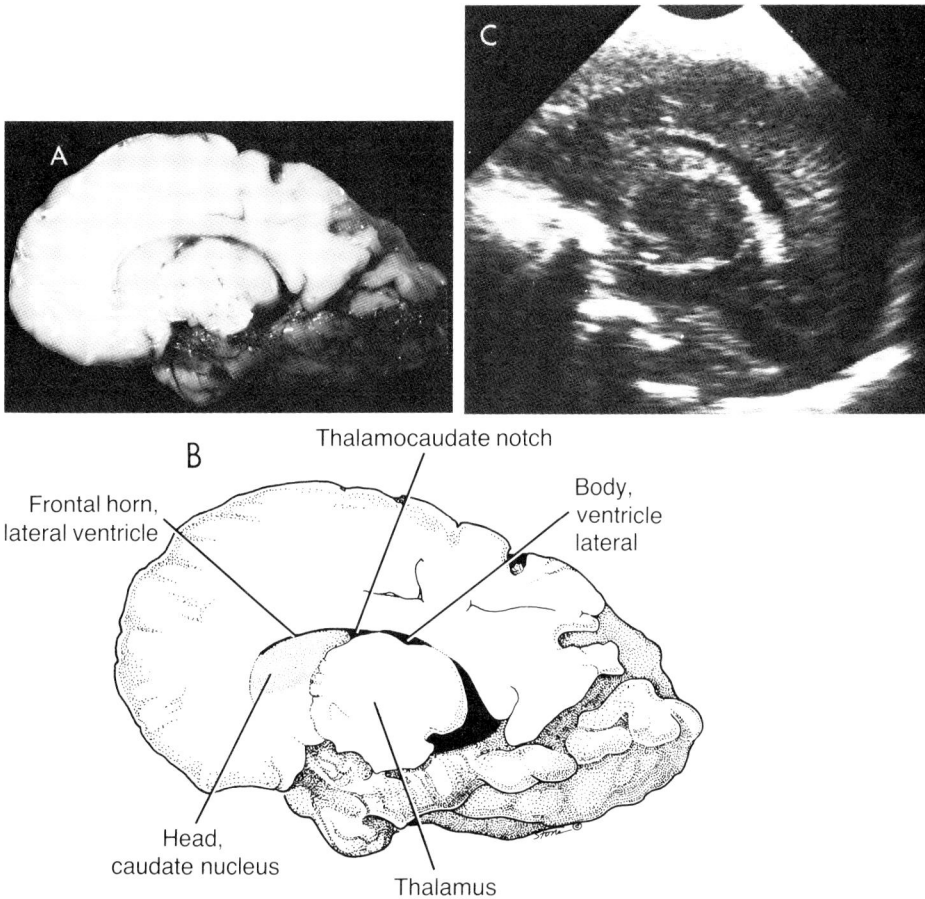

Fig 3–15.—Sagittal scan through the head of the caudate nucleus. **A,** Anatomical specimen; **B,** Anatomical drawing; **C,** Sonogram. (Part **A,** from Shuman et al.[3] Reproduced with permission.)

Midline Sagittal Scan

The cavum septi pellucidi is seen as a comma-shaped fluid-filled structure superior and anterior to the third ventricle (Fig 3–14). The third ventricle can be seen with its recesses, especially if the ventricle is slightly dilated. The massa intermedia is often seen as a soft tissue structure bridging the ventricle. The echogenic choroid plexus of the third ventricle forms its superior margin. The cerebral peduncles, pons, and medulla are seen as hypoechoic structures extending caudally into the posterior fossa. The roof of the fourth ventricle and the echogenic cerebellar vermis are also present on this section.

Sagittal Scan Through the Head of the Caudate Nucleus

By angling the transducer laterally with posterior divergence of the plane of section, an image may be obtained which cuts through the frontal horn and body of the lateral ventricle (Fig 3–15). The echogenic head of the caudate nucleus is seen anteriorly. Posteriorly the thalamus is seen beneath the body of the lateral ventricle. Between these the thalamocaudate notch is seen as an echogenic area containing the anterior extent of the choroid plexus. The bony margins of the middle cranial fossa are also seen.

Sagittal Scan Through the Body of the Lateral Ventricle

By angling the transducer still more laterally, a plane through the body and occipital and temporal horns of the lateral ventricle may be obtained (Fig 3–16). The choroid plexus is seen as an echogenic structure coursing around the thalamus with its greatest thickness in the region of the trigone of the lateral ventricle. The tentorium and echogenic cerebellar hemisphere may also be noted on this section.

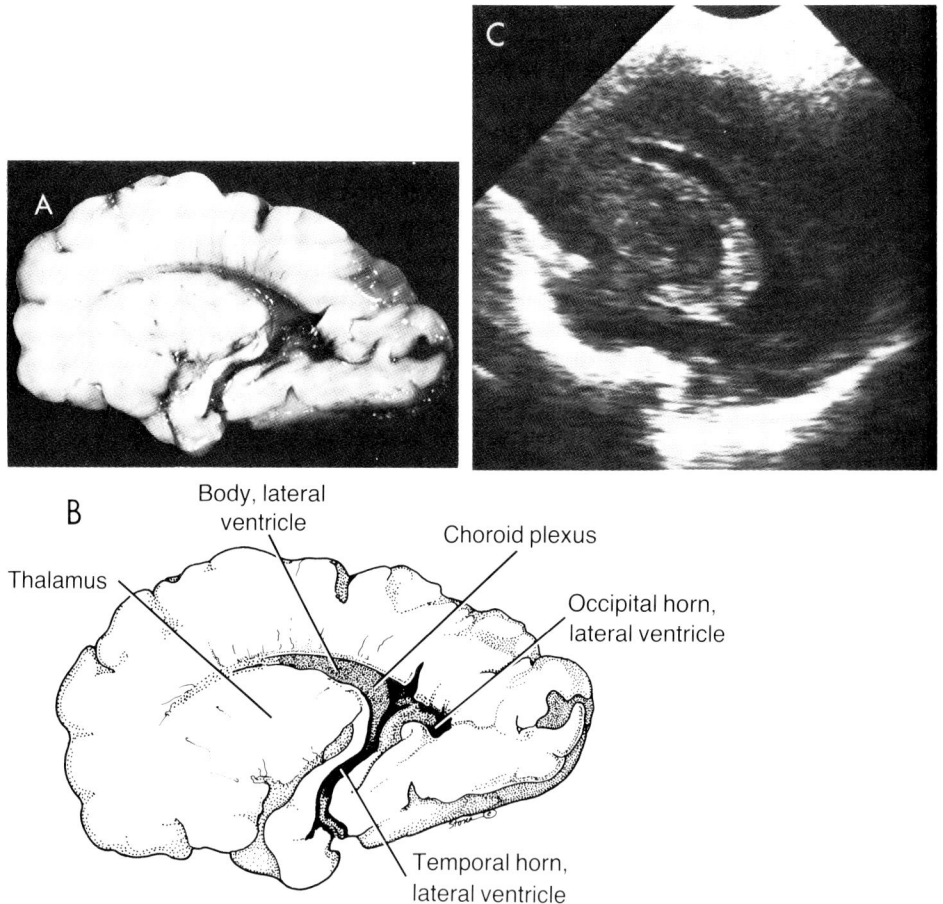

Fig 3–16.—Sagittal scan through the body of the lateral ventricle. A, Anatomical specimen; B, Anatomical drawing; C, Sonogram. (Part A, from Shuman et al.[3] Reproduced with permission.)

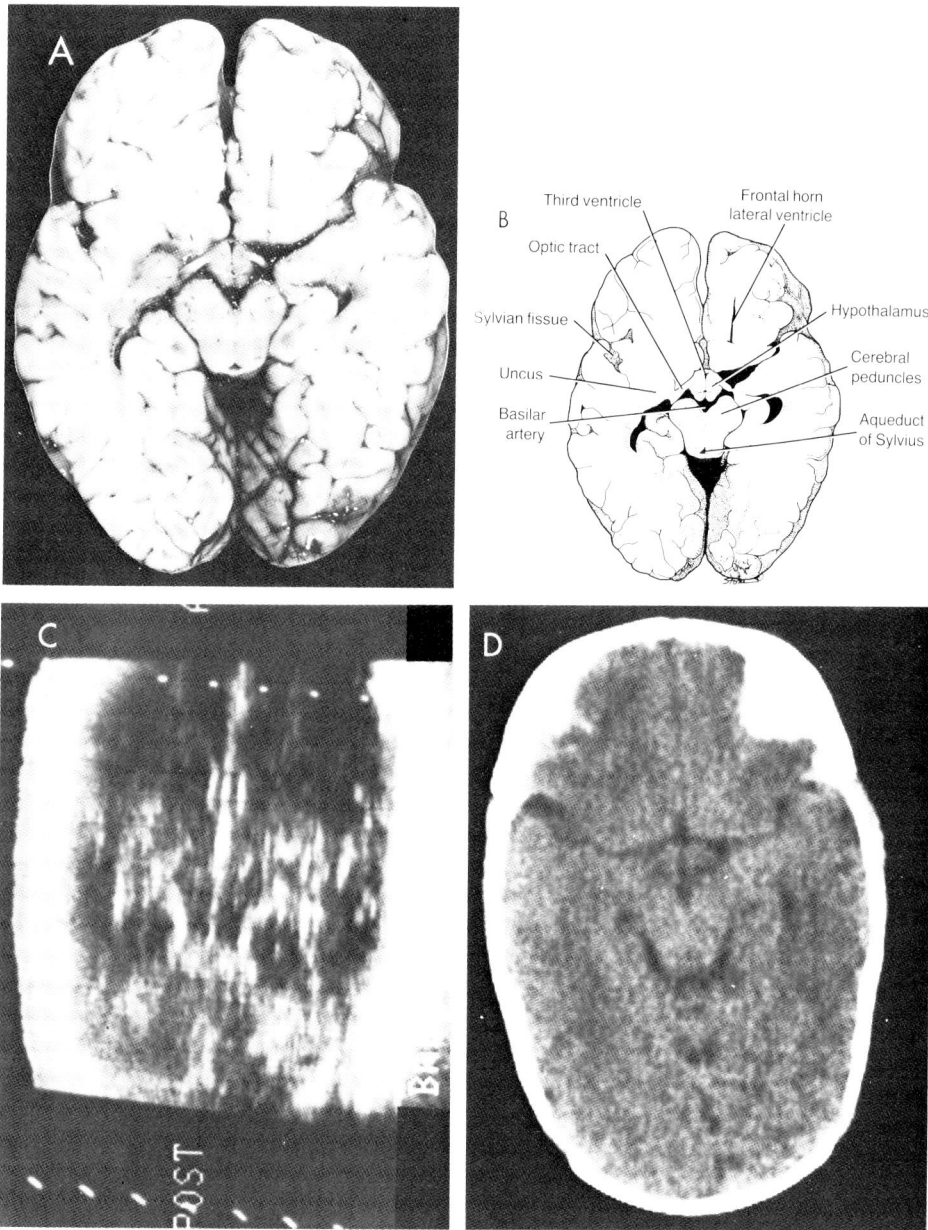

Fig 3–17.—Axial scan through cerebral peduncle. **A,** Anatomical specimen; **B,** Anatomical drawing; **C,** Sonogram; **D,** CT scan. (Part **A,** from Mack and Alvord[5]; reproduced with permission. Part **C,** from Shuman et al.[3]; reproduced with permission.)

AXIAL SECTIONS

Axial images were first obtained using a water path scanner angled 0–10 degrees to the anatomical baseline.[2] Similar images were then obtained using a commercially available articulated arm device.[1] Mechanical sector scanners can produce images in this projection as well, and while such images are more limited in scope, diagnostic information, especially concerning the size and contents of the lateral ventricles, may be obtained.[5] In the following section, axial sonograms are compared with CT images and pathologic specimens at similar anatomical levels.

Axial Scan at the Level of the Upper Midbrain and Cerebral Peduncles

The midbrain at the level of the cerebral peduncles is seen as a central, heart-shaped structure of relatively low echogenicity (Fig 3–17). The basilar artery pulsates in the interpeduncular cistern and the posterior cerebral arteries course around the midbrain. The aqueduct of Sylvius appears as an echogenic structure in the posterior aspect of the midbrain. The inferior, anterior extent of the third ventricle, surrounded by the hypothalmus, is seen anterior to the basilar artery. The sylvian fissures are also visualized on this plane of section with characteristic middle cerebral artery pulsations. Prominent subarachnoid spaces are noted on CT.

Axial Scan at the Level of the Thalami

At this level, the frontal horns of the lateral ventricles are seen anteriorly and the trigone and occipital horns are seen posteriorly (Fig 3–18). The third ventricle is seen as a narrow structure separating the oval thalami. Between the frontal horns of the lateral ventricles, the cavum septi pellucidi and the columns of the fornix are present. Lateral to the frontal horn is the head of the caudate nucleus. Characteristic areas of low attenuation are seen in the periventricular frontal regions bilaterally on CT.

Axial Scan at the Level of the Body of the Lateral Ventricles

At this level, the lateral margins of the body of the lateral ventricles are visualized sonographically as posteriorly diverging lines (Fig 3–19). The medial margin of the ventricle and the centrally placed corpus callosum are not seen on sonograms because of their steep angulation with respect to the transducer. On CT, areas of low attenuation corresponding to periventricular white matter are noted anteriorly and posteriorly.

ACKNOWLEDGMENTS

The authors wish to express their appreciation to Ms. Heidrun Eberhardt, R.D.M.S., for technical assistance, Ms. Jacqueline Gilliam for secretarial support,

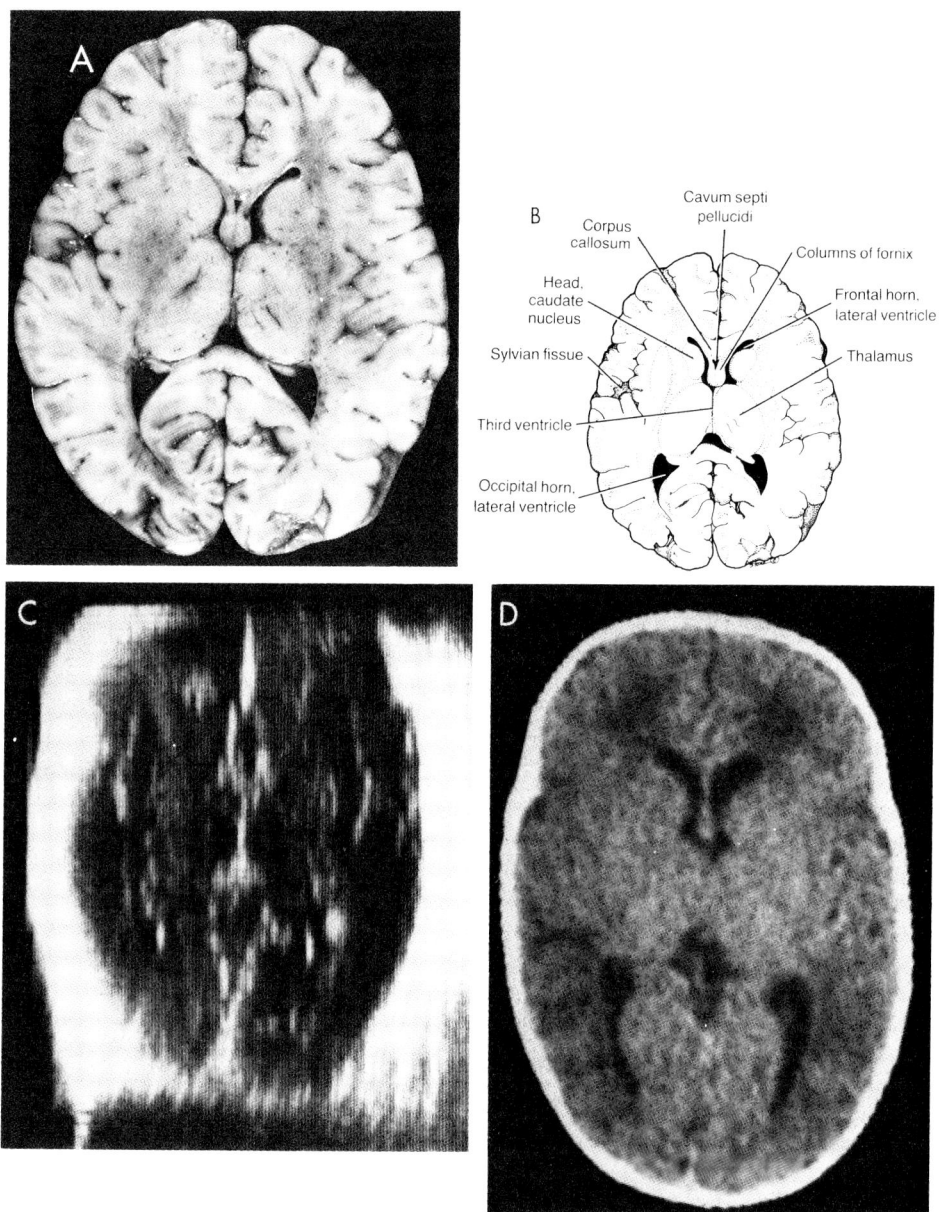

Fig 3–18.—Axial scan at the level of the thalami. **A,** Anatomical specimen; **B,** Anatomical drawing; **C,** Sonogram; **D,** CT scan. (Part **A,** from Mack and Alvord[5]; reproduced with permission. Part **C,** from Shuman et al.[3]; reproduced with permission.)

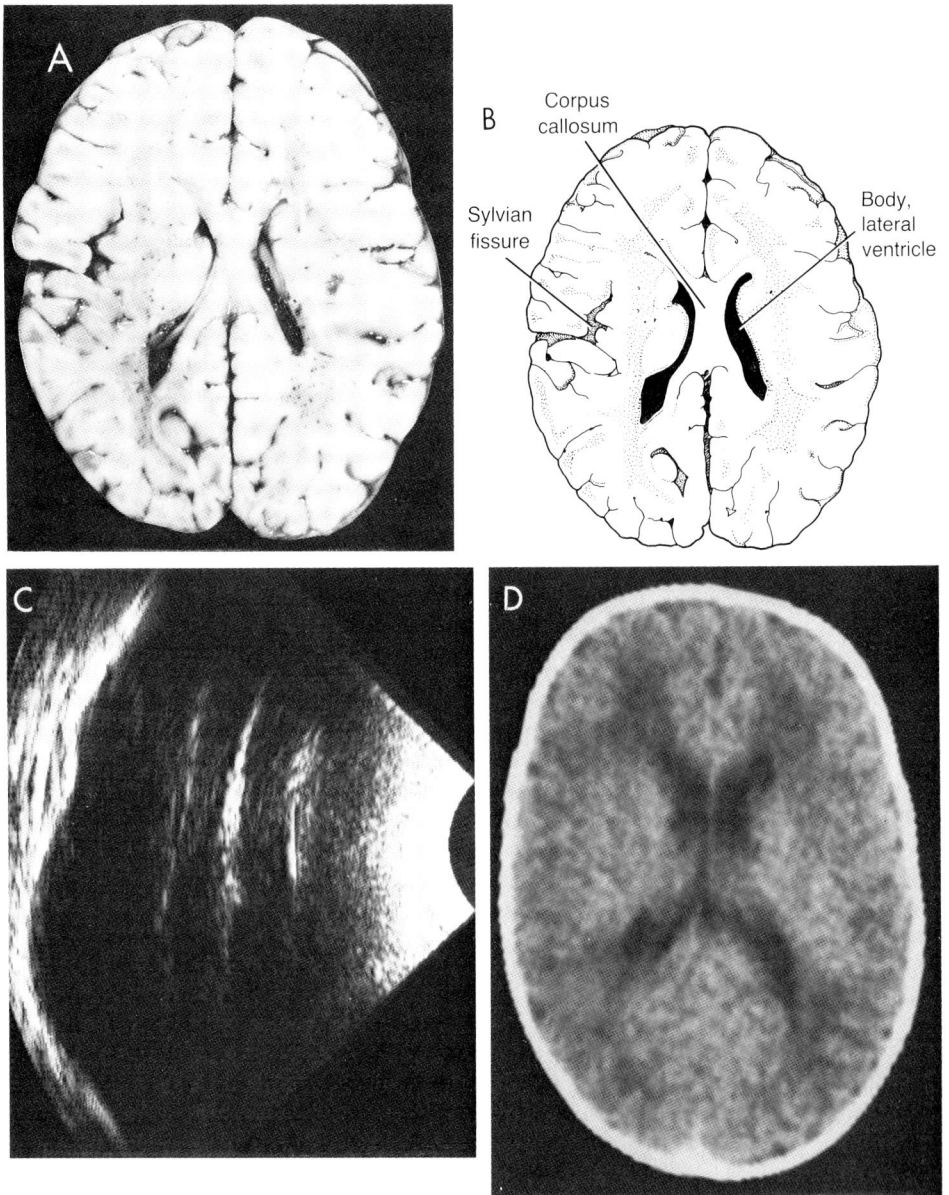

Fig 3–19.—Axial scan at the level of the body of the lateral ventricle. **A,** Anatomical specimen; **B,** Anatomical drawing; **C,** Sonogram; **D,** CT scan. (Part **A,** from Mack and Alvord[5]; reproduced with permission. Part **C,** from Shuman et al.[3]; reproduced with permission.)

and Mr. Joel Weinstein of ATL, Division of Squibb Medical Systems, for technical support.

REFERENCES

1. Johnson M.L., Mack L.A., Rumack C.M., et al: B-mode encephalography in the normal and high risk infant. *AJR* 133:375-381, 1979.
2. Kosseff G., Garrett W.L., Radavanovich G.: Ultrasonic atlas of the normal brain of infants. *Ultrasound Med. Biol.* 1:259-266, 1974.
3. Shuman W.P., Rogers J.V., Mack L.A., et al.: Real-time sonographic sector scanning of the neonatal cranium: Technique and normal anatomy. *AJNR* 2:349-356, 1981.
4. Pigalas A., Thompson J.R., Grube G.L., Normal infant brain anatomy: Correlated real-time sonograms and brain specimens. *AJR* 137:815, 1981.
5. Mack L.A., Alvord E.C., Jr.: Neonatal cranial ultrasound: Normal appearances. *Semin. Ultrasound* 3:216-230, 1982.
6. Shaw C.M., Alvord E.C. Jr.: Cava septi pellucidi et Vergae: Their normal and pathological states. *Brain* 92:213-224, 1969.
7. Pape K.E., Wigglesworth J.S.: Haemorrhage, ischaemia and the perinatal brain. *Clin. Dev. Med.* 69/70:11-39, 1979.
8. DeReuck J.L.: The significance of arterial angioarchitecture in perinatal cerebral damage. *Acta Neurol. Belg.* 1977;65-94.
9. Rakic P.: in Buchwald A., Brazier M.A.B. (eds.): *Brain Mechanisms in Mental Retardation.* New York, Academic Press, 1975, pp. 3-39.
10. Amundsen P., Newton T.H.: Subarachnoid cisterns in radiology of the skull and brain, in Newton T.H., Potts D.G. (eds.): *Radiology of the Skull and Brain: Ventricles and Cisterns.* St. Louis, C.V. Mosby Co., 1978, vol. 4, p. 3629.

CHAPTER 4

Fetal Intracranial Diagnosis

Dolores H. Pretorius, M.D.*
Michael L. Johnson, M.D.
Carol M. Rumack, M.D.

FETAL INTRACRANIAL STRUCTURES can be evaluated very effectively with high-resolution real-time ultrasound equipment and should be routinely evaluated in obstetric ultrasound studies to exclude intracranial pathology. The ability to study the fetal brain has given our patients a new group of management options. Intrauterine ventricular amniotic shunting, termination of pregnancy, or early delivery can each be considered in terms of the best possible outcome for the fetus. The main concern of the physician is to have a firm diagnosis, and with that goal in mind, the next most immediate concerns are a clear understanding of the normal fetal brain, knowledge of abnormalities most likely to be encountered, and knowledge of the potential pitfalls of each scanning technique. This chapter describes both normal and abnormal fetal brain studies.

NORMAL FETAL BRAIN

The normal fetal brain should be scanned in three projections: axial, coronal, and sagittal. At least two projections are necessary to arrive at a three-dimensional representation of normal anatomy and pathology. In utero, the fetus is not always in an optimal position for imaging in all three projections, and therefore two or even one projection may be all that is available. The basic anatomy can be easily identified on any of the three projections, although one particular projection may be more useful for identifying a particular structure. The developing brain changes during gestation, particularly in regard to ventricular size, so a detailed knowledge of normal structures is essential.

Axial scans are optimal for evaluating the fetal brain, and this is the projection normally used to establish the biparietal diameter (BPD) to determine gestational age. The BPD may be large or small because of intracranial pathology rather than

*Assistant Professor of Radiology, Division of Diagnostic Ultrasound, University of Colorado Health Sciences Center

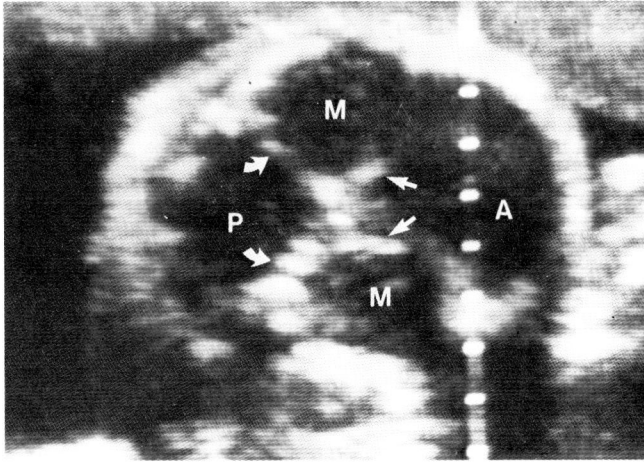

Fig 4–1.—Axial scan at the base of the skull in 28 weeks' gestation fetus with occiput to left of image. The anterior *(A)*, middle *(M)*, and posterior *(P)* fossae are marked. The petrous ridges *(curved arrows)* are visible posteriorly and the sphenoid wings *(straight arrows)* anteriorly. (From Johnson et al.[2] Reproduced with permission.)

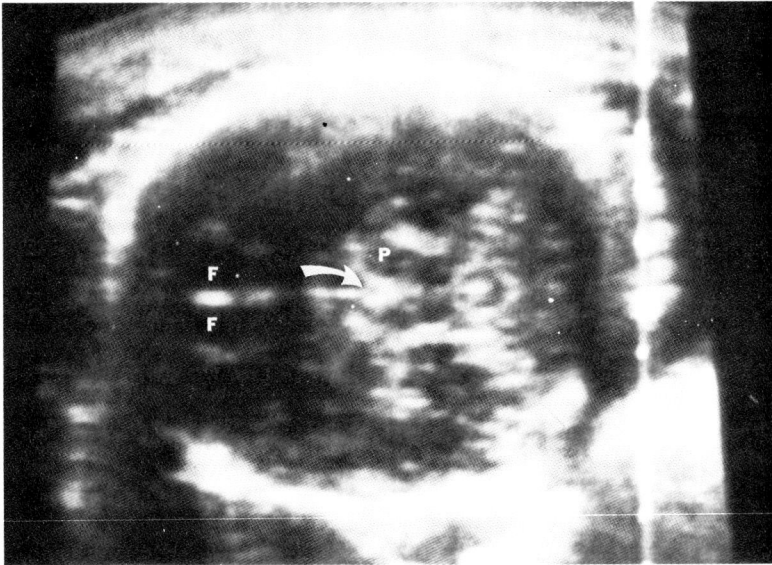

Fig 4–2.—Axial scan at the level of the cerebral peduncles *(P)* in 32-week-gestation fetus. The occiput is to the right of image. The basilar artery *(arrow)* pulsates in the interpeduncular cistern. The frontal horns *(F)* of the lateral ventricles are seen anteriorly. (From Johnson et al.[2] Reproduced with permission.)

because the fetus is off dates. Therefore, the intracranial structures also must be evaluated as well. The basic anatomy will be explained first as it appears in the axial projection, and then as it appears in the sagittal and coronal views.

AXIAL ANATOMY.—Axial scans near the base of the skull allow identification of the posterior, middle, and anterior fossae (Fig 4–1). Moving slightly rostrally, the circle of Willis may be seen with the pulsating internal carotid arteries, anterior cerebral arteries, and middle cerebral arteries. The basilar artery can be identified in the interpeduncular cistern anterior to the cerebral peduncles. The posterior cerebral arteries can be seen as they course around the midbrain in the ambient cistern. The cerebral peduncles are identified as heart-shaped, paired structures which are hypoechoic and have dense surrounding echoes (Fig 4–2). As noted previously, the basilar artery pulsates in the notch of the cerebral peduncles; it also marks the frontal aspect of the cranium in relation to the peduncles.

Moving rostrally, the next set of paired structures seen on an axial scan are the thalami. The thalami, like the peduncles, are hypoechoic; however, there are fewer surrounding high-density echoes, and the thalami are slightly larger and more rounded anteriorly. The level at which the thalami appear symmetric on an axial scan marks the best level for BPD measurement of the fetal cranium and thus is an

TABLE 4–1.—BIPARIETAL DIAMETER
(IN CENTIMETERS) FOR GESTATIONAL AGE
(IN WEEKS)*

BPD	AGE	BPD	AGE	BPD	AGE
1.9	11	4.5	20	6.9	28
2.0	11½	4.6	20½	7.0	28½
2.1	12	4.7	20½	7.1	29
2.2	12½	4.8	21	7.3	29½
2.3	12½	4.9	21½	7.4	30
2.4	13	5.0	21½	7.5	30½
2.5	13½	5.1	22	7.6	31
2.6	13½	5.2	22½	7.7	31½
2.7	14	5.3	22½	7.8	32
2.8	14½	5.4	23	7.9	32½
2.9	14½	5.5	23½	8.0	33
3.0	15	5.6	23½	8.2	33½
3.1	15½	5.7	24	8.3	34
3.2	15½	5.8	24	8.4	34½
3.3	16	5.9	24	8.5	35
3.4	16½	6.0	24½	8.6	35½
3.5	16½	6.1	25	8.8	36
3.6	17	6.2	25	8.9	36½
3.7	17½	6.3	25½	9.0	37
3.8	18	6.4	26	9.1	37½
4.0	18½	6.5	26	9.2	38
4.2	19	6.6	26½	9.3	38½
4.3	19½	6.7	27	9.4	39
4.4	19½	6.8	27½	9.6	39½
				9.7	40

*Reproduced by permission of John Hobbins, M.D., Yale University School of Medicine, New Haven (unpublished data).

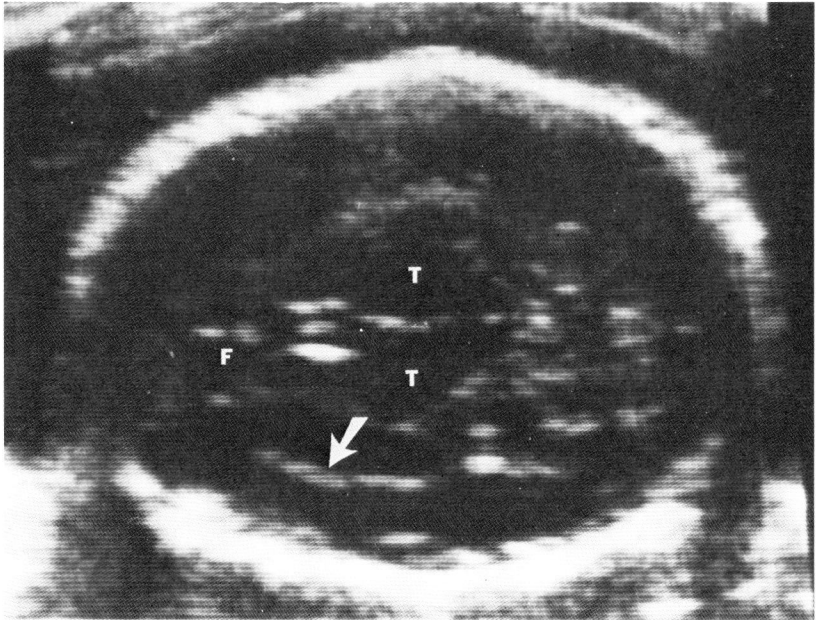

Fig 4–3.—Axial scan at the level of the thalami *(T)* in a 31-week-gestation fetus. The frontal horns *(F)* of the lateral ventricles are seen anteriorly. The middle cerebral artery pulsates in the sylvian fissure *(curved arrow)*. (From Johnson et al.[2] Reproduced with permission.)

excellent measure of gestational dating (Table 4–1). At the level of the thalami, the frontal horns of the lateral ventricles can be seen anteriorly and the sylvian fissures containing the pulsating middle cerebral arteries can be seen laterally (Fig 4–3). The cavum septi pellucidi is a hypoechoic structure in the midline located between the frontal horns of the lateral ventricles and the anterior border of the thalami. It is present in all fetuses and should not be mistaken for the third ventricle.

The lateral ventricles are visualized in a more rostral scan (Fig 4–4). The lateral walls of the lateral ventricles appear as parallel lines, equidistant from the inter-hemispheric fissure on the axial scan. The medial walls of the lateral ventricles are occasionally seen. An echogenic structure is usually identified within the lateral ventricles; this is the choroid plexus. The choroid plexus is particularly prominent earlier in gestation and becomes less prominent as the fetus reaches term; this prominence (at 8–22 weeks' gestational age) may be due to glycogen deposits within the choroid, which may provide an extra energy source for brain development.[1] A range of normal ventricular measurements has been identified by several authors by relating ventricular width to hemispheric width.[2-4] In our department, the lateral ventricular width (LVW) is measured from the middle of the midline echo to the lateral wall of the lateral ventricle at the point where the ventricular walls parallel the midline. The hemispheric width (HW) is measured on the same scan from the middle of the midline echo to the inner table of the calvarium. The LVW/HW ratio (LVR) is calculated and compared with a table of values developed in our

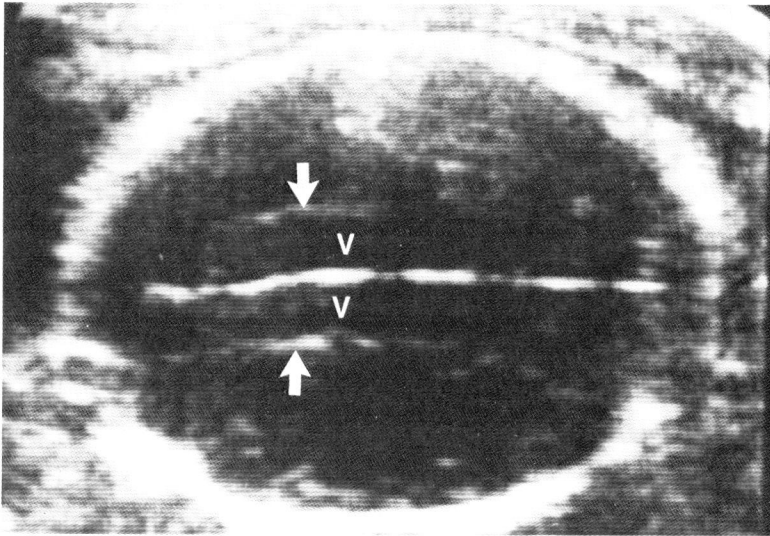

Fig 4–4.—Axial scan at the level of the lateral ventricles *(V)* in term fetus. *Arrows* point to lateral walls of the lateral ventricles.

department from a study of 196 normal fetuses (Fig 4–5).[2] The LVR is much higher early in pregnancy, reaching as high as 71% at 15 weeks' gestational age, then gradually decreases to 33% at term. The decrease in the LVR is consistent with the rapid growth of the cerebral hemispheres that occurs as pregnancy proceeds. This growth relationship has been observed by neuroanatomists.[5, 6]

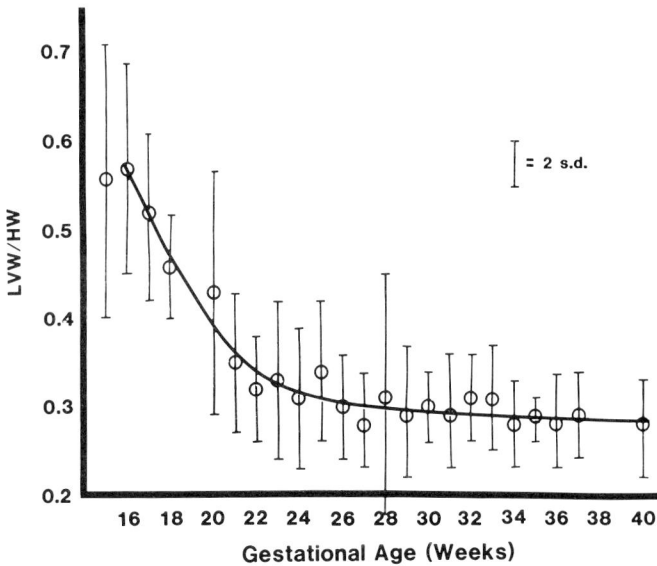

Fig 4–5.—Normal fetal lateral ventricular width to hemispheric width ratios. (From Meier P.R., Clewell W.H., Johnson M.L.: Amelioration of fetal hydrocephalus by placement of a ventriculo-amniotic shunt in utero, unpublished manuscript. Reproduced with permission.)

Another echogenic line is seen parallel to the fetal skull in approximately 80% of fetuses less than 30 weeks' gestational age. The space between the echogenic line and the skull is thought to represent the normal subarachnoid space. It becomes smaller as the fetus approaches term, being identified on ultrasound in only 3% of term infants.[7]

SAGITTAL ANATOMY.—Sagittal scans permit a three-dimensional appreciation of fetal brain anatomy. In the midline, the cavum septi pellucidi may be seen coursing posteriorly into the cavum Vergae. The corpus callosum courses parallel to the superior aspect of the cavum septi pellucidi and appears as a hypoechoic thin structure. Its upper margin is demarcated by the strongly echogenic pericallosal arteries. In a sagittal scan slightly off midline (Fig 4–6), the cerebellum appears to be more echogenic than the cerebral cortex. Cerebellar growth occurs later, and thus the fetal cerebellum is relatively small, compared to its size in the adult.[8] The third ventricle may be traced into the aqueduct of Sylvius and then into the fourth ventricle. The cisterna magna in the posterior aspect of the posterior fossa may appear particularly prominent.

The development of sulci in the brain parenchyma occurs at a standard rate throughout pregnancy; identification of these structures at particular times permits a rough estimation of gestational age. The cingulate sulcus lies superior to the corpus callosum on the midline sagittal scan. On sagittal scans angled lateral to the midline, the lateral ventricles may be identified, with the dense, echogenic, choroid plexus within them. A continuum of the lateral ventricular system from the frontal horns to the atrium, the temporal horns, and the occipital horns may be noted.

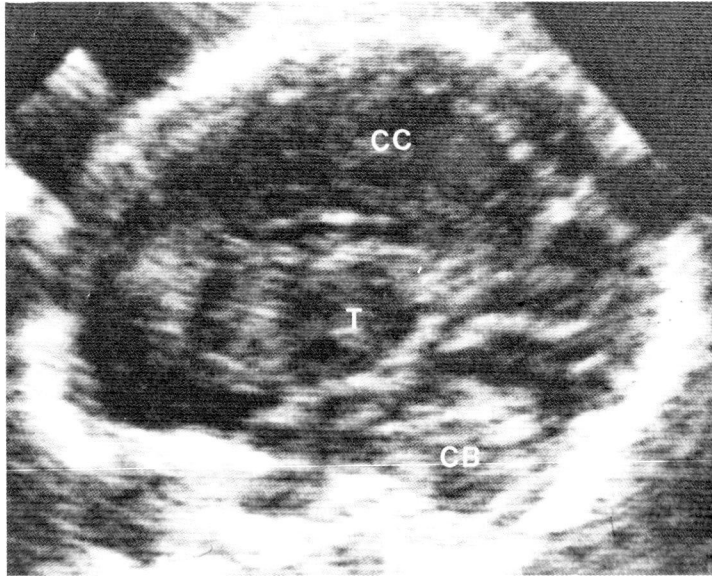

Fig 4–6.—Sagittal scan slightly lateral to the midline in 30-week-gestation fetus. The cerebral cortex (CC), thalamus (T), and cerebellum (CB) are visible.

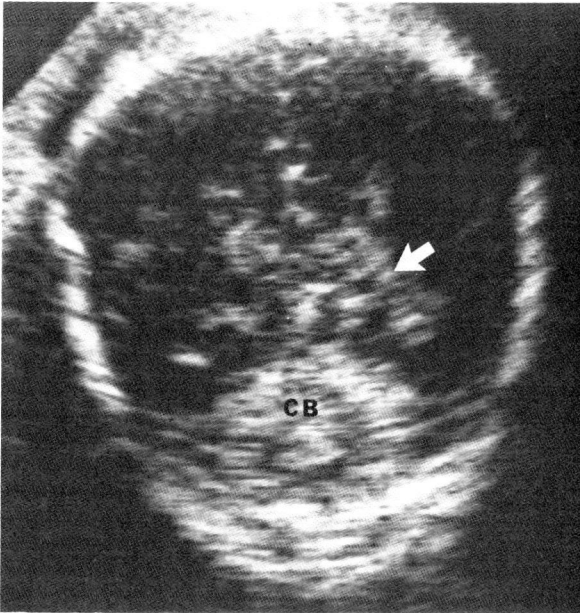

Fig 4–7.—Coronal scan through the level of the trigone demonstrating the choroid plexus *(arrow)* in a fetus at 30 weeks. The cerebellum *(CB)* is normally denser than the cerebral hemispheres.

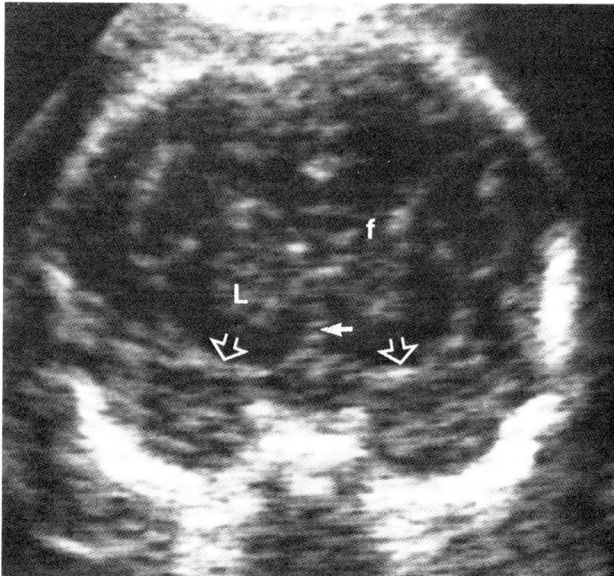

Fig 4–8.—Coronal scan through the frontal horns *(f)* in a 30-week-gestation fetus. Hypoechoic brain is seen surrounding the lentiform nucleus *(L)*. The anterior cerebral arteries *(closed arrow)* and middle cerebral arteries *(open arrow)* are noted.

CORONAL ANATOMY.—Coronal views of the fetal cranium may be useful when axial views are difficult to obtain because of fetal head position or when axial sections have a significant amount of noise or artifact in the near field, which may include the up side of the fetal brain. The BPD may be approximated by using the cerebral peduncles on a coronal scan. The temporal horns of the ventricular system may appear prominent, with hypoechoic matter surrounding them; the hypoechoic area is normal brain parenchyma. This appearance resembles owl's eyes. Moving anteriorly, the choroid plexus may be seen in a batwing appearance; the cerebellum is dense and is located inferiorly (Fig 4–7). Even more anteriorly, the frontal horns may be seen as small, slitlike structures and the lentiform nucleus may be seen surrounded by hypoechoic brain parenchyma (Fig 4–8). More detailed descriptions of the sagittal and coronal views are presented in the chapter on normal neonatal brain anatomy.

The cranium should be evaluated for normal calcification, shape, and continuity. Except in regions of anatomical sutures, a high-density echo surrounds the fetal brain beginning early in pregnancy; the fetal skull can be identified as early as 8–9 weeks' gestational age, although BPD measurement generally is not attempted until aproximately 12 weeks' gestational age. The fetal skull is oval in shape. After 26 weeks, fetal molding may occur, leading to a longer and more flattened shape. Molding leads to underestimation of the BPD and occurs most frequently in breech or transverse lie presentations.

Artifacts

Artifacts may complicate the evaluation of the fetal cranium. The near field of the scans may be filled with noise suggesting asymmetric disease when in fact the abnormality is present bilaterally (Fig 4–9). The cresent sign, seen as a hypoechoic area in the hemisphere farthest from the transducer, is present in most obstetric scanning and can be misinterpreted as hydrocephalus or a mass lesion (Fig 4–10).[9] Reverberation echoes may be misinterpreted as ventricular walls displaced laterally (Fig 4–11). An engaged fetal head late in pregnancy may be difficult to image and may be misinterpreted as being absent; the suspicion of possible anencephaly should signal the sonographer to ask for assistance in lifting the fetal head out of the pelvis to definitively see the fetal cranium. These artifacts should be identified as nonpathologic entities.

DEVELOPMENTAL ABNORMALITIES

Hydrocephalus

Hydrocephalus is dilatation of the ventricular system secondary to an increase in cerebral spinal fluid (CSF) pressure. Congenital hydrocephalus occurs with an incidence of 0.5–1.8 per 1,000 births.[10] There is an increased risk of CNS abnormalities in children of a family in which a previous child was born with hydrocephalus or another neural tube defect. Specifically, the risk of hydrocephalus recurring in a family is 2%; the risk of aqueductal stenosis recurring is 5.7%.[11]

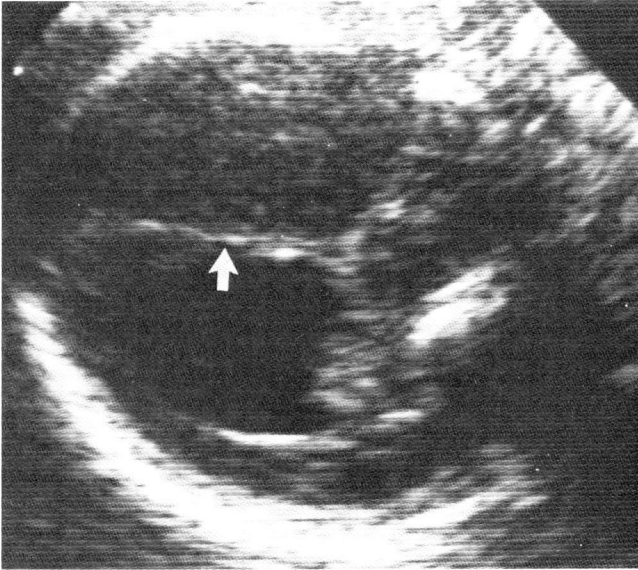

Fig 4–9.—Coronal scan in a 24-week-gestation fetus with hydrocephalus demonstrating noise in the near field obscuring the dilated ventricle closest to the transducer. The midline echo *(arrow)* is not displaced.

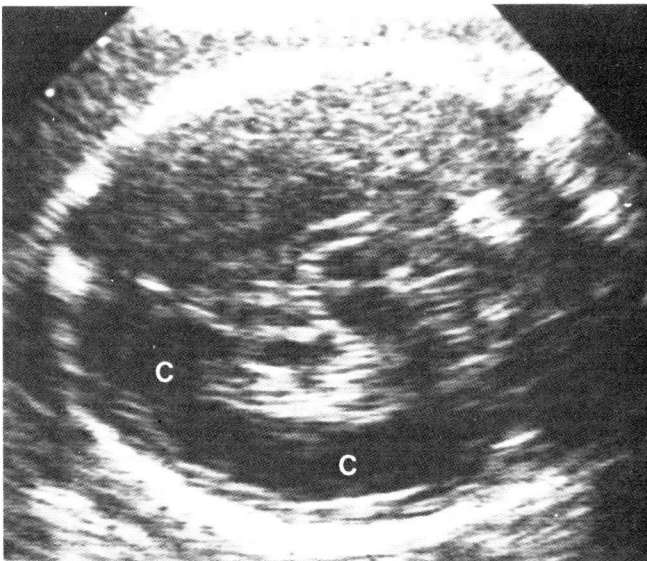

Fig 4–10.—Axial scan in a 25-week-gestation fetus showing the cresent sign. The hypoechoic area in the hemisphere farthest from the transducer *(C)* resembles a cresent and should not be misinterpreted as hydrocephalus.

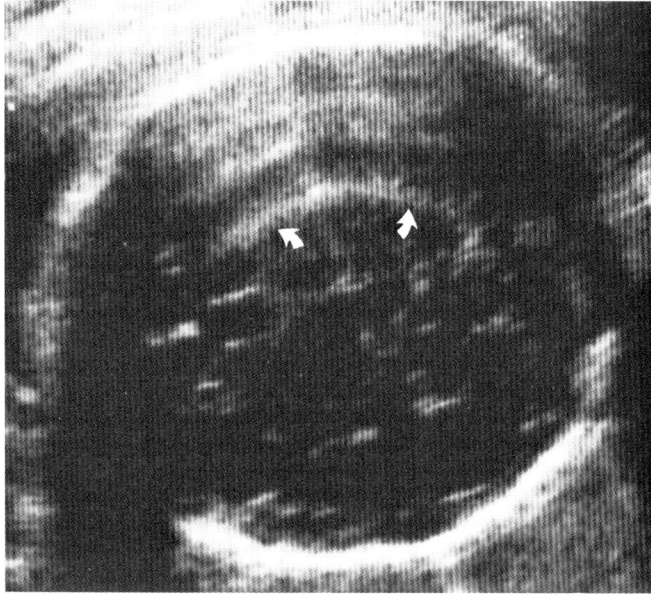

Fig 4–11.—Axial scan of a fetus at 30 weeks' gestation demonstrating reverberation artifact. The curved artifact *(arrows)* is caused by reverberation off the curved skull; it may be mistaken for hydrocephalus.

The causes of hydrocephalus in utero are multiple. There are definite X-linked and autosomal recessive causes of aqueductal stenosis, the X-linked cases comprising 2% of cases of congenital hydrocephalus. Familial inheritance of spina bifida and meningomyeloceles with associated hydrocephalus is well documented. These cases account for only a small percentage of the children born with hydrocephalus. Isolated chromosomal abnormalities, including trisomy 13 and 18, chromosomal deletions, and translocations, have also been reported with hydrocephalus.[11,12]

In the study by Bay et al. of 56 children with congenital hydrocephalus, the etiology of hydrocephalus was aqueductal stenosis in 20 children, communicating hydrocephalus in 10, Dandy-Walker malformation in 5, and unknown in 21.[11] In another study by McCullough and Balzer-Martin of 37 children treated for congenital hydrocephalus the etiology was aqueductal stenosis in 18, communicating hydrocephalus in 9, Dandy-Walker malformation in 6, and Arnold-Chiari myelodysplasia in 4.[13] Intrauterine infection has also been implicated as a cause of hydrocephalus.

The exact effect of hydrocephalus in utero on the developing brain is not entirely understood. Some histologic studies have implicated acute and severe hydrocephalus in causing brain damage.[14,15] Apparently, the increased CSF pressure leads to flattening of the ependyma and interruption of the CSF-brain barrier. The CSF is then able to penetrate the periventricular white matter. Demyelination secondary to axonal degeneration results in irreversible tissue damage. The exact point at which tissue damage becomes irreversible is unknown. Treatment of hydrocephalus

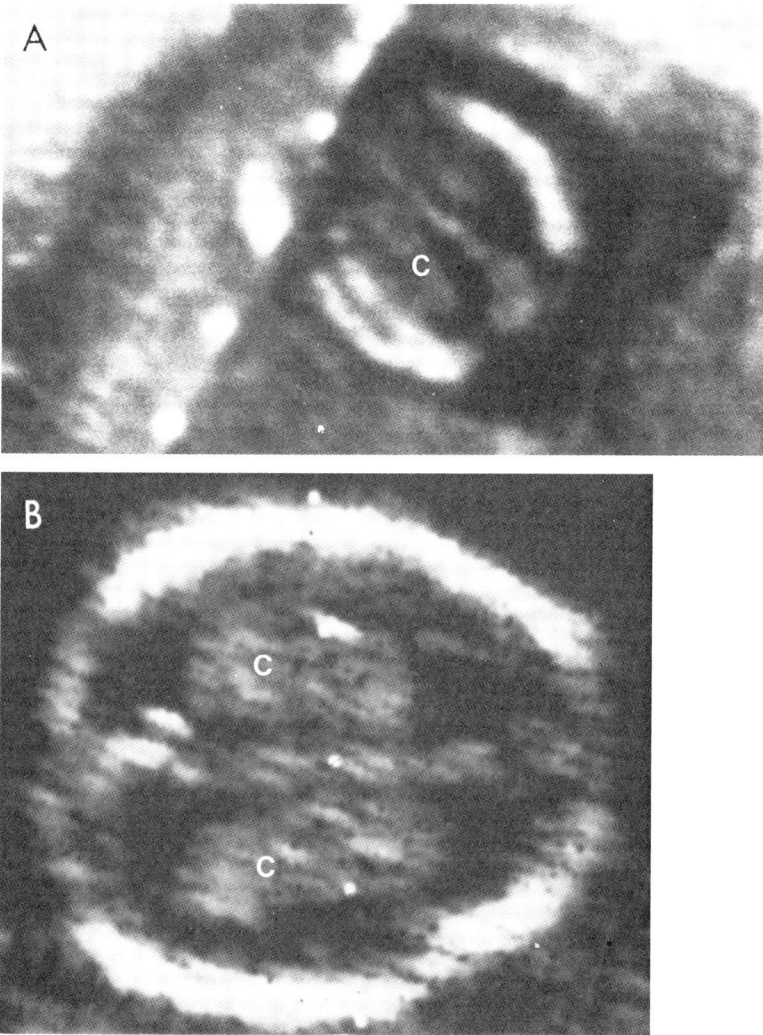

Fig 4–12.—Axial scans showing choroid plexus *(C)* filling the lateral ventricles at 13.5 weeks **(A)** and 16 weeks **(B)** in normal fetuses.

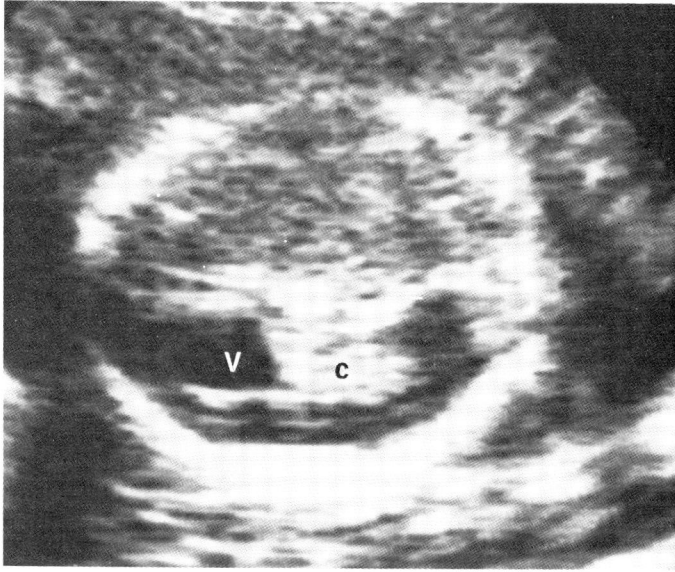

Fig 4–13.—Axial scan in a fetus of 18 weeks' gestation. The choroid plexus *(C)* is seen adjacent to CSF in the lateral ventricle *(V)* of a normal fetus. Note also obliteration of the near field by noise.

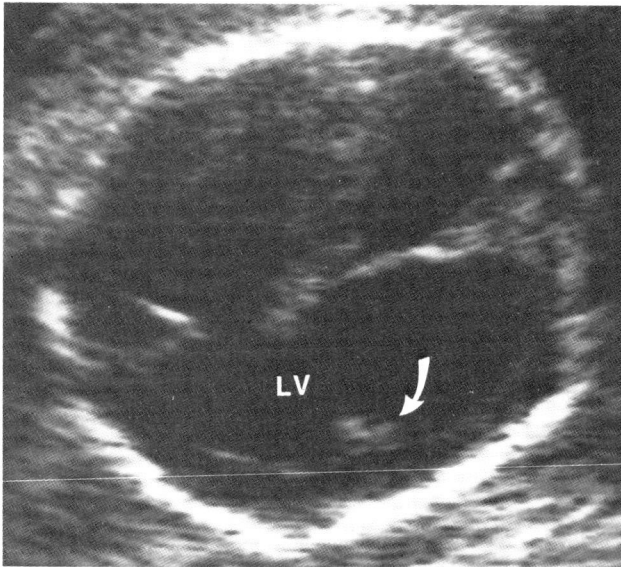

Fig 4–14.—Hydrocephalus in a fetus at 23 weeks' gestation. Choroid plexus *(arrow)* lies dependently in the large lateral ventricles *(LV)*.

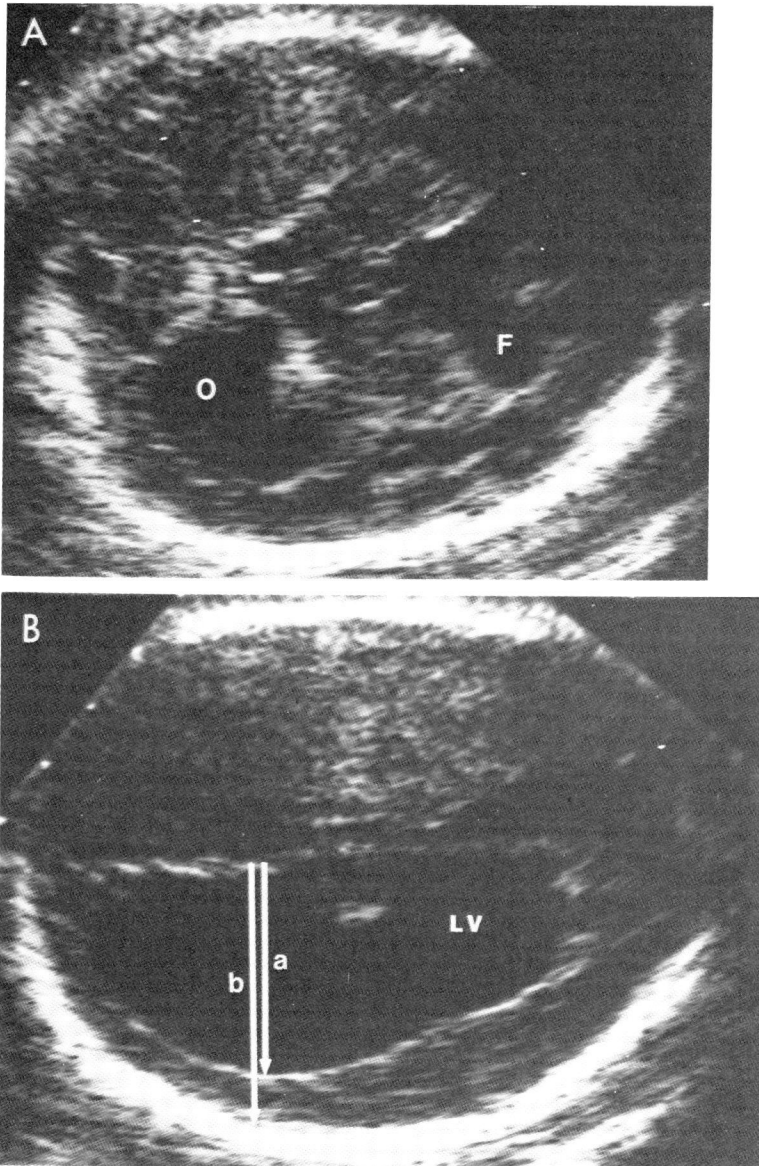

Fig 4–15.—Hydrocephalus in a fetus at 29 weeks' gestation. **A,** The frontal horns *(F)* and occipital horns *(O)* are dilated. **B,** The lateral ventricles *(LV)* are markedly dilated. The mantle thickness is only 8 mm. The lateral ventricular width *(a)* is 30 mm and the hemispheric width *(b)* is 38 mm. The lateral ventricular ratio *(LVR)* is .79.

before this point is reached could certainly improve the mental outcome of these children.

The diagnosis of fetal hydrocephalus can be made by ultrasound. In the first trimester, the choroid plexus nearly fills the entire volume of the lateral ventricle (Fig 4–12). At this time the LVR may reach as high as 71%. Hydrocephalus is suspected when CSF is seen in the lateral ventricle totally surrounding and compressing the choroid plexus. The LVR is then measured and compared with normal values (see Fig 4–5). In the second and third trimesters, the CSF normally appears beside the choroid plexus (Fig 4–13). The LVR should be determined if hydrocephalus is suspected (Fig 4–14). Fiske and Filly have found that medial displacement of the medial wall of the lateral ventricle can be helpful in diagnosing early hydrocephalus.[16]

Brain mantle thickness is measured on an axial scan, adjacent to the lateral ventricle in the region of the parietal lobe (Fig 4–15). It should be determined whether the hydrocephalus is symmetric or asymmetric (Fig 4–16). This may require manipulation of the fetus or a second ultrasound examination if the hemisphere in the near field cannot be evaluated. Other ultrasound parameters which the sonographer should observe are brain texture and periventricular echogenicity. A dense brain texture may suggest intrauterine infection. High-density echoes in the region of the ependyma of the ventricle may also be a sign of infection.

Associated ultrasound findings seen with hydrocephalus include (1) polyhydramnios and oligohydramnios, (2) abnormal fetal lie, (3) hepatomegaly and fetal ascites with associated infection, (4) meningomyelocele associated with Chiari type II malformation, and (5) other intracranial abnormalities, such as Dandy-Walker cyst, en-

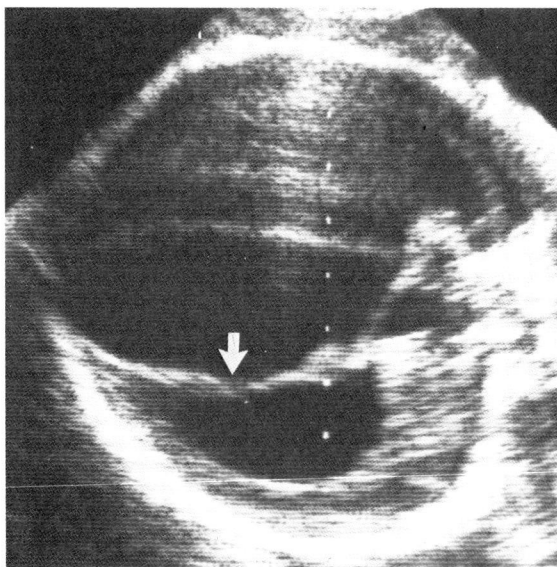

Fig 4–16.—Asymmetric hydrocephalus is demonstrated in a 28-week-gestation fetus. The midline *(arrow)* is deviated downward. Note noise and reverberation artifacts in near ventricle.

cephalocele, and intracranial tumor. In all cases of fetal hydrocephalus, the fetal spine should be examined in a longitudinal and transverse projection to exclude spinal dysraphism. In addition, each fetal organ system should be examined completely. Serial scans should be performed to determine whether the hydrocephalus is progressive.

Obstetric Management of the Fetus With Hydrocephalus

There are multiple options for therapy in the fetus diagnosed with hydrocephalus by ultrasound: (1) termination of pregnancy if the diagnosis is made during the time period allowable for legal abortion, (2) early delivery after administration of corticosteriods to accelerate lung maturity, with placement of a ventricular shunt after delivery, (3) small needle ventricular decompression at vaginal delivery, usually leading to a nonviable fetus, or possibly (4) intrauterine ventricular shunting to the amniotic cavity. Amniocentesis may be indicated in particular circumstances to exclude viral infection or chromosomal abnormalities.

Many factors influence the selection of therapy by the parents and physician. First, religious or moral constraints prohibit many parents from choosing abortion. Second, some parents are relieved to be allowed the option of abortion or decompression. Third, an associated abnormality may influence the parents to take a less aggressive approach to fetal therapy. Fourth, a mantle thickness of less than 5 mm may suggest a less successful outcome of aggressive therapy, and so physicians may not encourage that course of action. Fifth, hydrocephalus associated with infection may lead the physicians to encourage a less aggressive therapy due to the poor prognosis of these infants. Also, these fetuses probably do better by remaining in utero than by being delivered early. Sixth, the gestational age at diagnosis may influence whether an intrauterine shunt should be placed. Since intrauterine shunting has been performed only on an investigational basis in our department, certain requirements must be fulfilled before this experimental procedure can be considered.

Therapy for the fetus with hydrocephalus should involve a group of physicians and social workers who are willing to talk to the parents and allow personal, moral, and ethical factors to influence the therapy. A multispecialty group is essential and should include experts in ultrasound, pediatrics, obstetrics, and neurosurgery. A multispecialty group is essential to weigh fully all of the factors. At the time of diagnosis, a physician should be available to explain the abnormality and answer parents' questions. Therapeutic options should be discussed openly. A second or third examination may be required to ascertain that the ventricles are definitely enlarging. The parents may need significant emotional support.

Before intrauterine diagnosis of hydrocephalus was possible, children with congenital hydrocephalus were treated with ventriculoatrial or ventriculoperitoneal shunts following delivery or they were not treated neurosurgically because of multiple associated congenital abnormalities and severe hydrocephalus leading to a poor prognosis. In a study by McCullogh and Balzer-Martin of overt neonatal hydrocephalus,[13] 11 patients were untreated and 37 patients were treated with shunting. In the patients treated with shunting the survival was 85% (mean length of follow-up was 6 years); normal intelligence was demonstrated in 53%. The amount of brain mass (volumetric determination of brain size)[17] present at diagnosis was a predictor of intelligence. An IQ below normal was predominantly associated with aqueductal stenosis and Dandy-Walker malformation and a smaller brain mass. The patients who were untreated all died between 2 weeks and 16 months after birth.[13]

Ventriculoamniotic shunting was developed in an attempt to arrest progressive brain damage in fetuses with hydrocephalus.[18] Since many children diagnosed with fetal hydrocephalus have a poor prognosis, it is hoped that an invasive, intrauterine procedure might significantly alter their outcome in regard to brain development. We have placed intrauterine ventriculoamniotic shunts in fetuses only after they have satisfied our research protocol guidelines, which have been approved by our human research committee. The fetus must be less than 30 weeks' gestational age. We feel that fetuses older than 30 weeks should be given corticosteroids to accelerate lung maturity and then should be delivered vaginally at 32 weeks and be given a ventricular shunt after birth. The fetuses must show evidence of progressive ventriculomegaly on serial ultrasound examinations. The parents must be informed of the various options available and must be fully aware of the experimental nature of fetal surgery; they must then accept the risks of this surgery. We do not shunt fetuses with associated congenital abnormalities or verified infection. To date we have performed intrauterine shunts on four fetuses (Fig 4–17). All four babies are alive with ventriculoperitoneal shunts in place that were placed after delivery.

The shunt used in the ventriculoamniotic technique is designed with a one-way valve to allow CSF to drain into the amniotic cavity and to prevent amniotic fluid

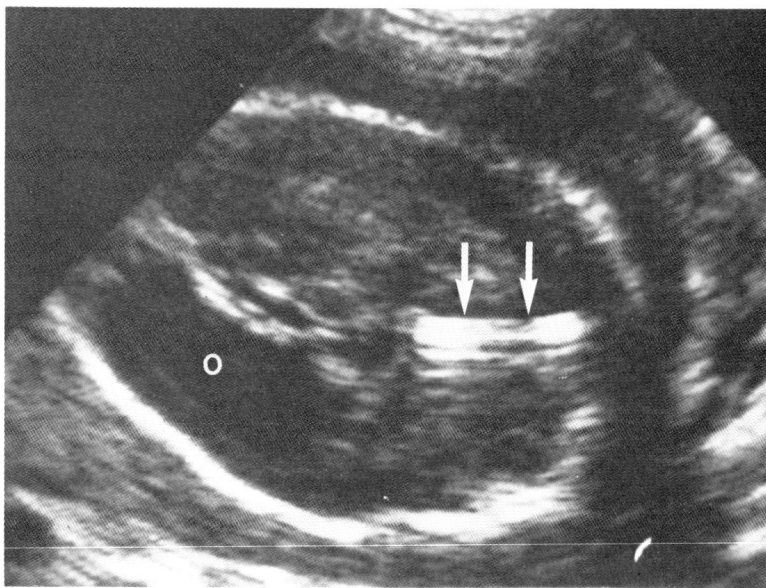

Fig 4–17.—Intraoperative sonogram showing shunt *(arrows)* in the anterior horn of the left lateral ventricle in a 23-week-gestation fetus. The right occipital horn *(O)* is also dilated. (From Clewell et al.[18] Reproduced with permission.)

from flowing into the ventricles. Even though amniotic fluid is sterile, particulate matter in the fluid could theoretically cause ventriculitis. Fetal activity may be a factor in dislodging a shunt from the fetal ventricles. For this reason, two anchors are placed in the shunt to minimize the possibility of dislodgment.

The prognosis in terms of brain development and mental intelligence is not yet available. Infants that we have not treated with intrauterine shunting have had a dismal prognosis. We hope that intrauterine shunting and early delivery with subsequent ventricular peritoneal shunting will result in a better prognosis for these children. Our early results are encouraging and exciting. It is important, however, not to lose sight of the fact that fetal surgery is still an experimental technique.

Anencephaly

Anencephaly is the most common congenital defect involving the central nervous system (CNS), the incidence being 1 in 1,000 births. The recurrence risk of a family having another anencephalic child is approximately 4%. For this reason, subsequent pregnancies should be screened with serum α-fetoprotein (AFP) levels, ultrasound examination, and perhaps amniotic AFP levels.

Anencephaly denotes the absence of the cranial vault and functioning brain tissue. Portions of the midbrain and brain stem may be present. Anencephaly occurs at approximately 2–3 weeks' gestational age, when the cephalic end of the neural tube fails to close. The open neural tube defect is covered by a membrane of angiomatous stroma rather than by skin or bone. The basal portions of the frontal, parietal, and occipital bones may be present.

Ultrasonographically, a fetal head should be identified at the latest by 15 weeks' gestational age. In anencephaly, major portions of the cranium and intracranial structures are absent; the fetal orbits and the skull base should be identified, however (Fig 4–18). The diagnosis is often made when attempting to determine a BPD for gestational dating. The uterus frequently appears to be large for gestational age. Polyhydramnios is present in approximately 40%–50% of anencephalic pregnancies which is apparently the result of failure of normal fetal swallowing of amniotic fluid. The polyhydramnios generally does not occur until 26–28 weeks' gestational age. Oligohydramnios is also associated with anencephaly, although its incidence is much lower than that of polyhydramnios. Although anencephalic fetuses are generally viable at diagnosis, the abnormality is fatal, and all such fetuses die before or soon after delivery.

Multiple congenital anomalies are associated with anencephaly. Approximately 50% of anencephalic fetuses have a spinal defect, including spina bifida, meningocele, and meningomyelocele. Other congenital abnormalities include cleft palate, umbilical hernia, and equinovarus.

A false diagnosis of anencephaly may be made when the fetal head is engaged in the pelvis. Before this diagnosis is finalized, a careful search should be made in the pelvis, and if there is any doubt a plain abdominal film may be necessary for confirmation before deciding on definite management.

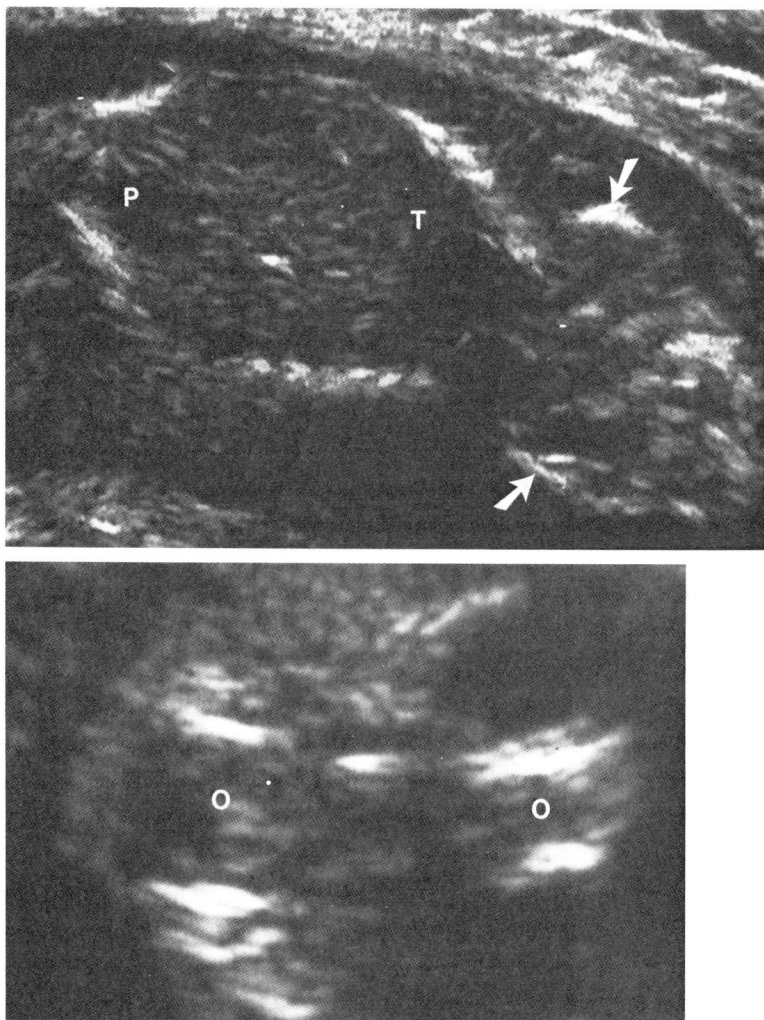

Fig 4–18.—Anencephaly in fetus of 24 weeks' gestation. **A,** the fetus is in vertex presentation with only the base of the skull seen *(arrows)*. The thorax *(T)* and pelvis *(P)* are marked. **B,** the fetal orbits *(O)* are seen. (Courtesy of Dr. George Leopold, University of California, San Diego.)

Encephalocele-Meningomyelocele

The primary defect of an encephalocele is failure of the surface ectoderm to separate from the neuroectoderm, leading to a bony defect in the calvarium.[16] This disorder of closure may cause only the meninges (meningomyelocele) or both meninges and brain tissue (encephalomeningocele) to herniate outside the cranium. The incidence of encephalomeningocele is 1 in 2,500. Encephalomeningoceles occur most frequently in the occiput in the midline (75%); they occur with equal frequency in the frontoethmoid and the parietal regions (12%–13%).[16] Encephalomeningoceles may present with polyhydramnios on ultrasound. A mass is seen extending from the calvarium. The mass may contain only fluid and a meningocele will be suspected. If the mass contains brain tissue an encephalomeningocele is diagnosed (Fig 4–19). A cranial defect may be identified occasionally. The cranial cavity will appear small if a significant portion of neural tissue is in the encephalocele sac. As with all neural tube closure defects, hydrocephalus is a frequent association.

Lumbar meningomyeloceles are associated with multiple cranial defects, including the Chiari type II malformation, in over 90% of the infants.[19-21] When the diagnosis of intrauterine hydrocephalus is made, a careful search must be made for a meningomyelocele, encephalocele, or myeloschisis. The shape of the ventricular system may be the first clue to a Chiari malformation and neural tube defect (Fig 4–20). Myeloschisis may be missed because it is simply an open spinal canal with flared posterior elements. No mass is present adjacent to the spine. The hydrocephalus seen in the Chiari malformation is specific and may be diagnosed with

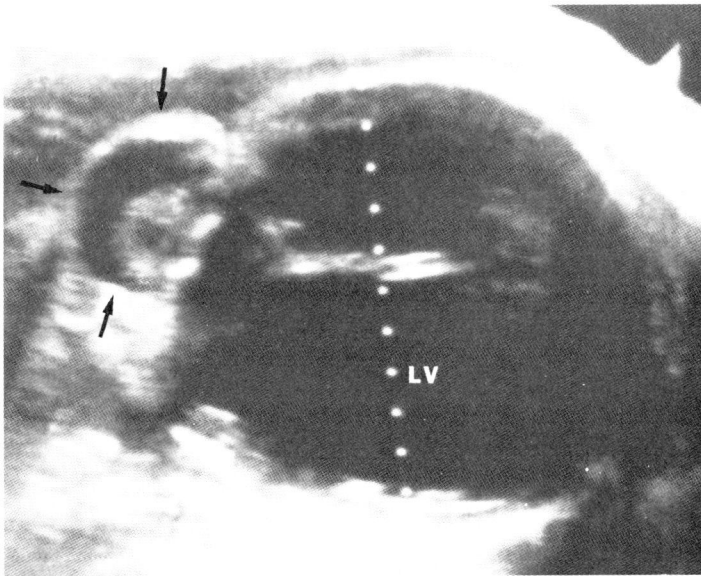

Fig 4–19.—Encephalocele in a term fetus. Dilated lateral ventricles *(LV)* and encephalocele *(arrows)* are seen. Brain tissue is seen within the encephalocele. (From Johnson et al.[2] Reproduced with permission.)

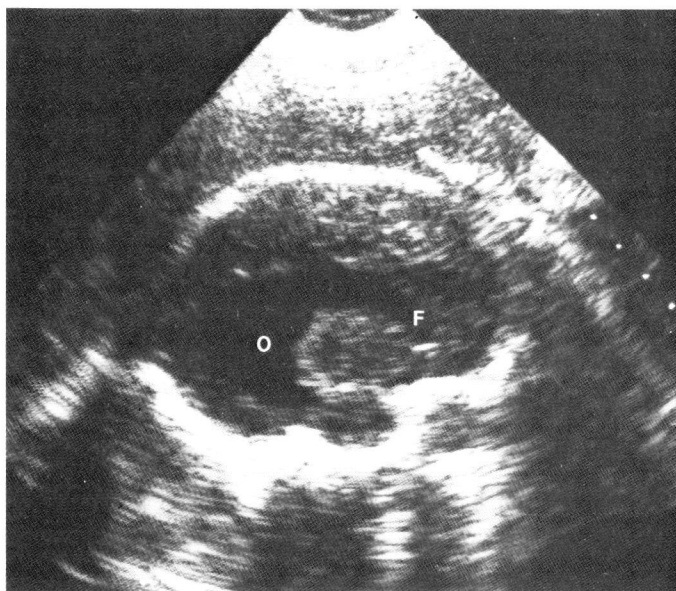

Fig 4–20.—Chiari II malformation in utero at 32 weeks' gestation. Sagittal scan of fetus with meningomyelocele shows characteristic pointing of the frontal horn *(F)* with much larger occipital horn *(O)*. (Courtesy of Dr. Gloria Komppa, Fitzsimmons Army Medical Center, Denver.)

ultrasound. Specifically, the findings are dilated ventricles, anterior pointing of the frontal horns on axial scans and inferior pointing on coronal scans, dilated third ventricle with prominent massa intermedia, and a small posterior fossa with downward displacement, leading to a "tentorial-cerebellar" pseudomass appearance (Fig 4–21).[21]

Dandy-Walker Syndrome

The Dandy-Walker Syndrome is characterized by (1) a posterior fossa cyst continuous with the fourth ventricle, (2) posterior fossa enlargement, with elevation of the torcula and tentorium, and (3) cerebellar vermian dysgenesis.[22] The associated congenital anomalies include hydrocephalus, atresia of the foramina of Magendie and Luschka, absence of the corpus callosum, and occipital meningoencephalocele. The etiology of this defect is thought to be failure of the velum medullare to regress, combined with cerebellar cleft malformation and failure of the cerebellar commissure to develop into vermis.[23]

On ultrasound examination, the fetal cranium has a large cystic structure in the posterior fossa which is the fourth ventricle; there is associated enlargement of the posterior fossa. The cerebellar hemispheres may be seen separated and flattened adjacent to the tentorium anterolaterally. Hydrocephalus is almost always present and may also be identified on ultrasound when it complicates this abnormality (Fig 4–22). These fetuses may have polyhydramnios, probably secondary to abnormal

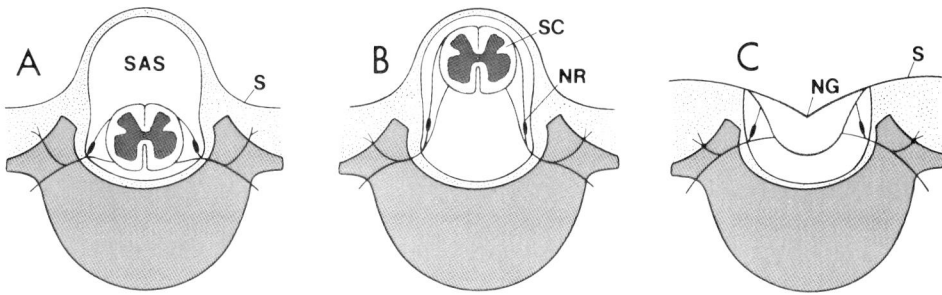

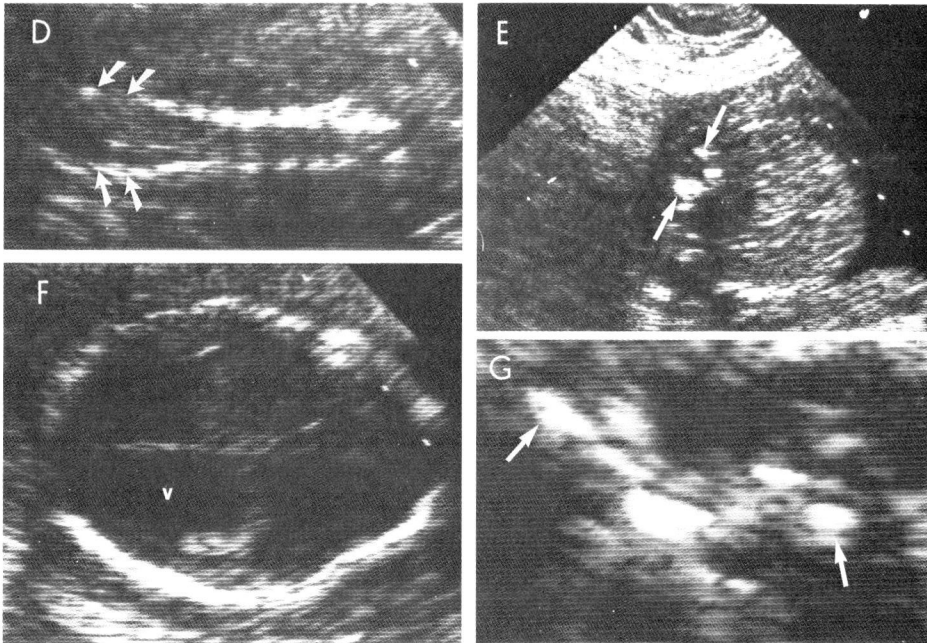

Fig 4–21.—Dysraphism. **A,** Meningocele; **B,** Meningomyelocele; **C,** Myeloschisis. (Diagrams modified from Fitzgerald M.J.T.: *Human Embryology.* New York, Harper & Row, 1978.) **D,** 18-week-fetus with widened spinal canal in lumbar region *(arrows)* on longitudinal scan. **E,** Transverse scan in lumbar region showing splaying of the posterior elements *(arrows)* of fetus in **D. F,** Hydrocephalus with dependent choroid in lateral ventricle *(V)* in same fetus as in **D. G,** Marked splaying of the posterior elements *(arrows)* of the lumbar spine in a term fetus with myeloschisis. This is the most severe form of dysraphism as diagrammed in **C.** This neural tube defect is not associated with a mass.

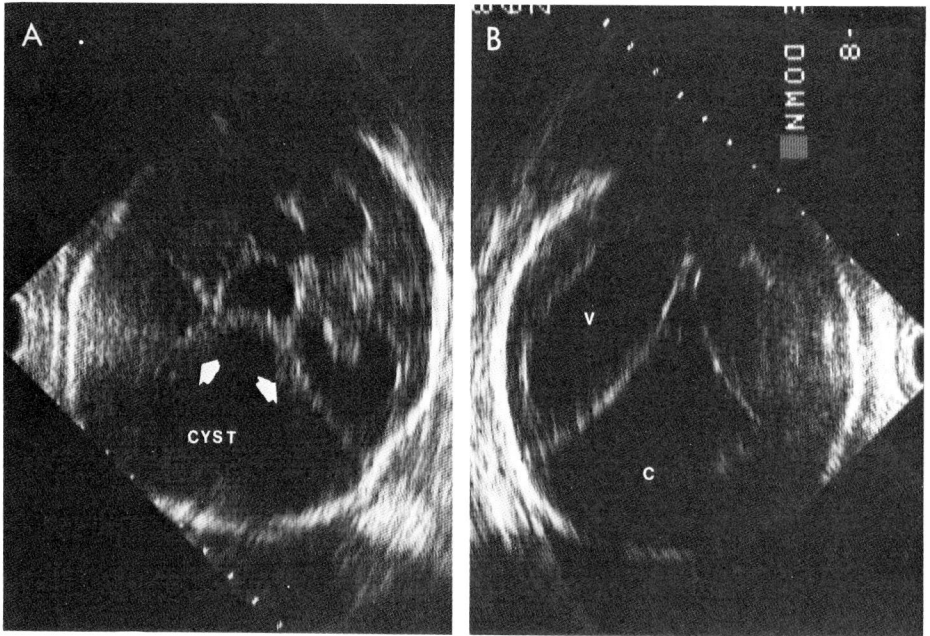

Fig 4–22.—Dandy-Walker malformation in a 30-week-gestation fetus. **A,** The cerebellar hemispheres *(arrows)* lie separated adjacent to the tentorium. The cyst (dilated 4th ventricle) is seen in the posterior fossa *(CYST)*. **B,** Hydrocephalus with large ventricles *(V)* has resulted from the posterior fossa cyst *(C)*.

swallowing resulting from brain stem or vagal nerve compression. A Dandy-Walker cyst is differentiated from a subarachnoid cyst in that it is continuous with the fourth ventricle and a subarachnoid cyst is not. Differentiation may be difficult on in utero ultrasound and neonatal evaluation may be required.

Holoprosencephaly

Holoprosencephaly is a developmental abnormality of forebrain diverticulation leading to a single large midline ventricle.[24] The spectrum of holoprosencephaly includes the most severe form, alobar holoprosencephaly, a less severe form, semilobar holoprosencephaly, and the mildest form, lobar holoprosencephaly. In alobar holoprosencephaly there is a large single cavity with only a minimal amount of cerebral tissue seen peripherally and fused thalami. Affected children have a specific facial syndrome that includes orbital hypotelorism and other craniofacial abnormalities. They rarely survive past the first year of life. In semilobar holoprosencephaly more cerebral tissue is present and occipital lobe tissue may be present. These children have associated facial anomalies (cleft palate and lip) but no specific syndrome. In lobar holoprosencephaly there is more cleavage of the prosencephalon but it remains incomplete. There is a deficiency of cerebral tissue, resulting in slightly dilated lateral ventricles, flattening of the roof and squaring of the frontal horns, and absence of the cavum septi pellucidi.[24] These children do not generally have facial abnormalities and they do survive to adulthood, with associated mental retardation. On ultrasound, the fetal cranium appears as a large cystic space with a mantle of cerebral tissue peripherally. Absence of the falx and corpus callosum with fused thalami allows differentiation from a large subarachnoid cyst or a posterior fossa cyst.

Craniosynostosis

Craniosynostosis, or premature fusion of the cranial sutures, may be complete or partial. Premature fusion of the lambdoidal, coronal, and sagittal sutures may lead to microcephaly and secondary microencephaly (small brain). Affected children may eventually have mental retardation and cranial nerve dysfunction unless they are treated surgically. Kleeblattschädel, or cloverleaf skull, is caused by premature fusion of only the coronal and lambdoidal sutures. The open sagittal suture permits growth only in one direction, leading to the trilobed "cloverleaf" skull appearance. These children have other associated congenital abnormalities and generally die in infancy; the children who survive are severely retarded. An in utero description of kleeblattschädel showed findings of delayed growth of the BPD early in pregnancy; late in the third trimester the cloverleaf head appeared as a cystic mass and was misinterpreted as a probable meningoencephalocele.[25] True microcephaly is due to failure of brain growth with normal suture development.

Other Skull Lesions

Fetal demise may lead to overlapping cranial bones, well known in the radiologic literature as Spalding's sign. The overlapping bones may be seen several days after fetal death. This sign may also be seen ultrasonographically in association with absent fetal heart movement and oligohydramnios, but it is a late ultrasonographic sign of fetal death (Fig 4–23).

Inadequate cranial calcification may develop due to skeletal syndromes, including achondrogenesis, osteogenesis imperfecta, and hypophosphatasia (Fig 4–24). On ultrasound the cranium looks fuzzy and poorly defined. The falx may appear particularly prominent in hypophosphatasia due to the poor mineralization of the cranium.[26] Theoretically, the falx could appear prominent in achondrogenesis and osteogenesis imperfecta as well. Other findings seen on ultrasound, including short limbs and abnormal vertebral body ossification, will help diagnose achondrogenesis and hypophosphatasia. Bent limbs representing in utero fractures assist in diagnosing osteogenesis imperfecta. Detailed bony anatomy requires a radiograph.

Microcephaly

Microcephaly is a small head which is more than 3 standard deviations below the mean. It is thought to occur in an incidence of $1:6,200$ to $1:8,500$; however, studies from which these figures were derived did not include stillborns or children who had died early in childhood.[27] The causes of microcephaly include (1) craniosynostosis (discussed above); (2) inherited factors, in association with the Meckel-Gruber

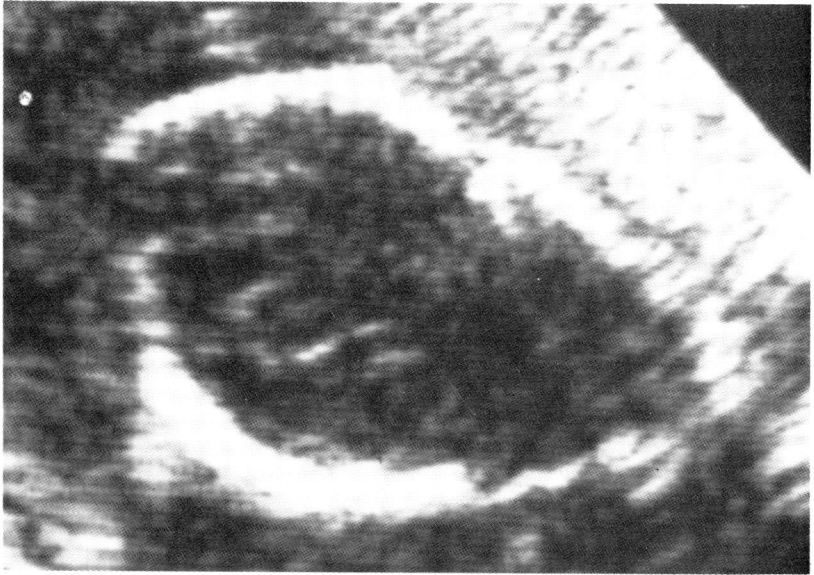

Fig 4–23.—Fetal demise in an 18-week-gestation fetus. Overlapping sutures of the collapsed fetal skull are apparent.

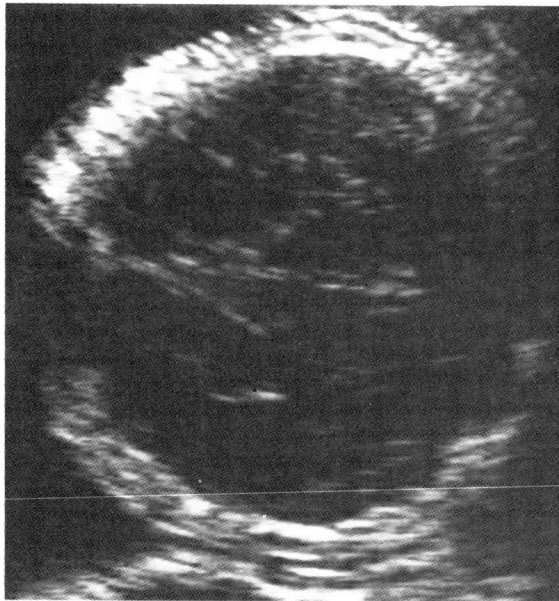

Fig 4–24.—Achondrogenesis in a 20-week-gestation fetus. The cranium is poorly calcified and there is associated mild skin thickening.

syndrome (polydactyly, encephalocele, polycystic kidney disease, and microcephaly); (3) chromosomal abnormalities, including trisomies, and (4) exposure to environmental teratogens including radiation and viruses.[16,28] The clinical findings of microcephaly are important only when the brain itself is small and unable to grow normally. Mental retardation is the major sequela. The findings that are seen on in utero ultrasound examination are (1) a BPD smaller than usual for gestational age as determined by menstrual dates or other parameters (e.g., femur length); (2) abdomen-head size discrepancy, with the head being smaller than expected; (3) poor growth of the fetal cranium on serial examinations, with normal abdominal growth; and (4) abnormal intracranial architecture from intrauterine infection. Totally normal brain parenchyma is also compatible with this abnormality. Intrauterine growth retardation (IUGR) affects fetal weight first, then fetal length, and finally brain growth.

DESTRUCTIVE LESIONS

Hydranencephaly

Hydranencephaly is a congenital deformity defined as total or near-total absence of the cerebral hemispheres with normally developed meninges and skull. The cerebellum and the midbrain, including the basal ganglia, are intact. The etiology of hydranencephaly is unknown. The two most common hypotheses for the etiology

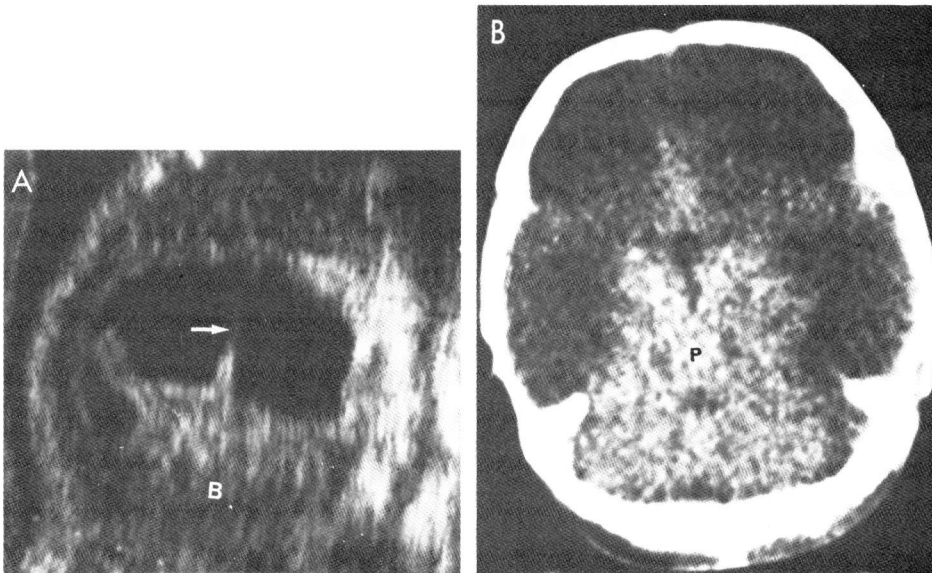

Fig 4–25.—Hydranencephaly in a 29-week-gestation fetus. **A,** A coronal scan shows absent frontal hemispheres, a normal falx *(arrow),* and brain tissue at base of skull *(B).* **B,** After delivery, a CT scan shows a cystic area in the region of the cerebral hemispheres and a normal posterior fossa *(P).*

are (1) intrauterine infarction secondary to bilateral internal carotid artery occlusion and (2) primary agenesis of the neural wall.[29] The former hypothesis is most widely accepted. Angiographic studies have supported this hypothesis by demonstrating hypoplasia of the supraclinoid internal carotid arteries with normal posterior cerebral and vertebrobasilar arteries.

The sonographic appearance of hydranencephaly includes (1) macrocephaly, (2) large echo-free areas within the cranial vault surrounding the midbrain and basal ganglia, (3) occasionally no midline echo (absent falx), (4) variable presence of the third ventricle, (5) a tentorium separating a relatively normal posterior fossa from the anterior and middle cranial fossae, (6) polyhydramnios, and (7) occasionally small portions of the occipital lobes supplied by the posterior cerebral arteries and the subfrontal cortex supplied by small vessels in the anterior cerebral distribution (Fig 4–25).[16,30]

The differential diagnosis of hydranencephaly includes massive subdural effusions, hydrocephalus, and alobar holoprosencephaly.[30] The prognosis for affected children is dismal, with few children living several years.

Infection

In utero infection may lead to abnormal fetal brain architecture. The ultrasound findings include hydrocephalus, increased echogenicity in the ependyma of the ventricular system, and variably increased and decreased echogenicity of the brain parenchyma. Intracranial calcifications associated with infection have been demonstrated by ultrasound both in utero and in the neonate (Fig 4–26). The calcium is seen only as bright echoes; shadowing is rarely seen.[31,32]

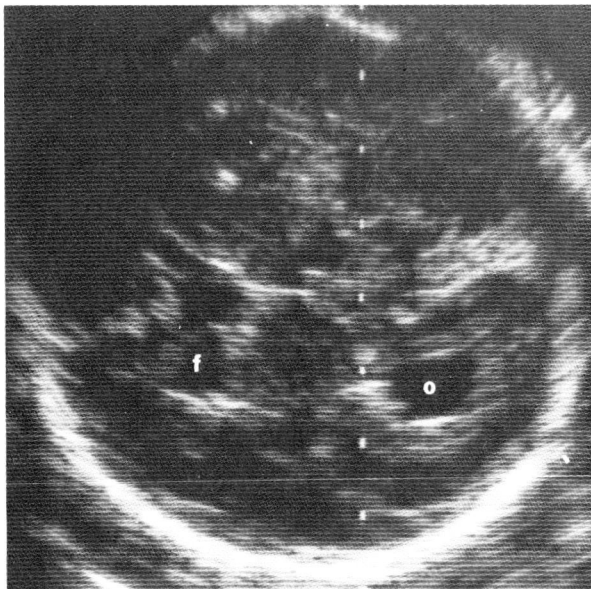

Fig 4–26.—Cytomegalovirus infection in a fetus at 31 weeks' gestation. The bright ependymal echoes around the frontal horns (f) and occipital horns (o) suggest infection. There is distortion of the parenchymal architecture.

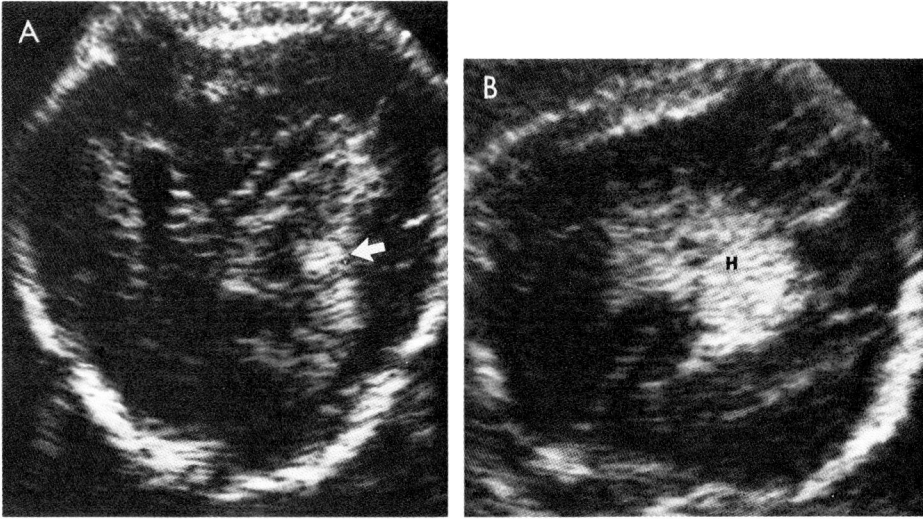

Fig 4–27.—Intracranial hemorrhage diagnosed in a nonviable fetus at 27 weeks' gestation. **A,** Coronal scan shows hemorrhage in the brain parenchyma *(arrow).* **B,** Oblique sagittal scan shows hemorrhage *(H)* in the periventricular white matter.

Intracranial Hemorrhage

Today, intracranial hemorrhage can be routinely diagnosed by ultrasound in premature infants. In utero diagnosis of intracranial hemorrhage has also been reported.[33] In the case studied by Kim and Elyaderani[33] and in a case from our laboratory, fetal death and hydrocephalus were concurrent with the finding of intracranial hemorrhage. The hemorrhage was identified as echogenic matter filling the lateral ventricle, and the LVR was increased (Fig 4–27). Ultrasonic demonstration of in utero hemorrhage should reveal the same findings as are seen in the neonate,[34] although only severe hemorrhage has been identified so far.

MASS LESIONS

Cerebral arachnoid cysts lie between layers of the arachnoid or between arachnoid and dura.[35] These cysts do not communicate with the ventricles or the arachnoid space, although their contents are very similar to CSF. The cysts are thought to be a product of maldevelopment of the leptomeninges; originally they probably did communicate within the subarachnoid space and did contain CSF, but over time the communications closed.[36] The prognosis for children with arachnoid cysts is variable, from normal with diagnosis late in life to mental retardation, seizures, and paresis. On ultrasound examination a fluid collection is identified in the cerebral cortex or in the posterior fossa.

Congenital teratomas have been identified in utero; however, the number of cases is so small that only a few case reports appear in the literature. The teratomas have been located intracranially (Fig 4–28),[37] cervically,[38] and arising from the oral

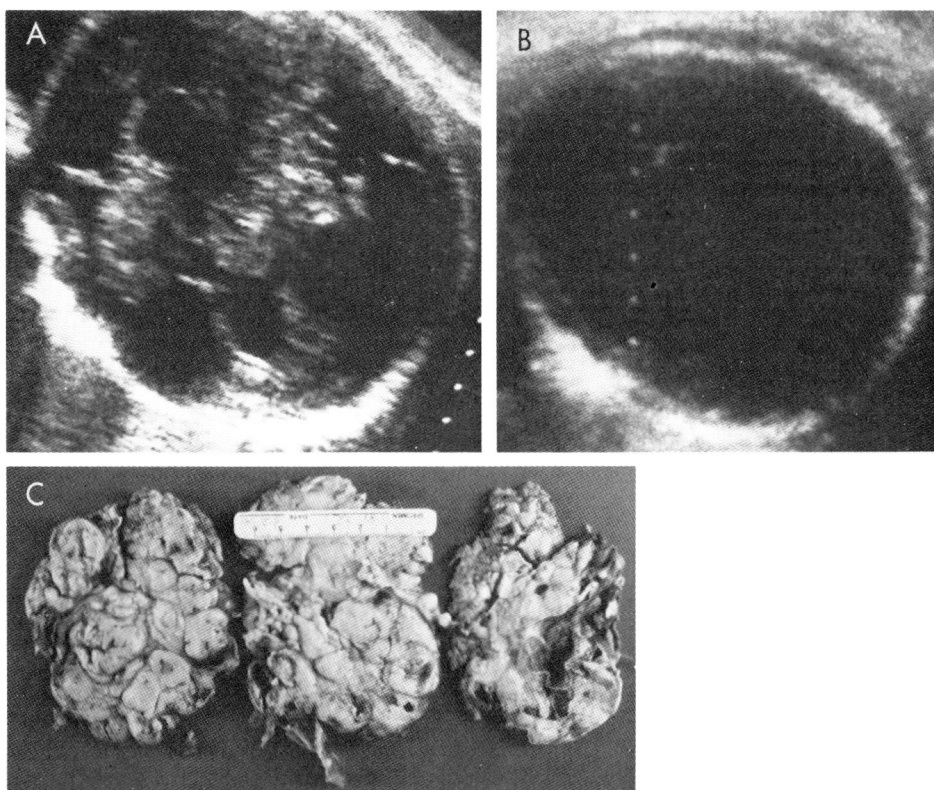

Fig 4–28.—Intracranial teratoma. **A,** A complex mass with both solid and cystic components nearly fills the cranium at the level of the third ventricle. **B,** Axial scan above the tumor demonstrates massive hydrocephalus. **C,** Pathologic specimen of teratoma with cystic and solid components.

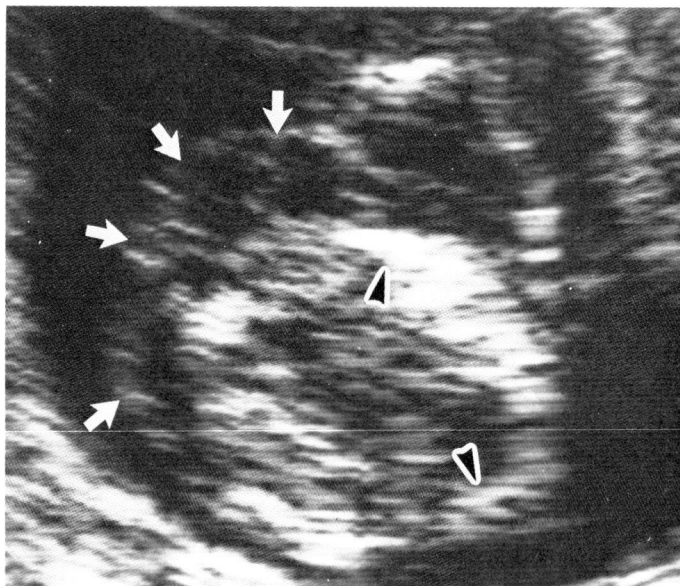

Fig 4–29.—Cystic hygroma in a 12-week-gestation fetus. A multicystic mass *(arrows)* is seen adjacent to the fetal head *(arrowheads)*. This fetus had XO chromosomes (Turner's syndrome).

cavity or pharynx (also referred to as epignathus).[39] The masses are complex but may be predominantly cystic, simulating a cystic hygroma. Calcifications may cause dense echoes with acoustic shadowing. Polyhydramnios was noted in all three of these case reports; however, the literature does not support such a high incidence of polyhydramnios associated with teratomas. Polyhydramnios certainly can be a significant factor in seeing mass lesions on ultrasound. Cystic hygromas (Fig 4–29), which form secondary to sequestration of lymphatic tissue or lymphatic obstruction, branchial cleft cysts resulting from failed closure of an embryonic cleft, and hemangiomas may simulate encephaloceles or teratomas.[40]

REFERENCES

1. Crade M., Patel J., McQuown D.: Sonographic imaging of the glycogen state of the fetal choroid plexus. *AJR* 137:489–491, 1981.
2. Johnson M.L., Dunne M.G., Mack L.A., et al.: Evaluation of fetal intracranial anatomy by static and real-time ultrasound. *JCU* 8:311–318, 1980.
3. Hadlock F.P., Deter R.L., Park S.K.: Real-time sonography: Ventricular and vascular anatomy of the fetal brain in utero. *AJR* 136:133–137, 1981.
4. Denkhaus H., Winsberg F.: Ultrasonic measurement of the fetal ventricular system. *Radiology* 131:781–787, 1979.
5. Lemire R.J., Loeser J.D., Leech R.W., et al.: *Normal and Abnormal Development of the Human Nervous System.* New York, Harper & Row, 1975, p. 95.
6. Moore K.L.: *The Developing Human: Clinically Oriented Embryology,* ed. 2. Philadelphia, W.B. Saunders Co., 1977.
7. Laing F.C., Stamler C.E., Jeffrey R.B.: Ultrasonography of the fetal subarachnoid space. *J. Ultrasound Med.* 2:29–32, 1983.
8. Birnholz J.C.: Newborn cerebellar size. *Pediatrics* 70:284–287, 1982.
9. Reuter K.L., D'Orsi C.J., Raptopoulos V.D., et al.: Sonographic pseudoasymmetry of the prenatal cerebral hemispheres. *J. Ultrasound Med.* 1:91–92, 1982.
10. Robertson R.D., Sarti D.A., Brown W.J., et al.: Congenital hydrocephalus in two pregnancies following the birth of a child with neural tube defect: Aetiology and management. *J. Med. Genet.* 18:105–107, 1981.
11. Bay C., Kerzin L., Hall B.: Recurrence risk in hydrocephalus. *Birth Defects Original Article Series* 15:95–105, 1979.
12. Habib Z.: Genetics and genetic counselling in neonatal hydrocephalus. *Obstet. Gynecol. Surv.* 36:529–534, 1981.
13. McCullough D.C., Balzer-Martin L.A.: Current prognosis in overt neonatal hydrocephalus. *J. Neurosurg.* 57:378–383, 1982.
14. Milhorat T.H., Clark R.G., Hammock M.K., et al.: Structural, ultrastructural and permeability changes in the ependyma and surrounding brain favoring equilibrium in progressive hydrocephalus. *Arch. Neurol.* 22:397–407, 1970.
15. Weller R.O., Shulman K.: Infantile hydrocephalus: Clinical, histological, and ultrastructural study of brain damage. *J. Neurosurg.* 36:255–265, 1972.
16. Fiske C.E., Filly R.A.: Ultrasound of the normal and abnormal fetal neural axis. *Radiol. Clin. North Am.* 20:285–296, 1982.
17. Shurtleff D.B., Foltz E.L., Loeser J.D.: Hydrocephalus. *Am. J. Dis. Child.* 125:688–693, 1973.
18. Clewell W.H., Johnson M.L., Meier P.R., et al.: A surgical approach to the treatment of fetal hydrocephalus. *N. Engl. J. Med.* 306:1320–1325, 1982.
19. Nadich T., Pudlowsk R.M., Nadich J.B., et al.: Computed tomographic signs of the Chiari II malformation: II. Midbrain and cerebellum. *Radiology* 134:65–71, 1980.
20. Zimmerman R.D., Brockbill D., Dennis M.W., et al.: Cranial CT findings in patients with myelomeningocele. *AJR* 132:623–629, 1979.

21. Babcock D.S., Han B.K.: Cranial sonographic findings in meningomyelocele. *AJR* 136:563–569, 1981.
22. Newman G.C., Buschi A.I., Sugg N.K., et al.: Dandy-Walker syndrome diagnosed in utero by ultrasonography. *Neurology* 32:180–184, 1982.
23. Benda C.E.: The Dandy-Walker syndrome or the so-called atresia of the foramen of Magendie. *J. Neuropathol. Exp. Neurol.* 13:14–29, 1954.
24. Byrd S.E., Harwood-Nash D.C., Fitz C.R., et al.: Computed tomography evaluation of holoprosencephaly in infants and children. *J. Comput. Assist. Tomogr.* 4:456–462, 1977.
25. Brahman S., Jenna R., Wittenauer H.J.: Sonographic in utero appearance of kleeblatts-chadel syndrome. *JCU* 7:481–484, 1979.
26. Laughlin C.L., Lee T.G.: The prominent falx cerebri: New ultrasonic observation in hypophosphatasia. *JCU* 10:37–38, 1982.
27. Kurtz A.B., Wapner R.J., Rubin C.S., et al.: Ultrasound criteria for in utero diagnosis of microcephaly. *JCU* 8:11–16, 1980.
28. Smith D.W.: *Recognizable Patterns of Human Malformations*, ed. 3. Philadelphia, W.B. Saunders Co., 1982, p. 617.
29. Yakovlev P.L., Wadsworth R.C.: Schizencephalies: A study of congenital clefts in the cerebral mantle: I. Clefts with fused lips. *J. Neuropathol. Exp. Neurol.* 5:16–130, 1946.
30. Dublin A.B., French B.N.: Diagnostic image evaluation of hydranencephaly and picto-rially similar entities, with emphasis on computed tomography. *Radiology* 137:81–91, 1980.
31. Dykes F.D., Ahmann P.A., Lazzara A.: Cranial ultrasound in the detection of intracra-nial calcifications. *J. Pediatr.* 100:406–408, 1982.
32. Graham D., Guidi S., Sanders R.C.: Sonographic features of in utero periventricular calcification due to cytomegalovirus infection. *J. Ultrasound Med.* 1:171–172, 1982.
33. Kim M.S., Elyaderani M.K.: Sonographic diagnosis of cerebroventricular hemorrhage in utero. *Radiology* 142:479–480, 1982.
34. Johnson M.L., Rumack C.M., Mannes E.J., et al.: Detection of neonatal intracranial hemorrhage utilizing real-time and static ultrasound. *JCU* 9:427–433, 1981.
35. Anderson F.M., Landing B.H.: Cerebral arachnoid cysts in infants. *J. Pediatr.* 69:88–96, 1966.
36. Starkman S.P., Brown T.C., Linell E.A.: Cerebral arachnoid cysts. *J. Neuropathol. Exp. Neurol.* 17:484, 1958.
37. Hoff N.R., Mackay I.M.: Prenatal ultrasound diagnosis of intracranial teratoma. *JCU* 8:247–249, 1980.
38. Rosenfeld C.R., Coln C.D., Duenhoelter J.H.: Fetal cervical teratoma as a cause of polyhydramnios. *Pediatrics* 64:176–179, 1979.
39. Kang W.K., Hissong S.L., Langer A.: Prenatal ultrasonic diagnosis of epignathus. *JCU* 6:330–331, 1978.
40. Morgan C.L., Haney A.H., Christokos A.: Antenatal detection of fetal structural defects with ultrasound. *JCU* 3:287–290, 1975.

CHAPTER 5

Congenital Brain Malformations

Carol M. Rumack, M.D.
Michael L. Johnson, M.D.

NORMAL BRAIN DEVELOPMENT

THERE ARE SPECIFIC STAGES in brain development which, when altered, may cause specific brain defects. The normal stages of cellular, tissue, and organ development have been described in detail by Volpe (Table 5–1).[1] The development of cells, or cytogenesis, cannot yet be studied with available radiologic techniques. The development of tissues, or histogenesis, can be studied in neoplasms and in some vascular lesions, which will be described later in this chapter. The development of organs, or organogenesis, is the blending of tissues into organs. Knowledge of the normal stages of organogenesis can help in understanding most congenital malformations.

Organogenesis

At 3–4 weeks' gestation, the flat sheet of cells called the neural tube curls and fuses dorsally (Fig 5–1).[2] The anterior and posterior ends (neuropores) close off, forming a hollow neural tube. At 5–6 weeks' gestation, diverticulation of the forebrain occurs. The cerebral vesicles enlarge and separate into the cerebral hemi-

TABLE 5–1.—STAGES OF BRAIN DEVELOPMENT*

CYTOGENESIS—Development of molecules into cells
HISTOGENESIS—Development of cells into tissues
ORGANOGENESIS—Development of tissues into organs
    Neural tube closure (dorsal induction; 3–4 weeks' gestation)
    Diverticulation (ventral induction; 5–6 weeks' gestation)
    Neuronal proliferation (2–4 months' gestation)
    Neuronal migration (3–6 months' gestation)
    Organization (6 months' gestation to years postnatally)
    Myelination (birth to years postnatally)

    *Modified from Volpe.[1] Stages overlap in time but may be individually abnormal.

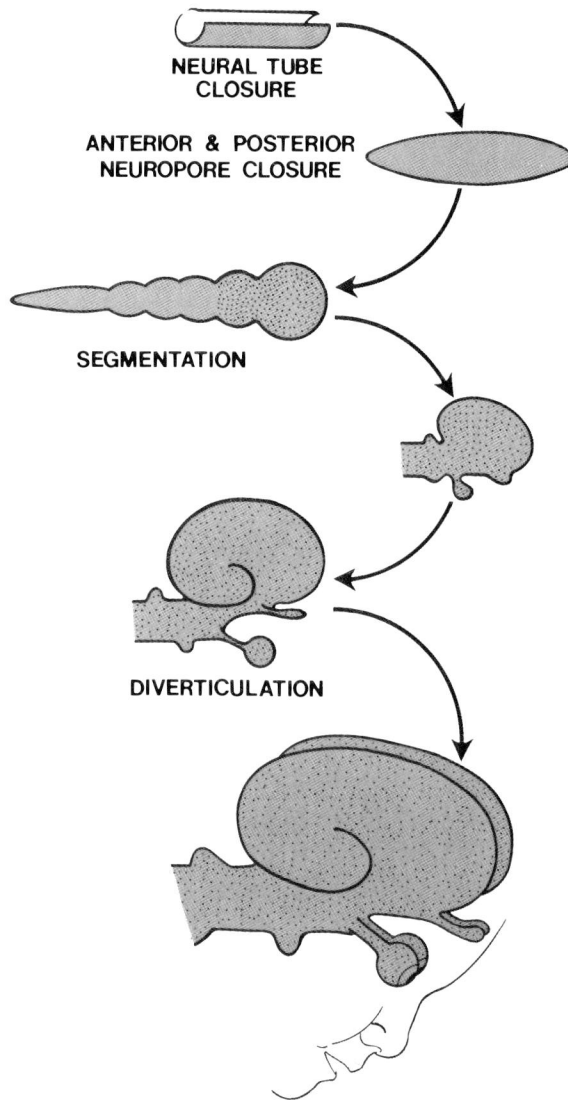

Fig 5–1.—Stages of organogenesis. The neural tube closes at 3–4 weeks' gestation, including closure of the anterior and posterior neuropores. At 5–6 weeks' gestation, the brain segments and then five types of diverticula form. First the paired olfactory tracts, optic tracts, and cerebral ventricles develop, then the unpaired pineal and pituitary. Neural proliferation, migration, organization, and myelination occur after these stages. (Modified from *DeMyer.*[2])

spheres. The paired optic tracts and olfactory tracts separate from the cerebral cortex. Two unpaired diverticula develop to form the posterior pituitary (ventrally) and the pineal (dorsally). At 2–6 months' gestation, neural proliferation and migration occur, resulting in multiple layers of cells. Organization of these tracts occurs throughout this time but is more developed from 6 months' gestation. Myelination occurs after birth so defects in myelination are not termed congenital.

TABLE 5–2.—CONGENITAL BRAIN MALFORMATIONS*

DISORDERS OF ORGANOGENESIS
  Disorders of neural tube closure
    Cranioschisis (dysraphism)
      Anencephaly
      Encephalocele
      Myelomeningocele
    Chiari malformation
    Dandy-Walker syndrome
    Agenesis of the corpus callosum
    Lipoma of the corpus callosum
    Teratoma
  Disorders of diverticulation
    Septo-optic dysplasia
    Lobar holoprosencephaly
    Alobar holoprosencephaly
    Aventricular cerebrum
  Disorders of proliferation
    Microcephaly
    Macrocephaly
      Cerebral gigantism—Soto's syndrome
      Megalocephaly—storage diseases
  Disorders of sulcation and migration
    Lissencephaly
    Schizencephaly
    Heterotopias
    Polymicrogyria
  Destructive lesions†
    Hydranencephaly
    Porencephaly
    Hypoxia
    Toxicoses
    Infectious diseases
      Rubella
      Cytomegalovirus
      Toxoplasmosis
      Herpes simplex
  Disorders of organization‡
  Disorders of myelination‡
DISORDERS OF HISTOGENESIS§
  Tuberous sclerosis
  Neurofibromatosis
  Sturge-Weber's disease (encephalotrigeminal angiomatosis)
  Neoplasia
  Vascular lesions

*Table modified from Harwood-Nash and Fitz.[3]
†See chapters on infarction and infection.
‡Not well described yet. May be defined by nuclear magnet resonance imaging.
§See chapters on tumors and vascular disorders.

DeMyer's elegant description and classification of pathogenesis are based on these stages and can be used to understand the logical pattern of the associated defects.[2] Harwood-Nash and Fitz have adapted the DeMyer classification to congenital brain anomalies which are recognized by present neuroradiologic techniques (Table 5–2).[3] Cerebral malformations are classified according to the types of errors that occur in cerebral development. The classification indicates the malformation, not the etiology. Thus, several mechanisms could cause the same malformation, and the final anatomical appearance will not necessarily reflect the original cause.

Developmental myelination defects have not been reported except in a recent description of cerebral white matter hypoplasia in 12 severely retarded patients studied by Chattha and Richardson.[4] With the development of nuclear magnetic resonance imaging, normal brain stages of myelination and myelination defects should be well defined and defects of organization may be recognizable. Organogenetic malformations are the main topic of this chapter; histogenetic malformations (abnormal tissue cell development) are discussed in other chapters.

DISORDERS OF NEURAL TUBE CLOSURE

Cranioschisis or Dysraphic Disorders

Cranioschisis or dysraphism means a splitting of the brain, but the condition is actually caused by failure of the neural tube to close. The level at which the neural tube closure defect occurs determines which anomaly will be present. Dysraphic disorders include anencephaly, encephalocele, and myelomeningocele. Anencephaly results from failure of development of the brain except for the base of the skull.[5] An encephalocele develops when the neural tube does not close, resulting in a skull defect, usually in the occipital region. The larger the amount of neural tissue that herniates out, the less the brain and skull will develop.

Below the brain, herniation of the meninges and cord into the neural tube defect is known as a meningomyelocele. If there is only a meningeal sac, it is a meningocele. If the entire neural tube is open with a flat, splayed cord and no actual sac, it is termed myeloschisis.[6]

Anencephaly, encephalocele, and meningomyelocele are usually easily diagnosed clinically in the newborn by direct observation and rarely require other studies. Computerized tomographic (CT) scans may be necessary to define the skull defect, to determine the presence or absence of neural tissue in the herniated contents, and to define intracranial anatomy.[7] It is possible to diagnose these entities in utero with ultrasound, and antenatal diagnosis can significantly alter maternal and fetal care. These lesions are discussed in the chapter on fetal intracranial diagnosis. Whenever a neural tube defect is present there is almost always a malformation of the brain known as the Arnold-Chiari or Chiari II malformation.

Chiari II Malformation

The Chiari malformation is actually a spectrum of anomalies having in common a basic malformation of the brain stem and cerebellum associated with dysraphism of the spinal cord and spine (Fig 5–2). There are many theories as to the exact mechanism, but it seems most sensible to assume that both anterior neural tube closure and posterior neural tube closure were affected simultaneously in the first 3–4 weeks of gestation. If a meningomyelocele is present, it is nearly pathogonomic for the Chiari II malformation, since 85%–90% of meningomyelocele patients have this anomaly.[8, 9] Variants of the Chiari II malformation are called type I and type III malformations. Type I denotes a variable downward displacement of the tonsils and cerebellum with a normal fourth ventricle and is usually diagnosed in adults without a meningomyelocele.[3] Type III denotes displacement of the medulla, fourth ventricle, and all of the cerebellum into an occipital and high cervical encephalomeningocele. These defects in organogenesis, known collectively as the Chiari II malformation, affect the entire brain and are discussed below by anatomical area of abnormality (Table 5–3).

VENTRICLES AND CISTERNS IN THE CHIARI II MALFORMATION.—Elongation and caudal displacement of the fourth ventricle are the key findings in Chiari II malformation.[10, 11] The fourth ventricle is frequently low and compressed due to herniation of the cerebellum into the foramen magnum.[9, 12] The third and lateral ventricles are usually enlarged, although the third ventricle may be narrowed by an

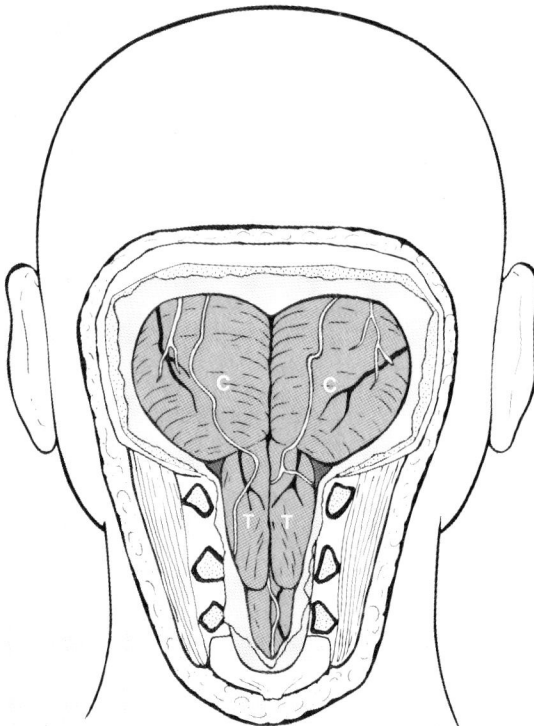

Fig 5–2.—Chiari II malformation. The cerebellar tonsils (T) and cerebellum (C) have herniated into the foramen magnum, causing compression of the fourth ventricle and hydrocephalus.

TABLE 5–3.—Disorders of Closure

CHIARI II MALFORMATION
Ventricles and cisterns
Elongation and caudal displacement of the fourth ventricle
Enlarged lateral ventricles with occipital larger than frontal horns
Anterior and inferior pointed frontal horns (batwing)
Third ventricle only slightly large due to encroachment of massa intermedia
Partial absence of the septum pellucidum
Prominent interhemispheric fissure
Prominent suprapineal recess
Poorly visualized basal cisterns
Midbrain and cerebellum
Enlarged massa intermedia (interthalamic adhesion)
Prominent anterior commissure
Beaking of the midbrain
Variable fusion of the colliculi
Tentorial-cerebellar pseudomass (especially with shunting)
Towering cerebellum
Bullet-shaped incisura
Skull and dura
Lacunar skull (lückenschädel)
Petrous pyramid scalloping
Enlarged foramen magnum
Hypoplastic falx
Hypoplastic tentorium with low insertion and V-shaped incisura
DANDY-WALKER SYNDROME
Enlarged fourth ventricle (Dandy-Walker cyst)
Hypoplastic cerebellar hemispheres
Hypoplastic or absent cerebellar vermis
High insertion of tentorium, torcular and transverse sinuses
Obliteration of basal cisterns by cyst
Hydrocephalus of third and lateral ventricles
AGENESIS OF CORPUS CALLOSUM
Elongation and superior displacement of the third ventricle
Widely separated lateral ventricles
No septum pellucidum
Peaked frontal horns
Medially concave occipital horns

enlarged massa intermedia. Hydrocephalus may not be apparent until repair of the meningomyelocele changes cerebrospinal fluid (CSF) dynamics, resulting in increasing ventricular size in the first few weeks of life.[13]

On ultrasound examination,[9] the cerebellum will appear low in the posterior fossa and the fourth ventricle will be barely visible, if at all (Fig 5–3). The massa intermedia obscures the third ventricle in the region below the the foramen magnum (Fig 5–4). The "batwing" appearance is particularly evident on sagittal and coronal views of the frontal horns (Fig 5–5).

The basal cisterns in the posterior fossa are frequently not visible, owing to compression by the cerebellum and brain stem structures. There may be prominence of the suprapineal recess and partial absence of the septum pellucidum. Frequently there is widening of the interhemispheric fissure, particularly after shunting.

On CT examination, the fourth ventricle is usually not visible and when present is usually small (Fig 5–6).[14] The third ventricle may be slightly enlarged but is

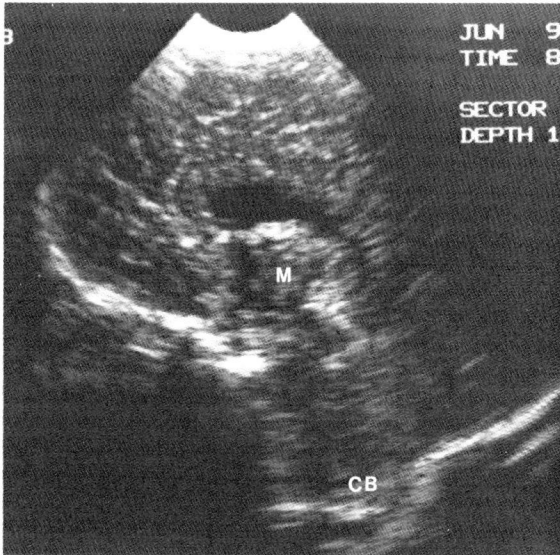

Fig 5–3.—Chiari II malformation. Sagittal ultrasound scan shows displacement of inferior cerebellum *(CB)* into foramen magnum. The fourth ventricle is difficult to visualize. The large massa intermedia *(M)* fills the posterior part of the third ventricle.

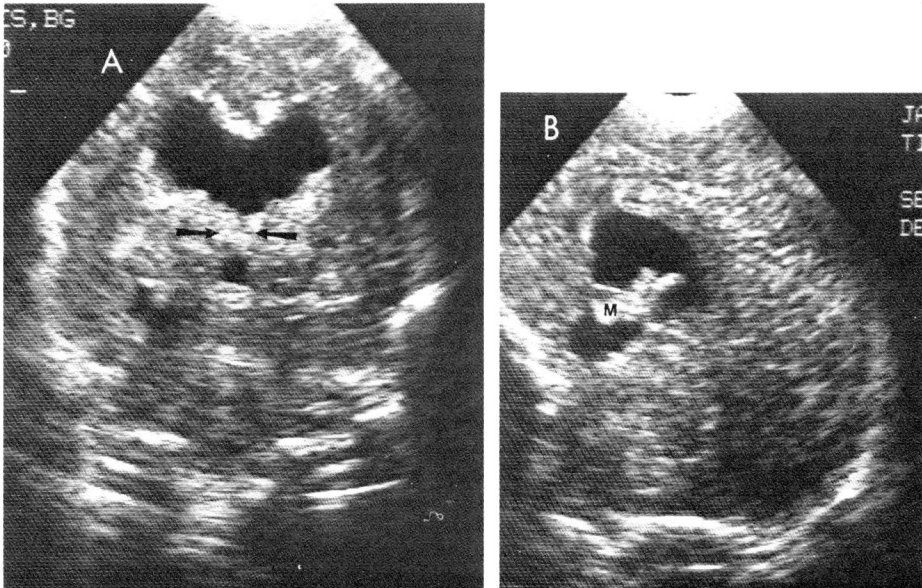

Fig 5–4.—**A,** Massa intermedia obscures top half of third ventricle *(arrows)* in coronal scan of patient with Chiari II malformation. **B,** Same patient with large massa intermedia *(M)* on sagittal ultrasound scan. The brain is highly echogenic owing to intrauterine infection.

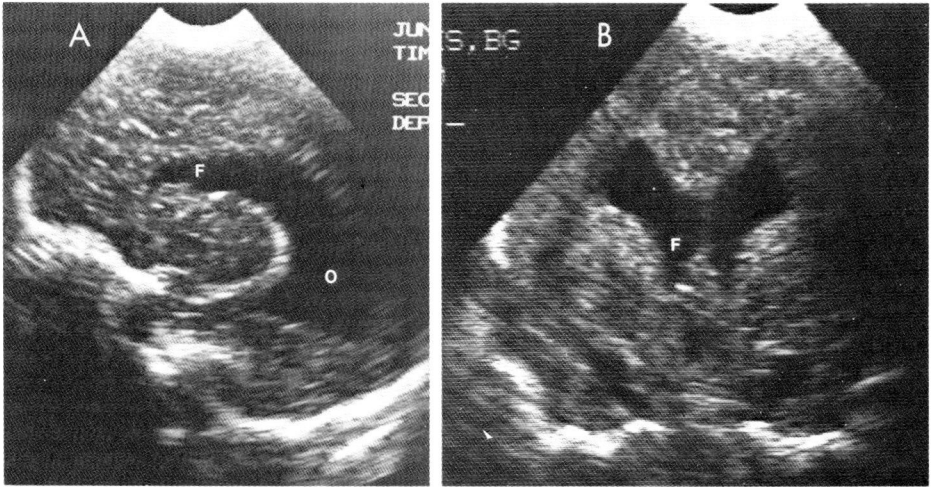

Fig 5–5.—A, "Batwing" or pointed frontal horns *(F)* with enlarged occipital horns *(O),* in newborn with Chiari II malformation on sagittal ultrasound scan. **B,** Coronal ultrasound scan demonstrating inferior pointing of frontal horns *(F)* in same patient.

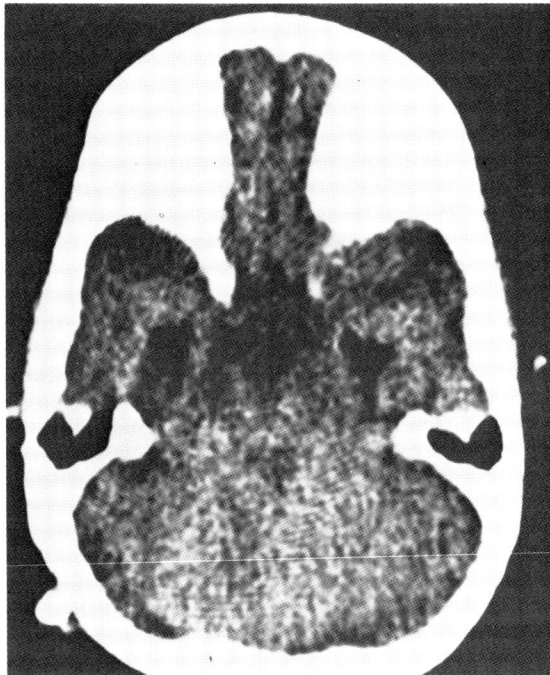

Fig 5–6.—Axial CT scan demonstrating absence of fourth ventricle in 7-month-old female infant shunted at birth for hydrocephalus and a Chiari II malformation.

partially obscured by the massa intermedia (Fig 5–7). The occipital horns are much larger than the frontal horns (Fig 5–8). The classic batwing appearance of the ventricles is caused by anterior and inferior pointing of the frontal horns (Fig 5–9).

MIDBRAIN AND CEREBELLUM IN THE CHIARI II MALFORMATION.—The massa intermedia or interthalamic adhesion is usually quite large and may tether the walls of the third ventricle. Typically there is beaking of the midbrain posteriorly (Fig 5–10).[15] More difficult to visualize with present techniques are an enlarged anterior commissure and fusion of the colliculi.

SKULL AND DURA IN CHIARI II MALFORMATION.—On CT examination, the skull is frequently scalloped on the inner table with bony ridges giving an appearance called a lacunar skull, or lückenschädel.[16] This will be particularly apparent with bone windows in the high frontal and parietal areas (Fig 5–11). The petrous pyramids are normally convex posteriorly and may be flattened or concave. The foramen magnum is typically enlarged in all directions and therefore round (Fig 5–12). Skull abnormalities can only be studied well with CT, since ultrasound shows only the inner surfaces of the skull.

On ultrasound and CT, the falx may be hypoplastic, allowing interdigitation of the cerebri sulci. The hypoplastic tentorium has a very large incisura and is attached low, resulting in a shallow posterior fossa (Fig 5–13).

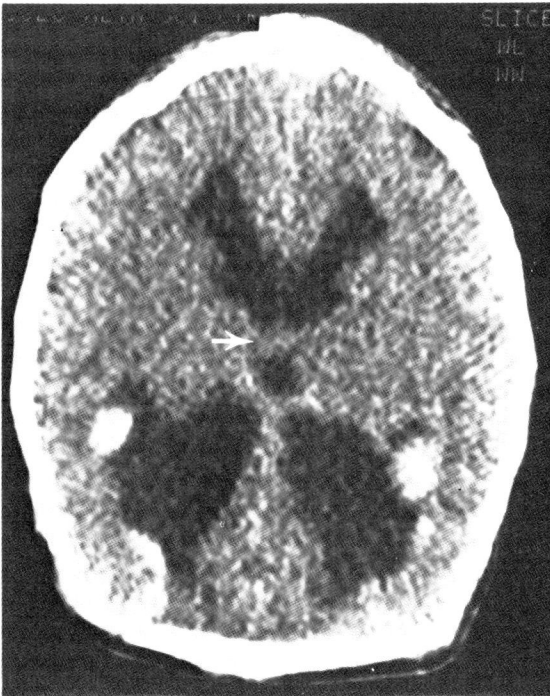

Fig 5–7.—Large massa intermedia *(arrow)* obscures anterior portion of third ventricle in term neonate with a Chiari II malformation. Note incidental subependymal hemorrhage.

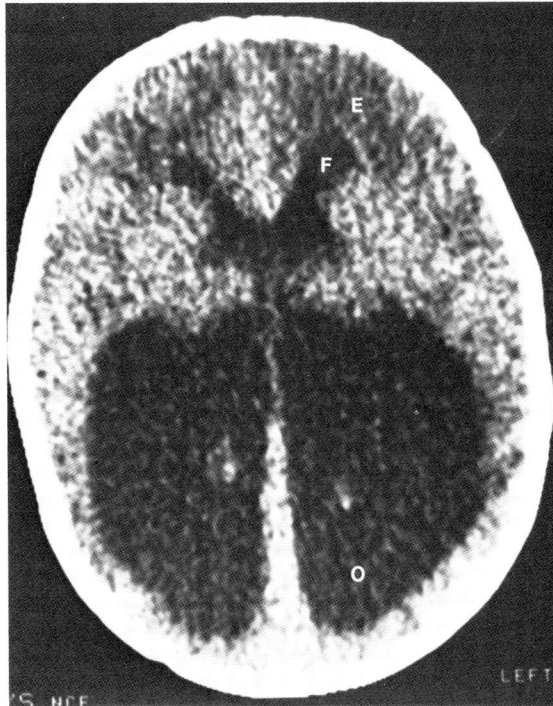

Fig 5–8.—Occipital horns *(O)* are typically much larger than the frontal horns *(F)* in the Chiari II malformation. This is an axial CT scan in a term neonate with periventricular edema *(E)* anterior to the frontal horns. Edema was secondary to pressure and disappeared after shunting.

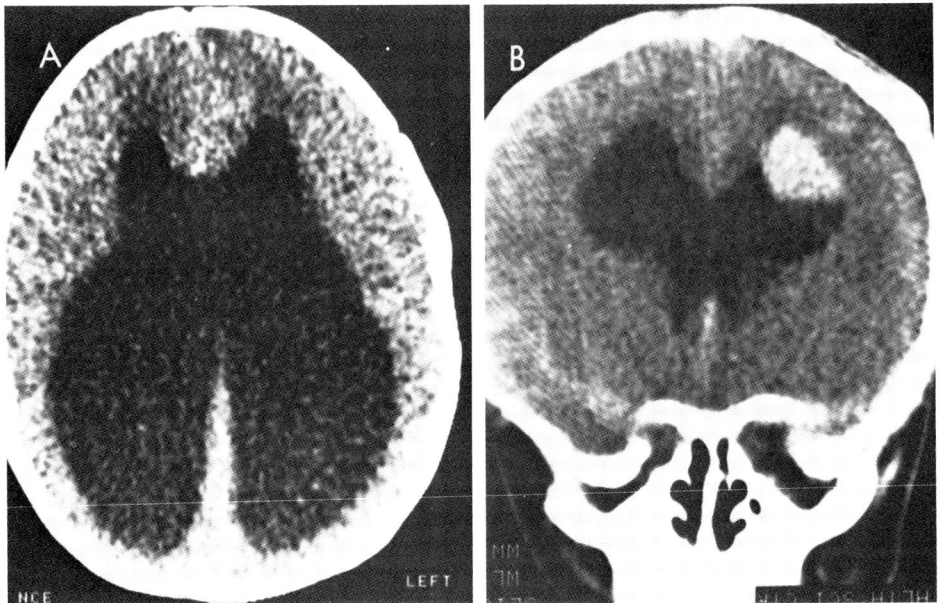

Fig 5–9.—**A,** Anterior pointing of frontal horns demonstrated on axial CT scan in term neonate with Chiari II malformation. **B,** Inferior pointing of frontal horns (same patient as in Fig 5–7). There is incidental subependymal hemorrhage.

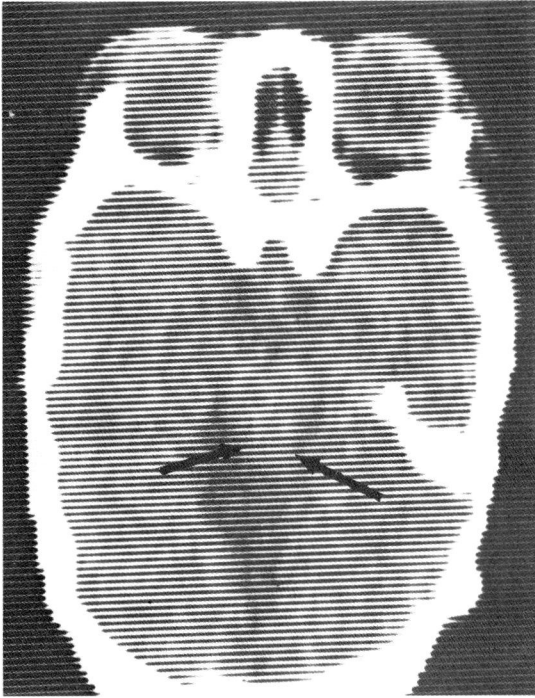

Fig 5-10.—Posterior beaking of the midbrain *(arrows)* on axial CT scan. The enlarged tentorial incisura is noted after shunting in this term neonate with a Chiari II malformation.

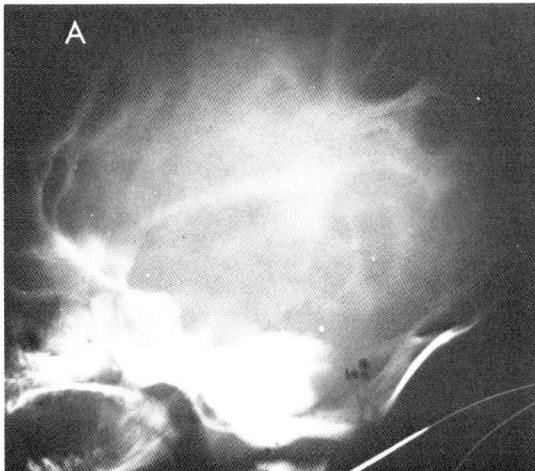

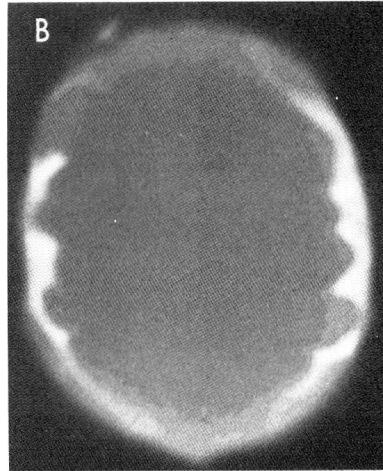

Fig 5-11.—Lacunar skull, or lükenschädel. **A,** Alternating lucent and dense areas of bone on lateral skull in newborn. **B,** Axial CT scan with bone window in same patient demonstrating scalloping of inner surface of the skull in the Chiari II malformation.

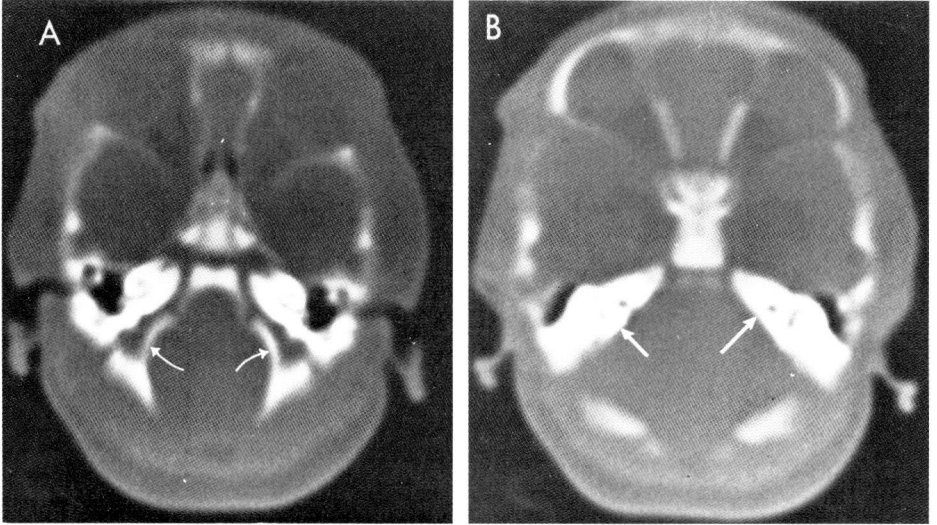

Fig 5–12.—A, Enlarged foramen magnum *(arrows)* in newborn with Chiari II malformation. **B,** Petrous pyramids are flattened posteriorly *(arrows)* in same patient.

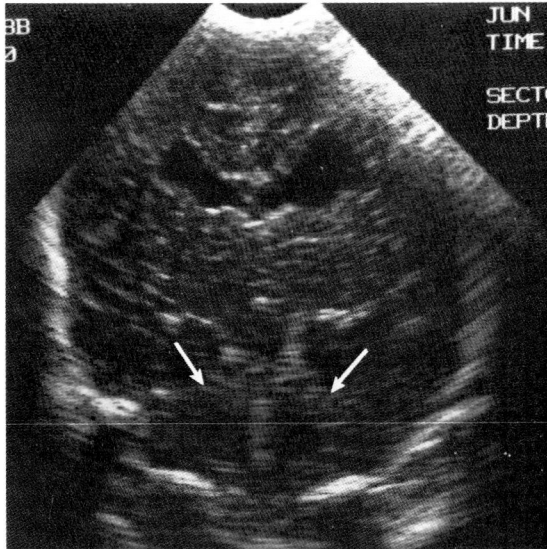

Fig 5–13.—Low placement of tentorium *(arrows)* in neonate with Chiari II malformation.

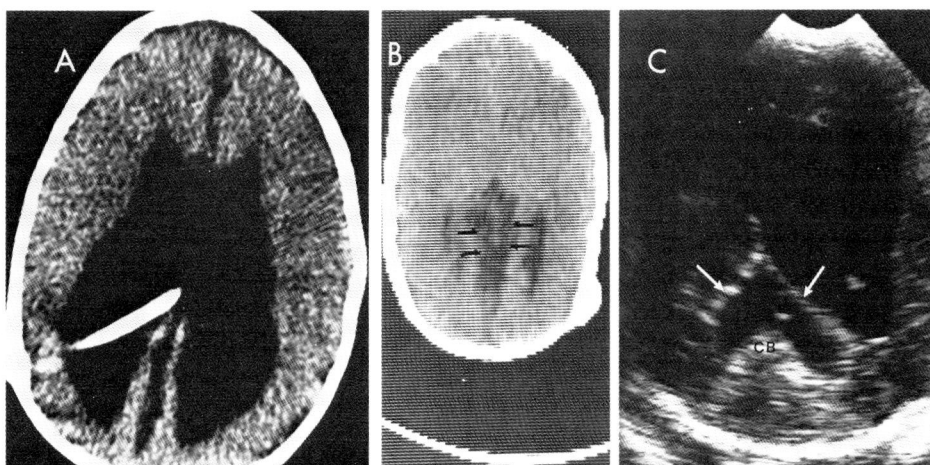

Fig 5–14.—Postshunt appearance of Chiari II malformation. **A,** Large interhemispheric fissure in 7-month-old infant with shunt malfunction and recurrent hydrocephalus on axial CT scans. **B,** Neonate at 1 week post shunt with tiny ventricles and large tentorial incisura *(arrows).* Cerebellum is visualized in the incisura on the axial CT scan. **C,** Coronal ultrasound of 1-week-old infant after shunt. Large supracerebellar cistern below tentorium *(arrows)* and cerebellum *(CB)* are prominent centrally. An axial section of this scan would show the same appearance as **B,** resulting in the posterior fossa pseudomass.

POSTSHUNTING CHIARI MALFORMATION.—After shunting for hydrocephalus, the ventricles are usually quite small, with a very prominent interhemispheric fissure and gyral interdigitations. The cerebellum may "tower" into the large, V-shaped midline cisterns behind the third ventricle caused by the confluence of the velum interpositum, superior vermian cistern, and enlarged ambient cisterns (Fig 5–14).

Dandy-Walker Syndrome

The key finding in the Dandy-Walker syndrome is the presence of a posterior fossa cyst, which is actually an extremely enlarged fourth ventricle due to atresia of the foramina of Magendie and Luschka with cerebellar vermian dysgenesis.[17, 18] In addition to the massively enlarged fourth ventricle, there is typically hydrocephalus involving the third and lateral ventricles. The basilar cisterns are obliterated by the cyst and the posterior fossa is markedly enlarged. The tentorium, straight sinus, and transverse sinuses are elevated. The cerebellar vermis may be absent or hypoplastic.

These findings are well demonstrated on either ultrasound or CT (Fig 5–15). Hydrocephalus presents at birth, causing an enlarged head in approximately half of patients with the Dandy-Walker syndrome.[17] These characteristic findings can be easily diagnosed in utero.

The skull findings consist mainly of enlargement of the posterior fossa. The transverse sinuses are typically superiorly displaced above the lambdoidal suture.

A Dandy-Walker variant is present if there is a large posterior fossa cyst, a par-

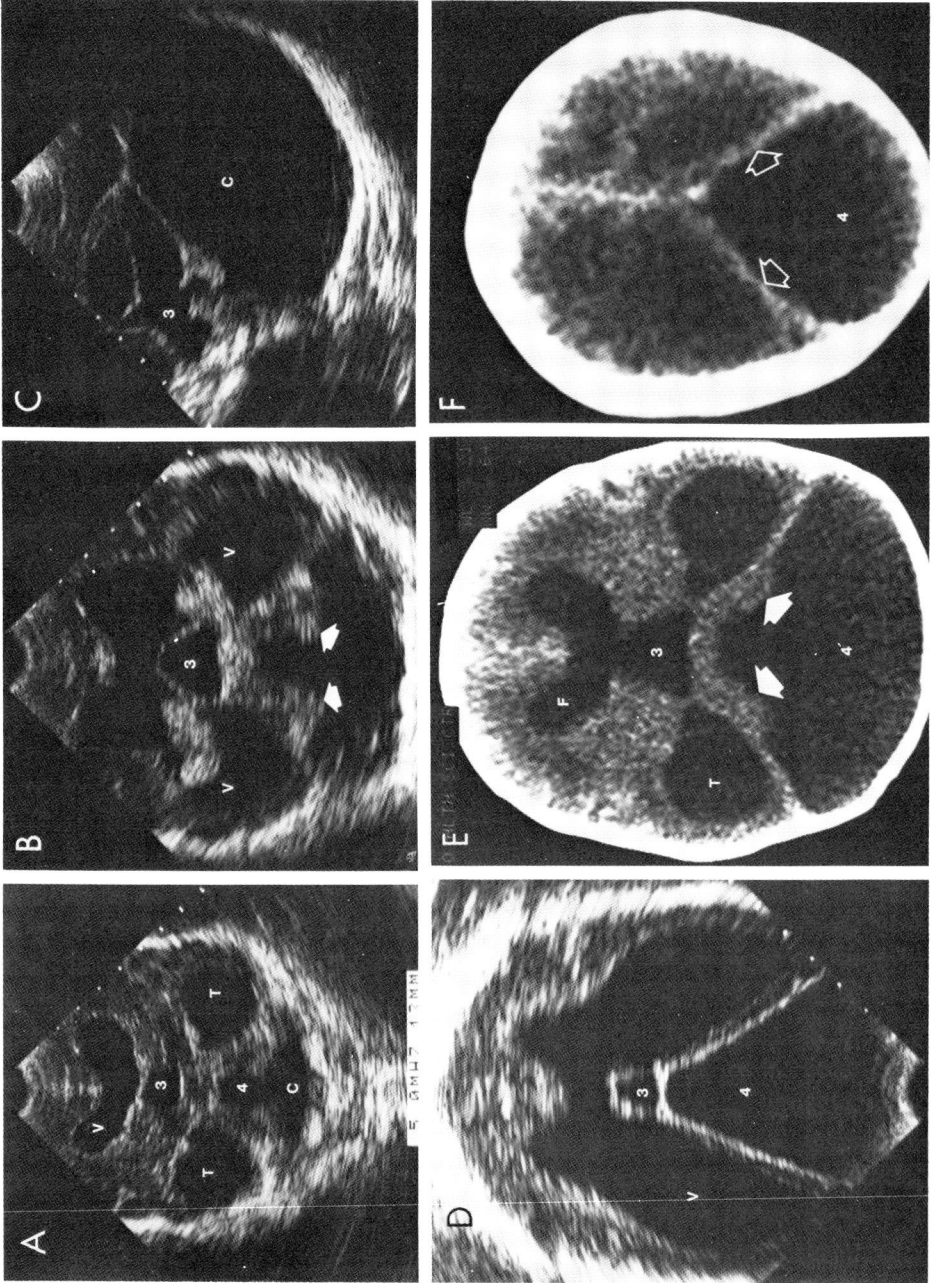

Fig 5–15.—See legend on facing page.

tially formed fourth ventricle with a large vallecula connecting the two areas. Cerebellar hypoplasia and absence of the inferior vermis occur. The posterior fossa is not as large as in the usual Dandy-Walker syndrome.

If there is a separate fourth ventricle, the condition represents only an arachnoid cyst or large cisterna magna, and not a Dandy-Walker malformation.

Agenesis of the Corpus Callosum

The corpus callosum is a commissure of fibers crossing between the cerebral cortex from both sides. If the initial connections between the commissural fibers are not made, there will be complete agenesis of the corpus callosum. Failure of neural tube closure leaves a longitudinal bundle of callosal fibers (Probst's bundle) passing on the superomedial aspect of the ventricles but not connected to each other.[5] There are many anomalies associated with this defect that occur at a similar time in development.[17] Shunting of hydrocephalus or an associated cyst may improve the prognosis, since agenesis of the corpus callosum alone may have a normal prognosis.[19]

RADIOLOGIC FINDINGS IN COMPLETE AGENESIS.—Ultrasound and CT demonstrate complete absence of the corpus callosum with resulting elongation and displacement of the third ventricle upward between the lateral ventricles. The lateral ventricles are widely separated and no septum pellucidum is present. The frontal horns are sharply peaked and indented medially by the longitudinal callosal bundles that failed to fuse in the middle (Fig 5–16). The occipital horns are larger and medially concave. The third ventricle may vary in size and in the amount of upward extension. Frequently there is an associated midline interhemispheric cystic lesion. This may be a separate arachnoid cyst or a communicating porencephalic cyst. It may be difficult to ascertain whether the cyst is separate from the ventricle without ventriculography.

On a sagittal ultrasound scan, the distortion of the sulcal pattern can be appreciated above the lateral ventricle. Instead of a cingulate sulcus parallel to the corpus callosum, the sulci are arranged more randomly and in fact may radiate in a circular pattern about the ventricle.

Fig 5–15.—Dandy-Walker Syndrome. Neonatal scans on day 1 after in utero diagnosis. **A,** Enlargement of the lateral ventricles (V) and third (3) and fourth (4) ventricles with inferior and posterior extension of the fourth ventricle forming a "cyst" (C). Temporal horns (T) are also dilated. **B,** More posterior coronal ultrasound scan at the level of the trigone where the echogenic choroid lies in each ventricle (V). The third ventricle (3) is enlarged. The fourth ventricular dilation outlines the hypoplastic cerebellum (arrows), which has not fused in the midline. **C,** Massive posterior fossa "cyst" (C) communicates by way of the aqueduct to the enlarged third (3) ventricle in this midline sagittal ultrasound scan. **D,** Axial ultrasound scan through the posterior fontanelle (note transducer artifact at bottom of scan) showing enlarged lateral ventricles (V) and third (3) and fourth (4) ventricles. **E, F,** Axial CT scans demonstrating enlarged third ventricle (3) and frontal (F) and temporal (T) horns of lateral ventricles. Note the small cerebellum in the posterior fossa (arrows). Massively enlarged, cystic fourth ventricle (4) extends superiorly to the vertex with the tentorium markedly elevated (open arrows, **F**).

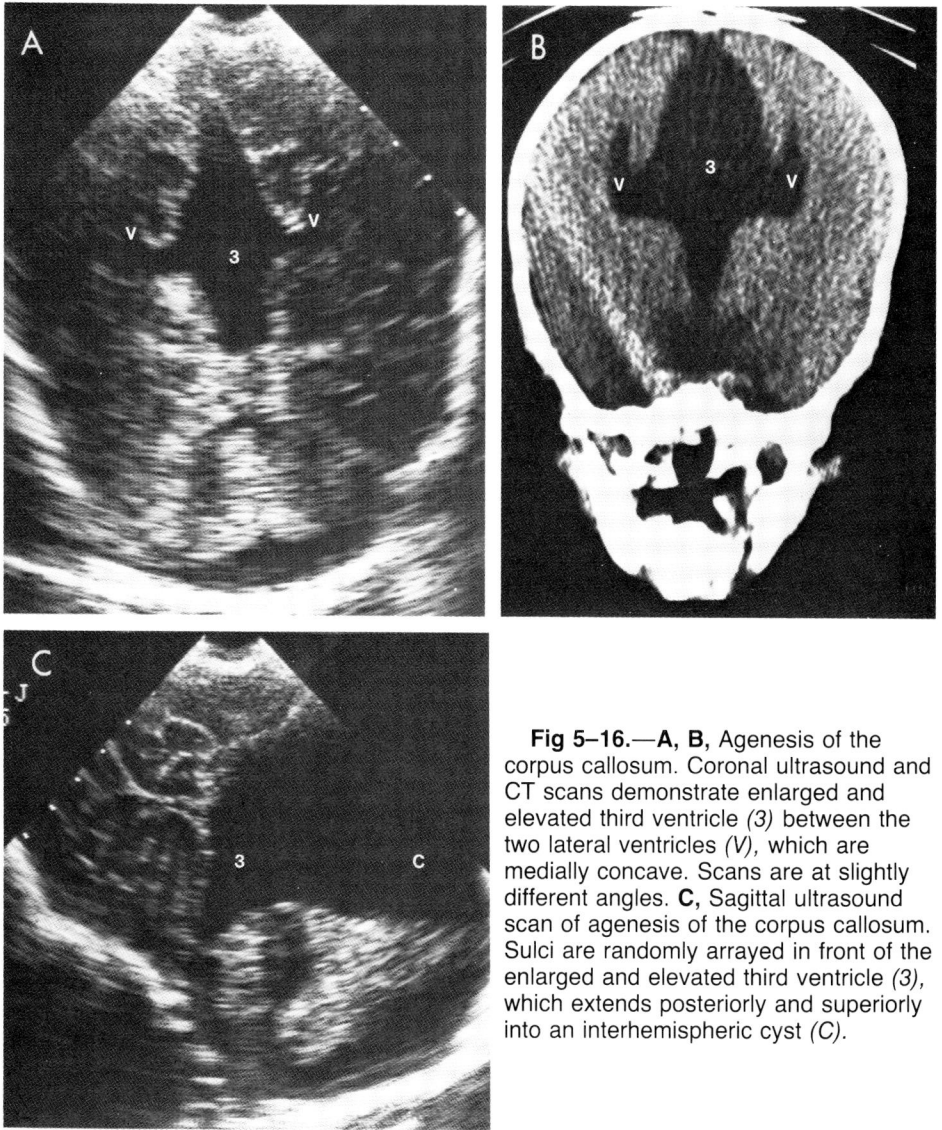

Fig 5–16.—A, B, Agenesis of the corpus callosum. Coronal ultrasound and CT scans demonstrate enlarged and elevated third ventricle *(3)* between the two lateral ventricles *(V),* which are medially concave. Scans are at slightly different angles. **C,** Sagittal ultrasound scan of agenesis of the corpus callosum. Sulci are randomly arrayed in front of the enlarged and elevated third ventricle *(3),* which extends posteriorly and superiorly into an interhemispheric cyst *(C).*

TABLE 5–4.—Disorders of Diverticulation

SEPTO-OPTIC DYSPLASIA
  Absent septum pellucidum→flat fused frontal horns
  Thin corpus callosum
  Dilated anterior third ventricle
  Hypoplasia of the infundibulum with large infundibular recess
  Small optic canals (primary optic nerve hypoplasia)
  Clinically blind, nystagmus, optic nerve hypoplasia
ALOBAR HOLOPROSENCEPHALY
  Single large midline ventricular cavity
  Relatively thin cerebral tissue
  Fused thalami
  Absent third ventricle
  Aqueduct small
  Relatively normal cerebellum and brain stem
  Associated midline craniofacial anomalies:
    Orbital hypotelorism
    Variable absence of nasal septum, median and bilateral cleft lip and
      palate, micrognathia, and trigonocephaly
    Rarely, cyclopia, ethmocephaly, cebocephaly
  Clinical microcephaly
SEMILOBAR OR LOBAR HOLOPROSENCEPHALY
  Single ventricle
  Fused thalami
  Moderate cerebral tissue
  Occipital horns partially formed
  Absent or small third ventricle
  Small or occluded aqueduct
  Absent septum pellucidum
  Squared frontal horns
    (lobar implies more complete formation of lobes and occipital horns
    and third ventricle)

RADIOLOGIC FINDINGS WITH PARTIAL ABSENCE.—Partial absence is probably caused by a destructive process after initial fusion occurs.[17] There may be partial anterior or posterior absence with associated absence of the septum pellucidum in the same area. Typically the third ventricle herniates upward only through the region where the corpus callosum is absent.

RADIOLOGIC FINDINGS IN LIPOMA.—A lipoma, though rare, may be identified in the corpus callosum at any age.[3] It is of low density on CT and high density on ultrasound. Absence of the corpus callosum is usually associated with the lipoma.

DISORDERS OF DIVERTICULATION

If any pathologic event interferes with diverticulation of the forebrain, the final results will lack one or more of the fundamental basic diverticuli, which include two olfactory tracts, two optic tracts, paired cerebral hemispheres, the pineal, and the pituitary (Table 5–4).[2]

At one end of the spectrum of malformations is arrhinencephaly, or absence of the olfactory tracts. This condition is of very little clinical significance and cannot be identified by any present neuroradiologic methods. Anophthalmia or an arrest in the growth of optic bulbs and nerves, is another simple disorder not requiring sophisticated techniques. It is very rare.

Septo-Optic Dysplasia

Septo-optic dysplasia is a developmental anomaly combining agenesis of the septum pellucidum and primitive optic ventricle with hypoplasia of the optic nerves, chiasm, and infundibilum.[20] Abnormalities of endocrine function apparently result from extension of the midline abnormalities into the hypothalamus and pituitary. Hypotonia and seizures are seen in the first few months of life. Blindness is suspected from wandering eye movements (nystagmus). Clinically, there are hypoplastic optic discs and frequently hypoglycemia, growth deficiency, and diabetes insipidus.[3, 20]

RADIOLOGIC FINDINGS IN SEPTO-OPTIC DYSPLASIA.—On ultrasound examination, the septum pellucidum is absent, and there is squaring or flattening of the frontal horns. Dilation of the optic recess and suprasellar and chiasmatic cisterns may be seen.

On CT examination, small optic nerves and optic canals are noted best on the direct coronal scan.[20–22] The septum pellucidum is absent, and there is dilation and flattening or squaring of the frontal horns (Fig 5–17). Enlargement of the infundibulum is sometimes seen with diabetes insipidus. The suprasellar and chiasmatic cisterns may be dilated. The optic recess of the anterior third ventricle may be enlarged in optic hypoplasia.

Holoprosencephaly

Holoprosencephaly is a malformation in which the first two segments of the brain (the prosencephalon) fail to undergo diverticulation (see Fig 5–1).[2] The most anterior telencephalon fails to form cerebral hemispheres, resulting in one central ventricle. The diencephalon fails to divide into two thalami, and there is at most a remnant of a third ventricle. As one would expect, the olfactory tracts are involved, usually being absent. These findings together are termed alobar holoprosencephaly when only one ventricle forms.

Clinically, there is a characteristic facies with midline hypotelorism, variable absence of the nasal septum, medial and bilateral cleft palate, micrognathia, and trigonocephaly.[23] Very rarely cyclopia, ethmocephaly, or cebocephaly is present. Alobar holoprosencephaly is the most severe malformation of this type, and most affected infants do not survive the first year of life.[24]

RADIOLOGIC FINDINGS IN ALOBAR HOLOPROSENCEPHALY.—On ultrasound examination, the single midline ventricle is noted, with absence of the interhemispheric fissure (Fig 5–18).[25] The moderately echogenic thalami can be distinctly identified as fused structures below and anterior to the fused densely echogenic

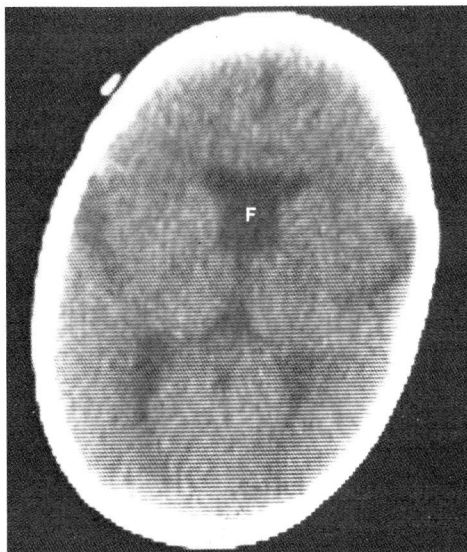

Fig 5–17.—Isolated septum pellucidum agenesis in a 3-year-old child with septo-optic dysplasia. CT scan shows classic squaring of the frontal horns *(F)*, which are continuous side to side, due to lack of the septum pellucidum.

choroid plexi. The single ventricle frequently extends posteriorly. There are no separate frontal, temporal, or occipital horns. The sulci (and the echogenic vessels within them) are markedly distorted as they cross the midline. There may be fewer sulci (pachygyri) than normal.

On CT examination, there is a single midline ventricle (see Fig 5–18), with absence of the falx. There is a relatively thin cerebral mantle which may cover the ventricle (ball shaped), or if it does not meet is termed a "pancake" or "cup type."[25] The thalami are fused in the midline. The third ventricle may be very small or absent. The aqueduct is small. The cerebellum and brain stem may be relatively normal.

RADIOLOGIC FINDINGS IN SEMILOBAR OR LOBAR HOLOPROSENCEPHALY.—Milder variants of alobar holoprosencephaly do occur and are well described by Byrd et al.[26] The radiologic findings depend on the extent of failure of diverticulation. Typically the occipital horns are least affected, with varying amounts of third ventricular and thalamic development. If the ventricles are fused anteriorly and separated posteriorly into occipital horns, lobar or semilobar holoprosencephaly will be diagnosed. All forms have in common at least absence of the septum pellucidum, so that septo-optic dysplasia is actually the mildest end of the spectrum. Milder facial anomalies are present, including cleft lip and palate.

Semilobar holoprosencephaly has a single ventricular cavity with fused and anteriorly rotated thalami.[27] The ventricle is smaller, there is more cerebral tissue, and occipital horns are present. The third ventricle is incorporated into the single ventricle but the aqueduct and fourth ventricle may be normal.

Lobar holoprosencephaly is the mildest form, being described even in mentally retarded adults. The frontal horns are squared with a flat roof and absent septum pellucidum. Variable amounts of cerebral lobes are developed.

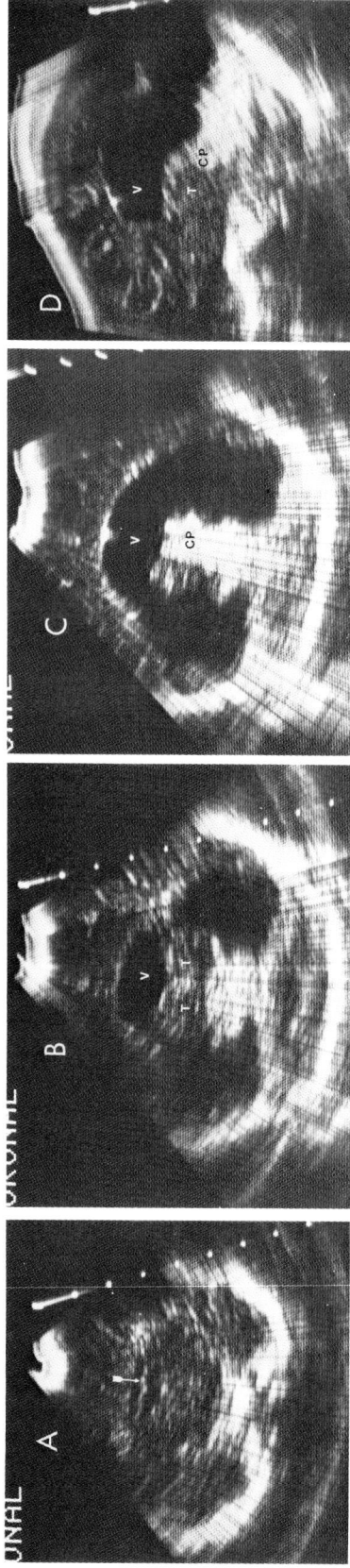

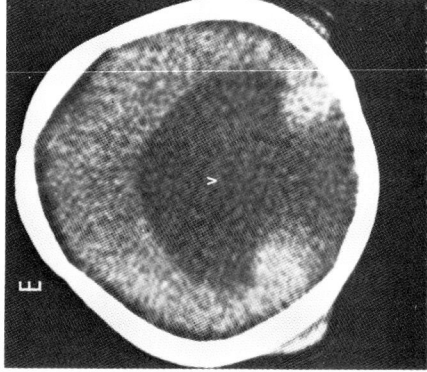

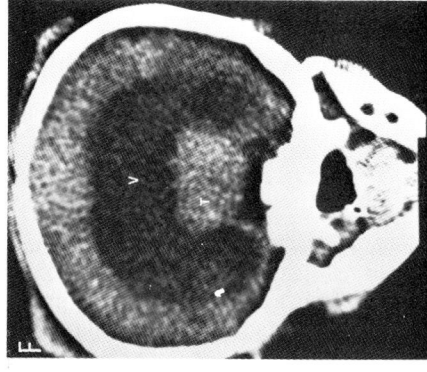

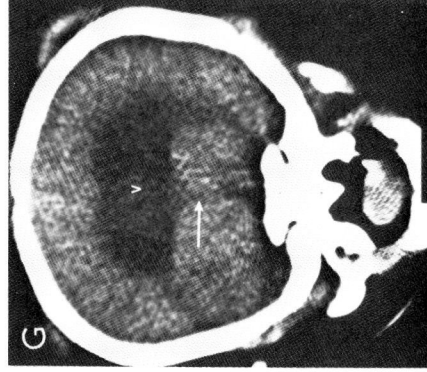

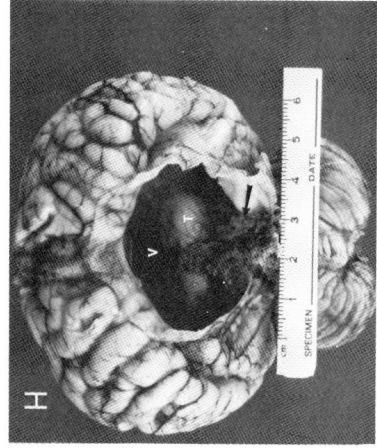

Fig 5–18.—See legend on facing page.

DISORDERS OF PROLIFERATION

Microcephaly

True microcephaly results in a normally formed but small brain. Although many of the chromosomal disorders result in microcephaly and the individual is mentally deficient, pathognomic changes in the brain are few.[17]

Macrocephaly

True macrocephaly occurs in cerebral gigantism, or Soto's syndrome, which is simply an enlarged brain.[17] Macrocephaly may be caused by an enlarged ventricular system, enlarged brain (in storage diseases), or a mass-occupying lesion such as porencephalic cyst, arachnoid cyst, or tumor. Cerebral gigantism or storage diseases are diagnosed by exclusion with present imaging techniques. Hydrocephalus, cysts, and tumors are discussed in other chapters.

DISORDERS OF SULCATION AND MIGRATION

Lissencephaly

Lissencephaly is thought to be caused by failure of neuronal migration at no later than the fourth fetal month, resulting in a four-layered cortex.[5, 28] The normal cortex contains six layers. Sulci and gyri usually develop with the last two layers and thus none or very few sulci form. The periventricular white matter is underdeveloped, resulting in persistence of the typically large fetal ventricles, termed colpocephaly.

The ultrasound appearance of lissencephaly may be less characteristic, since grey and white matter are less well differentiated. The abnormally large ventricles with very large sylvian fissures should be a diagnostic clue. As the neonatal brain matures, one normally sees the development of sulci and a decrease in the size of the sylvian fissure. These do not occur in lissencephaly.

Fig 5–18.—Alobar holoprosencephaly in 1-day-old infant. **A,** Coronal ultrasound scan in quite anterior projection demonstrates a sulcus *(arrow)* crossing the midline. **B,** Coronal ultrasound scan at the level of the fused, midline thalami *(T)* just cutting the midline ventricle *(V)*. **C,** Coronal ultrasound scan (more posterior) through single ventricle *(V)*, which is continuous over the echogenic fused choroid plexus *(CP)*. **D,** Sagittal ultrasound scan shows randomly oriented sulci due to associated agenesis of the corpus callusum. Single ventricle *(V)* starts above the thalami *(T)* and extends posteriorly over the choroid plexus *(CP)*. **E,** Axial CT scan demonstrates single "ball-shaped" ventricle *(V)* with relatively thick mantle. **F,** Coronal CT scan demonstrates single ventricle *(V)* above fused thalami *(T)*. **G,** Coronal CT scan demonstrates single ventricle *(V)* with a tiny remnant of a third ventricle *(arrow)*. **H,** Pathologic specimen of same patient with cerebellum pulled down below ruler to show single ventricle *(V)*, fused midline thalami *(T)*, and choroid plexus *(arrow)* (Reproduced with permission. Rumack et al.[25])

The CT appearance of lissencephaly is quite characteristic (Fig 5–19).[29–31] The brain is smooth with very wide sylvian fissures, thick grey matter, and thin white matter layers. Opercularization (covering of the insula by cerebral gyri) is deficient or absent. There may be a few coarse, shallow gyri. The ventricles are large, and the cavum septi pellucidi is usually present. Subarachnoid spaces are prominent. Midline calcification may occur near the septum pellucidum.

Schizencephaly

Schizencephaly is characterized by bilateral, roughly symmetric clefts. The clefts may extend to the ventricle and usually are separate from the sylvian fissure. The septum pellucidum and corpus callosum are often absent. The ventricles are often unusually shaped with squaring of the frontal horns due to associated partial absence of the septum pellucidum. The time of insult varies from 6 weeks' to as late as 6 months' gestation, when the optic nerves, septum pellucidum, and corpus callosum are forming.

CT findings include absence of the septum pellucidum with squared frontal horns, partial absence of the corpus callosum, and clefts extending toward the ventricle (Fig 5–20). The resulting appearance is a combination of porencephaly and septo-optic dysplasia. The insult occurs so early in gestation that the brain recovers

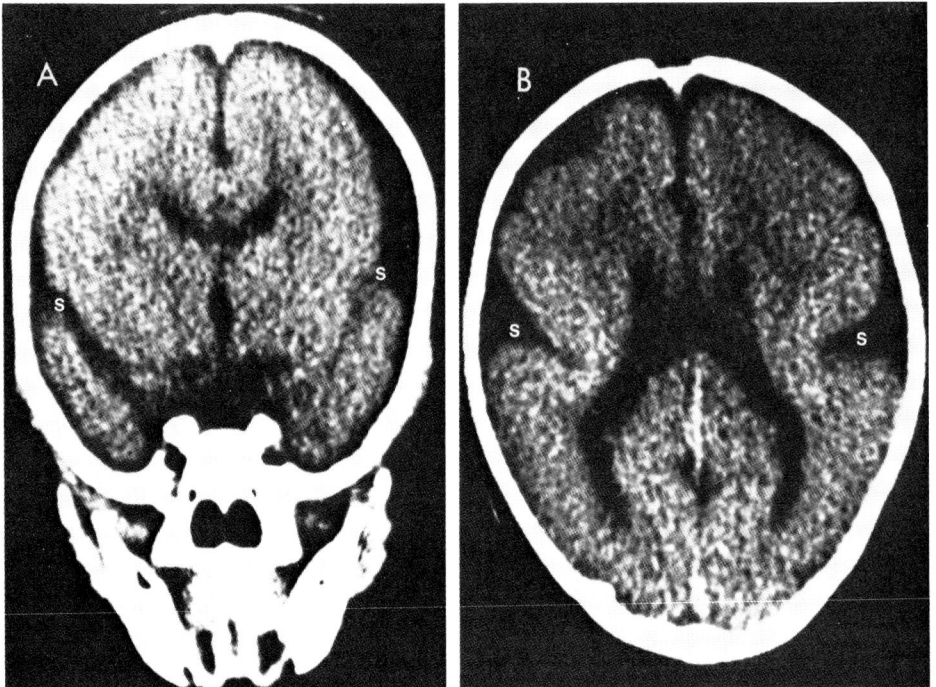

Fig 5–19.—Lissencephaly. Coronal **(A)** and axial **(B)** CT scans of 6-month-old infant with wide sylvian fissures *(S)*, thin white matter, and thick gray matter (From Johnson et al.[29] Reproduced with permission.)

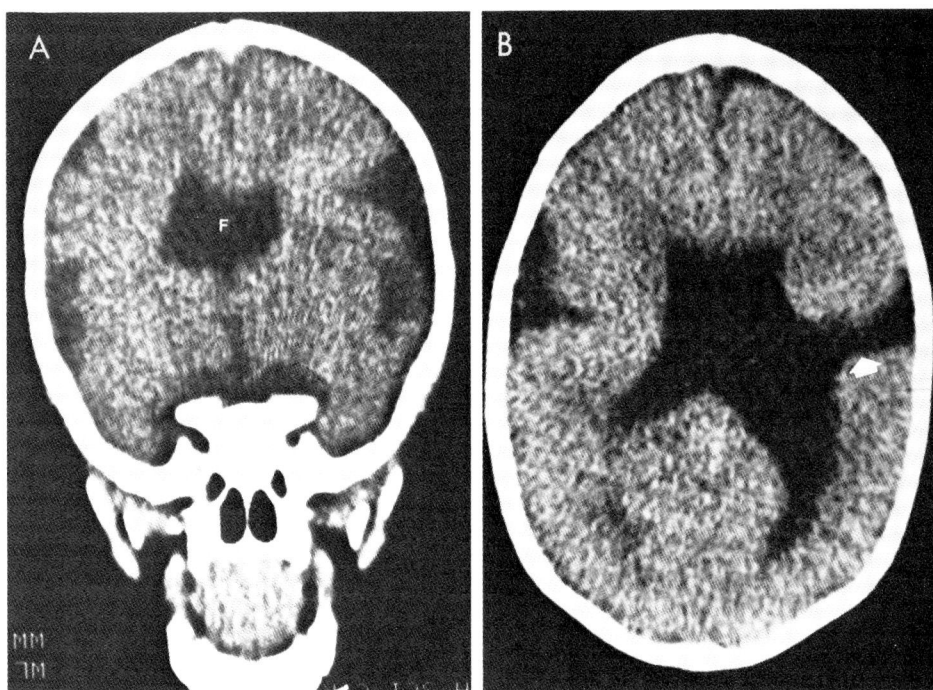

Fig 5–20.—Schizencephaly. Coronal **(A)** and axial **(B)** CT scans of 7-month-old infant with asymmetric clefts, one of which extends into the ventricular system *(arrow)*. Characteristic squared frontal horns *(F)* are due to absence of the septum pellucidum. (From Johnson et al.[29] Reproduced with permission.)

by maldevelopment rather than showing evidence of destruction and scarring. In fact, some authors consider schizencephaly, porencephaly, and hydranencephaly to be part of the same spectrum, depending on the time the insult occurs.[32] Yakovlev et al. originally proposed using schizencephaly to emphasize the fetal onset of this problem.[33] It really is a form of porencephaly occurring in utero.[3]

Heterotopias

Heterotopic grey matter appears as nodules bulging into the ventricles in many malformations.[5] They represent localized defects in neuronal migration (Fig 5–21). On CT or ultrasound, heterotopic nodules often have the same density and echogenicity as normal grey matter, although histologically they may lack the complete architecture.

Polymicrogyria

Polymicrogyria is an abnormal thickening of the cortex due to the piling up of many small gyri with fused surfaces.[5] It usually occurs near a local area of intrauterine insult, such as toxoplasmosis or cytomegalovirus infection. There is typically

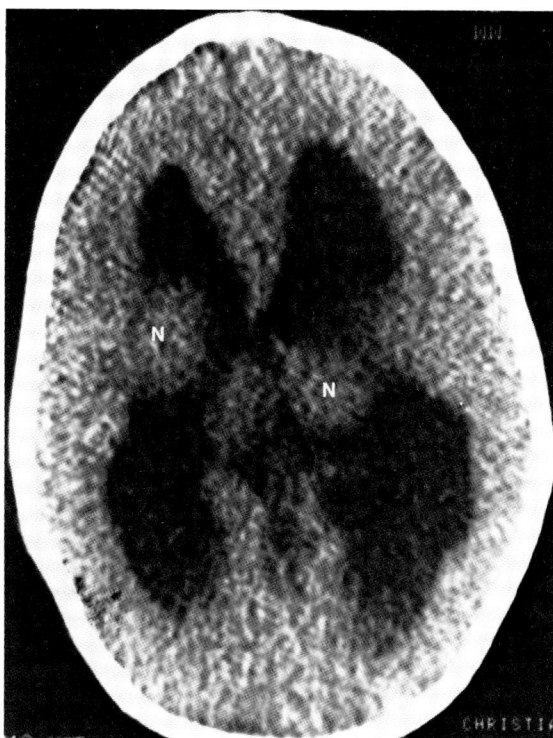

Fig 5–21.—Heterotopic gray matter. Nodules of gray matter *(N)* bulge into the ventricular system in a 3-month-infant with hydrocephalus.

a four-layered cortex in the affected region, indicating that the onset occurred no later than the sixth fetal month. Polymicrogyria may accompany the Chiari malformation, causing an increase in number and small-sized gyri.[3] The vasculature should reflect this difference, with a wavy appearance to the anterior and middle cerebral artery branches due to the poorly formed sulci.

Because polymicrogyria is best diagnosed histologically, CT or ultrasound may have difficulty with the diagnosis, but with improved detail, these areas may be more apparent. Careful attention to the detailed architecture of the vasculature is necessary to make this diagnosis on ultrasound.

REFERENCES

1. Volpe J.J.: Normal and abnormal human brain development. *Clin. Perinatol.* 4:3–30, 1977.
2. DeMyer W.: Classification of cerebral malformations. *Birth Defects* 7:78–93, 1971.
3. Harwood-Nash D.C., Fitz C.R.: *Neuroradiology in Infants and Children.* St. Louis, C.V. Mosby Co., 1976.
4. Volpe J.J.: Neurology of the newborn, in *Major Problems in Clinical Pediatrics.* Philadelphia, W.B. Saunders Co., 1981.
5. Friede R.L.: *Developmental Neuropathology.* New York, Springer-Verlag, 1975.
6. Fitzgerald M.J.T.: *Human Embryology.* New York, Harper & Row, 1978.
7. Byrd S.E., Harwood-Nash D.C., Fitz C.R., et al., Computed tomography in the evaluation of encephalocoeles in infants and children. *J. Comput. Assist. Tomogr.* 2:81–87, 1978.

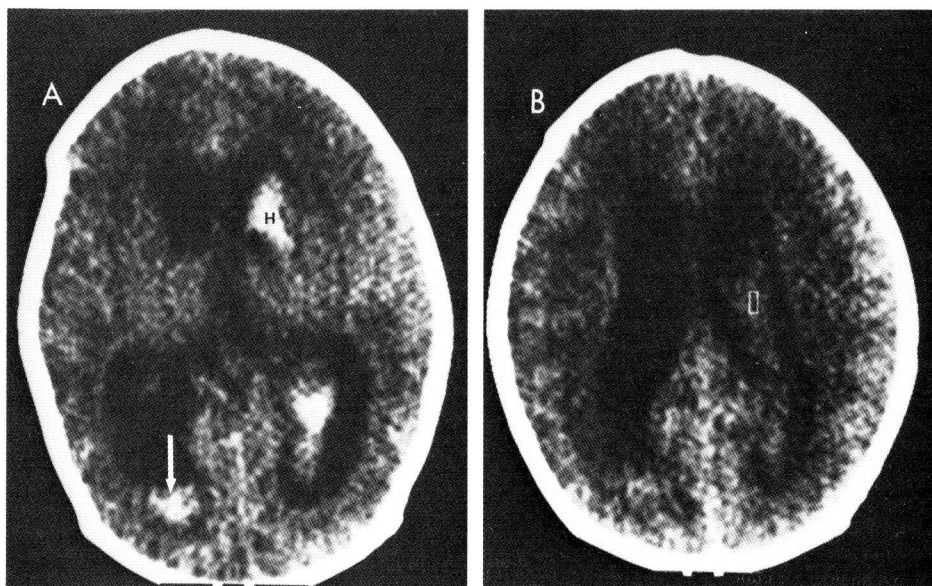

Fig 6–1.—A, Acute subependymal *(H)* and intraventricular hemorrhage *(arrow)* with hydrocephalus at 4 days of age. There is a classic subependymal hemorrhage near head of caudate nucleus. **B,** Chronic hemorrhage with isodense intraventricular clot *(cursor box)* is visible because it lies within the ventricle surrounded by CSF. These are axial CT scans.

On ultrasound, hemorrhage is initially highly echogenic without acoustic shadowing and becomes only slightly less echogenic without acoustic shadowing over the first week (Fig 6–2).[24] The architecture of the brain in the region of the hematoma is markedly distorted and remains echogenic for weeks to months, depending on the size of the lesion.

Care should be taken not to diagnose acute hemorrhage if the hematoma is not highly echogenic. Moderate echogenicity has been demonstrated in edema following infarction and meningitis.[25, 26] If a question arises, a CT examination should be done to confirm edema rather than mistakenly diagnose hemorrhage (see section on ischemic lesions).

Calcium deposition in the brain may be confused with acute hemorrhage on both CT and ultrasound. However, calcium usually has a CT density of 80–100 HU, whereas acute hemorrhage usually has a density of 60–80 HU. On ultrasound, calcium usually causes acoustic shadowing. Small flecks of calcium may cause an equivocal CT density and no shadowing on ultrasound. In this situation, ultrasound or CT scans repeated in 7–14 days will show a drop in the density if there is hemorrhage and no change in density if there is calcium present.[26]

Chronic or Old Hemorrhage

On CT, resolving hemorrhage becomes isodense with brain (30 HU) at 5–7 days (see Fig 6–1).[27, 28] A low-density area (0–10 HU) may develop, depending on the

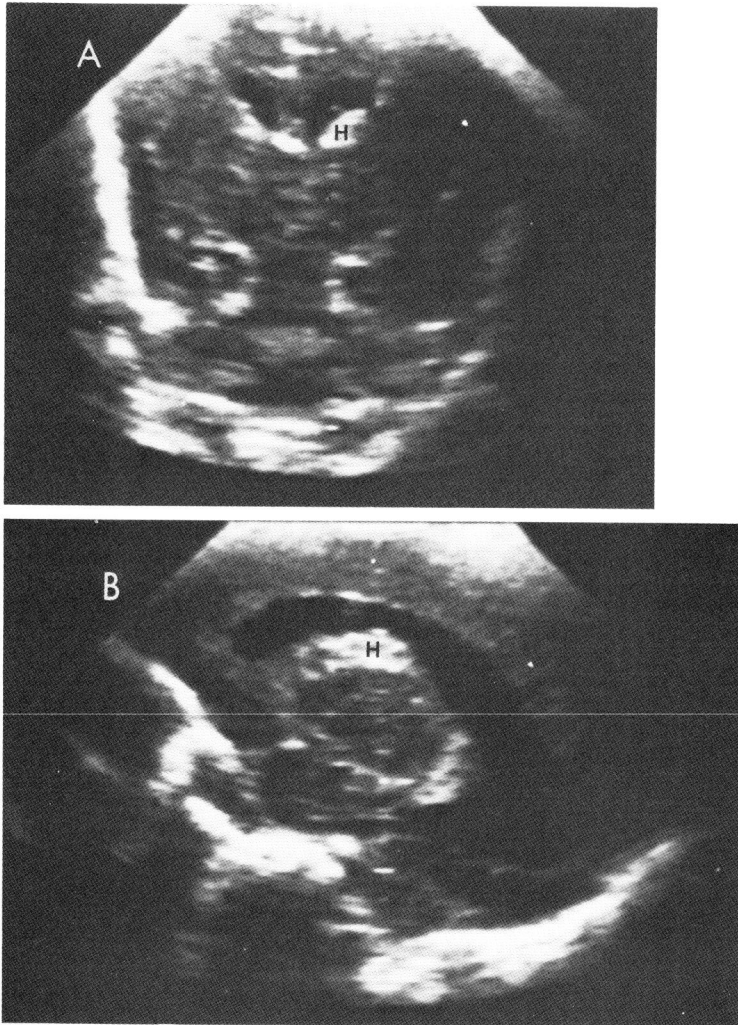

Fig 6–2.—Acute subependymal hemorrhage. Coronal **(A)** and sagittal **(B)** ultrasound scans show acute subependymal hematoma *(H)* at classic site above caudate nucleus at junction of caudate and choroid just behind the foramen of Monro. (From Johnson et al.[13] Reproduced with permission.)

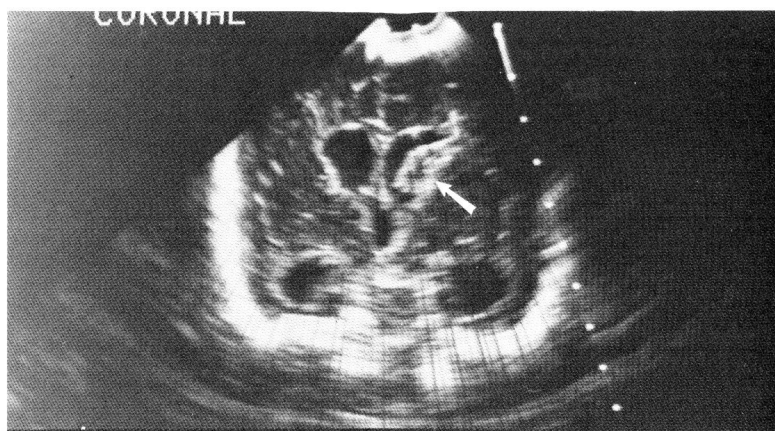

Fig 6–3.—Resolving subependymal hematoma *(arrow)* with central sonolucency about 2 weeks after the initial hemorrhage.

final amount of brain necrosis. The lesion may remain isodense with brain for many weeks before it is possible to determine the final amount of brain loss.

On ultrasound, resolving hemorrhage becomes gradually less echogenic over 1–2 weeks and a central area of sonolucency will develop (Fig 6–3).[29, 30] In large hemorrhages, the clot retracts from the edges and falls centrally. The hematoma may require several weeks to months for complete resorption.

With either modality, the initial and final results can be identified. There is, however, a period of 1–4 weeks when the CT study may be falsely negative but ICH can still be accurately diagnosed by ultrasound.

SUBEPENDYMAL GERMINAL MATRIX HEMORRHAGE

Subependymal germinal matrix hemorrhage occurs primarily in premature infants and is the commonest type of hemorrhage in that age group.[6]

The typical site of subependymal hemorrhage (SEH) is the germinal matrix, which lies above the caudate nucleus in the floor of the lateral ventricle and sweeps from the frontal horn posteriorly into the temporal horn.[31, 32] The most common site is at the junction of the caudate nucleus and choroid plexus. This is the region of the choroid attachment to the floor of the lateral ventricle, termed the telea choroidea, just behind the foramen of Monro.

On CT, SEH presents as a high-density lesion lying just lateral to the frontal horn (see Fig 6–1). The hematoma is typically in the head of the caudate, but it may extend or occur along the entire length of the caudate from the frontal to the temporal horn region. The hemorrhage may vary in size from 2 mm to 1 cm. Once it extends beyond the lateral edge of the frontal horn it is considered intraparenchymal. SEH may be unilateral or bilateral and may be asymmetric.

On ultrasound, SEH presents as a highly echogenic lesion adjacent to and inferior and lateral to the lateral ventricle (see Fig 6–2). With careful attention to

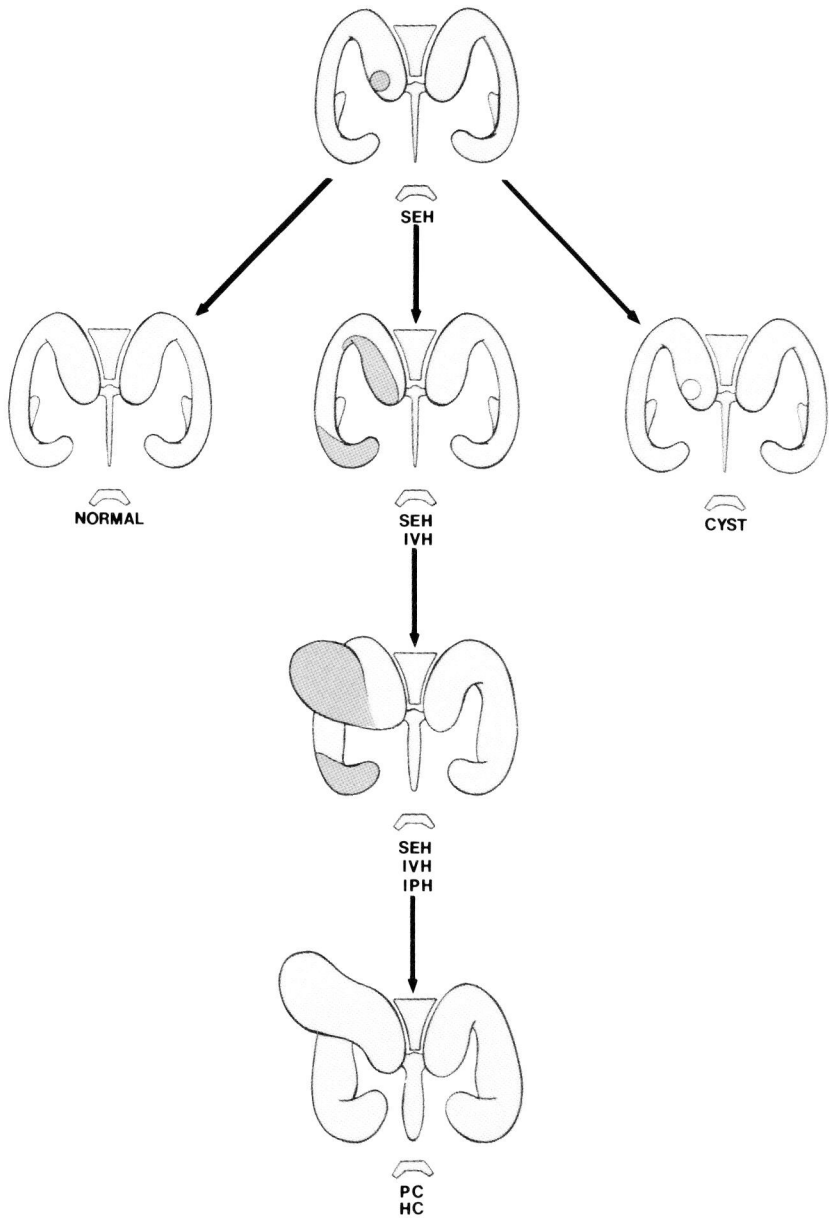

Fig 6–4.—Sequelae of subependymal hemorrhage. Subependymal hemorrhage *(SEH)* may resolve, leaving a normal scan; may resolve, leaving a small subependymal cyst; may progress, rupturing into the ventricle, causing interventricular hemorrhage *(IVH)*, or extending into the parenchyma, causing intraparenchymal hemorrhage *(IPH)*. Hydrocephalus *(HC)* and porencephaly *(PC)* are common sequelae of SEH.

symmetric scanning technique and proper gain settings, ultrasound is extremely accurate in detecting even small hematomas. If the bulging, highly echogenic hematoma is present on both coronal and saggital views, one can be certain of the diagnosis. Typically on the sagittal view, the SEH will appear as a bulge just anterior to the termination of the choroid. However, pitfalls do exist. The normal choroid plexus is extremely echogenic. It is broadest at the trigone and tapers as it courses anteriorly up to the foramen of Monro and inferomedially into the temporal horn: the choroid does not extend into the occipital horn. An oblique coronal scan may transect one frontal horn where there is no choroid and the opposite body with choroid, suggesting a subependymal hematoma. Real-time sweeps will prevent this error by visualizing the entire choroid from the posterior trigone region anteriorly through the telea choroidea. As the coronal sweep is made from posterior to anterior, the choroid should taper gradually and symmetrically.

As the hematoma resolves, small hemorrhages return to normal brain appearance on both ultrasound and CT (Fig 6–4). Whether small amounts of fibrosis persist in some of the survivors cannot be determined. In moderate-sized hemorrhages, a subependymal cyst may form which is best delineated on ultrasound (Fig 6–5) but may be averaged into the ventricular wall on CT. Subependymal cysts are common findings on autopsy after ICH.[33] The central material is gelatinous, with a sharply defined wall as permanent evidence of an earlier hemorrhage.

Progression of SEH may occur, leading to intraventricular or intraparenchymal hemorrhage, or both.

INTRAVENTRICULAR HEMORRHAGE

Extension From Subependymal Hemorrhage

Acute intraventricular hemorrhage (IVH) presents as high-density material within the ventricles. Cerebrospinal fluid (CSF)-blood levels or clots may be evident in the dependent portion of the ventricles, usually in the occipital horn (Fig 6–6). The lateral ventricle may be dilated and the third and fourth ventricles become more prominent because they are filled with blood. Acutely, small amounts of IVH are better appreciated on CT as thin CSF-fluid levels in the occipital horns. As the blood becomes isodense, the clot may be missed on CT but is clearly seen on ultrasound.[13] The ventricular wall usually is isodense with brain on CT after 1 week but may appear to have a subependymal halo which is either clot attached to the wall or a chemical ventriculitis in response to blood.[9]

On ultrasound, a very helpful indication of IVH is an irregular bulky choroid plexus, especially if there is a clot attached to the surface (Fig 6–7). Initially it may be difficult to identify a ventricle if it is filled with clot that obscures the ventricular margins (Fig 6–8). As the ventricle dilates, the diagnosis of IVH on ultrasound is easier, especially if dense echoes are seen within the occipital horn (Fig 6–9). Movement of clot within the ventricle is diagnostic of IVH but not usually seen. IVH often is difficult to diagnose with either modality after 2–3 weeks, because the ventricular clot resolves. However, the ventricular surfaces usually become very

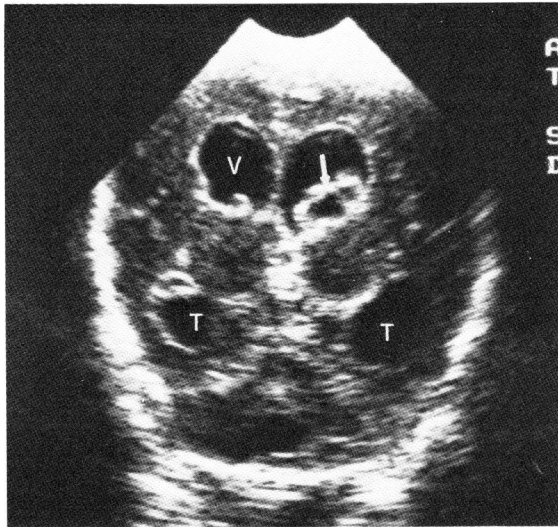

Fig 6–5.—Subependymal cyst *(arrow)* on coronal ultrasound scan in 1-month-old infant. Moderate hydrocephalus is present, with rounding of the bodies of the lateral ventricles *(V)* and temporal horns *(T)*.

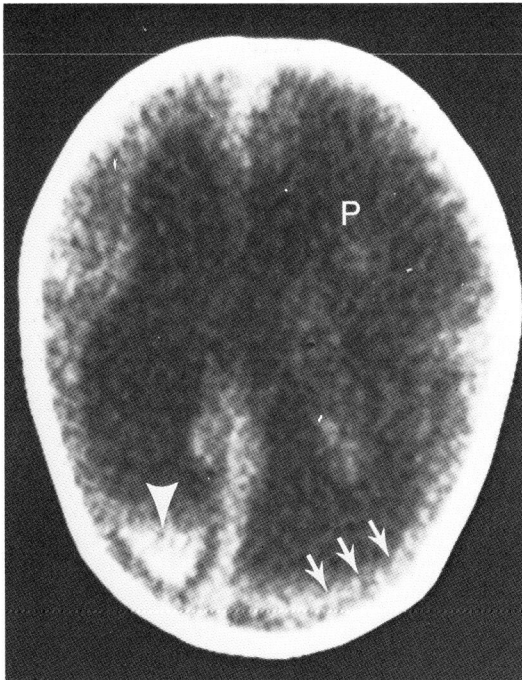

Fig 6–6.—Resolving intraventricular hemorrhage with CSF-blood fluid level *(arrowhead)*, hydrocephalus, and porencephaly *(P)* on axial CT scan. Subependymal halo around portions of the ventricle *(arrows)* may be clot attached to the wall or may be evidence of chemical ventriculitis.

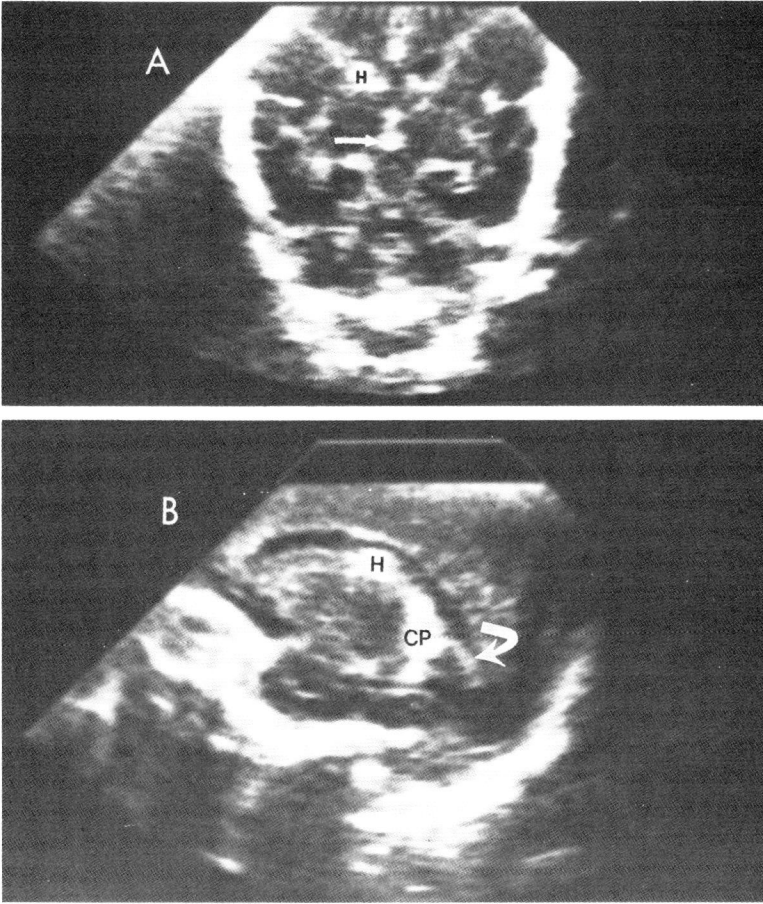

Fig 6–7.—Acute intraventricular hemorrhage on ultrasound. Coronal **(A)** and sagittal **(B)** scans demonstrate subependymal hematoma *(H)*, clot in the third ventricle *(arrow)*, and clot *(curved arrow)* attached to the choroid plexus *(CP)*. (From Rumack and Johnson.[127] Reproduced with permission.)

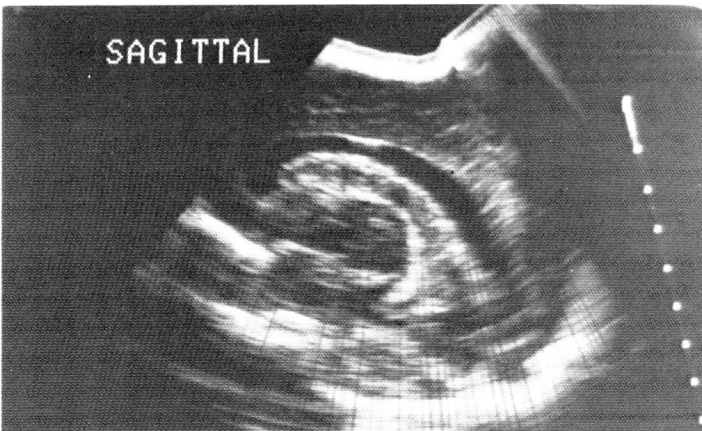

Fig 6–8.—Subacute intraventricular hemorrhage partially obscures the ventricular margins, especially the temporal and occipital horns. Sagittal ultrasound scan in 2-week-old infant.

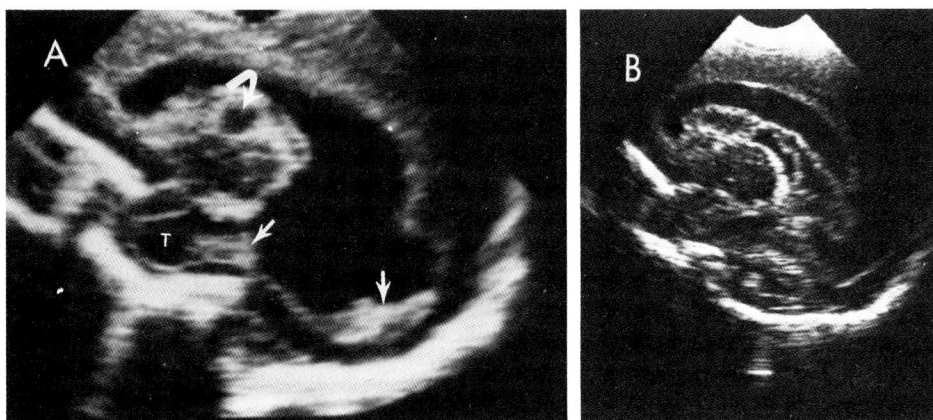

Fig 6–9.—Resolving intraventricular hemorrhage on sagittal ultrasound scans in two different patients. **A,** Echogenic debris in the occipital and temporal *(T)* horns *(arrows).* Subependymal cyst *(curved arrow)* fills region of caudate at 3 weeks of age. (From Johnson et al.[13] Reproduced with permission.) **B,** Intraventricular clot is becoming sonolucent centrally. It is attached in front of the choroid and extends back into the occipital horn. (From Rumack and Johnson.[127] Reproduced with permission.)

echogenic after IVH in what seems to be a chemical ventriculitis or cellular reaction to subependymal or intraventricular blood (see Figs 6–3, 6–5, and 6–9).[34]

Choroid Plexus Hemorrhage

If there is IVH without SEH, the ventricular hemorrhage probably originates in the choroid plexus, particularly in term infants, who rarely have SEH.[6] On CT, the choroid is isodense with brain so that choroid hemorrhage is easily identified, although it may simulate isolated IVH.

On ultrasound examination, the normal choroid may appear slightly bulbous posteriorly. The choroid is so echogenic that isolated choroid plexus hemorrhage is a difficult diagnosis. Clot that is attached to the choroid and extends posteriorly into the occipital horn can be a good diagnostic clue (see Figs 6–7,B and 6–9,B) and is usually associated with other debris from IVH. In small normal-sized ventricles, it can be difficult to differentiate IVH from choroid hemorrhage. SEH bulging into the ventricle may simulate IVH and ruptures more often into the ventricle than into the parenchyma.

INTRAPARENCHYMAL HEMORRHAGE

Extension From Subependymal Hemorrhage

Most parenchymal hemorrhages occur as a result of extension of subependymal hematomas. They may be seen on the first study or may develop progressively as an extension of SEH over the first few days of life.[35] Because most instances of

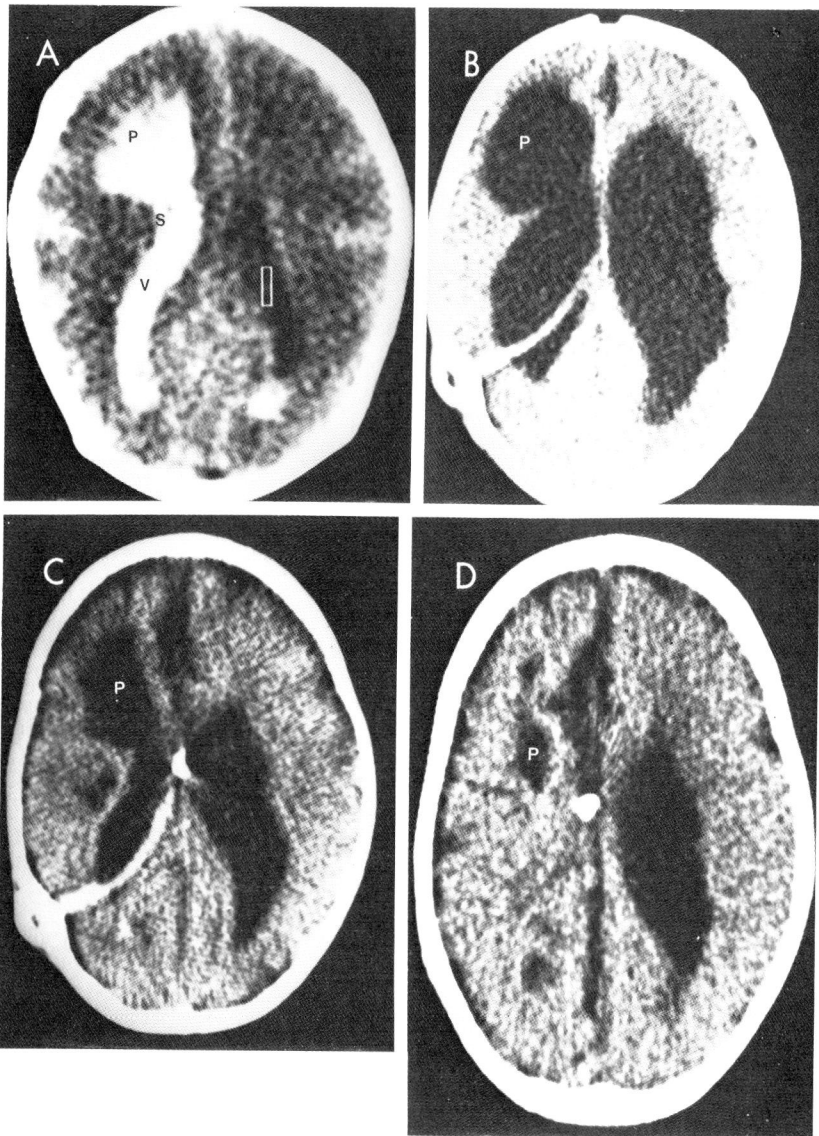

Fig 6–10.—Intraparenchymal hemorrhage and porencephaly on axial CT scans. **A,** Acute subependymal *(S),* intraparenchymal *(P),* and intraventricular *(V)* hemorrhage. **B,** Three months later, hemorrhage has resolved, leaving hydrocephalus and porencephaly *(P).* Intraventricular shunt was malfunctioning. **C,** One year later, ventricular size is closer to normal, especially on the side with the shunt, but the porencephalic area *(P)* is still fairly large. **D,** One and one-half years later, the porencephalic area *(P)* is barely visible; the left ventricle is still slightly enlarged. The right side of the brain is slightly larger than the left.

SEH originate adjacent to the frontal horn of the lateral ventricle, most parenchymal hemorrhages extend into the frontal or parietal lobes of the cerebral hemisphere.

On CT, parenchymal hemorrhage appears as a highly dense lesion which becomes isodense over 7–10 days (Fig 6–10). Isodense lesions progress to hypodensity and result in porencephaly (see Table 6–2).

During the process of resolution four specific stages can be identified on ultrasound (Fig 6–11).[13, 30] In the first stage the hematoma is homogeneous, highly echogenic, and has irregular margins. In the second stage, 1–2 weeks after the hemorrhage, the center of the hematoma becomes hypoechoic with a well-demarcated echogenic edge or "rind." By 2–4 weeks clot retraction can be seen around the demarcated hemorrhage, and the clot settles in a dependent fashion. Finally, by

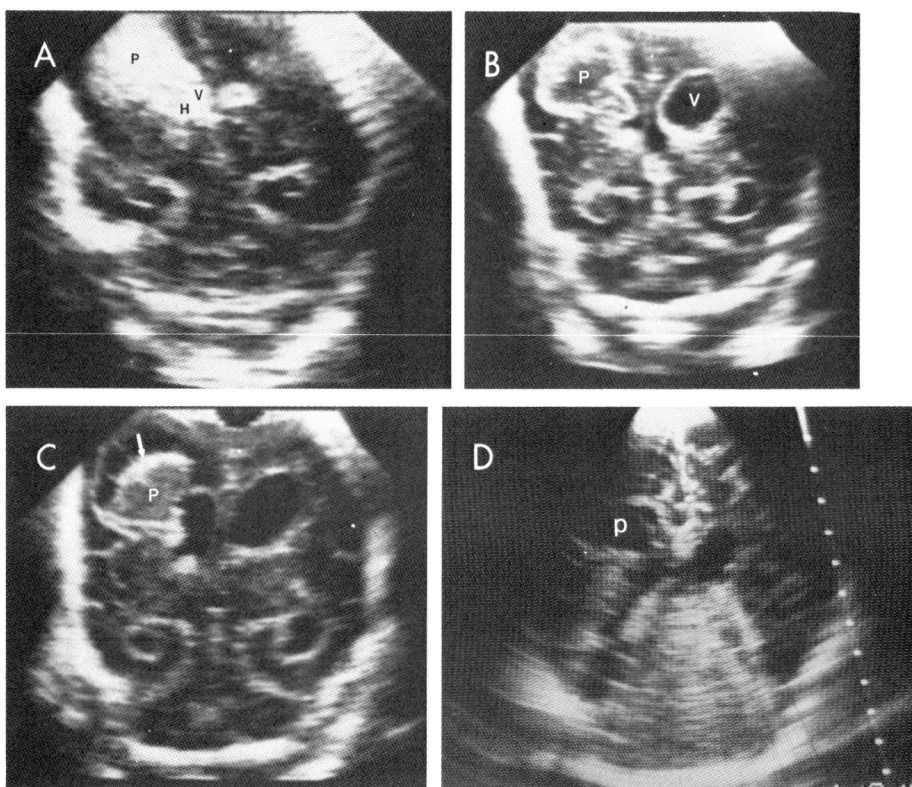

Fig 6–11.—Course of intraparenchymal hematoma resolution on ultrasound. **A,** Acute subependymal hematoma *(H)* has ruptured into the ventricle *(V)* and the parenchyma *(P).* **B,** Resolving hematoma at 2 weeks has developed central sonolucency and echogenic edges in the area of porencephaly *(P).* The body of the lateral ventricle *(V)* is dilated, with echogenic margins. **C,** Clot retraction *(arrow)* is occurring around the edges of the parenchymal hematoma *(P).* The ventricles have enlarged. (From Johnson et al.[13] Reproduced with permission.) **D,** Hematoma has completely resolved at 2 months, leaving porencephaly *(P).* Ventricular size has decreased after shunting.

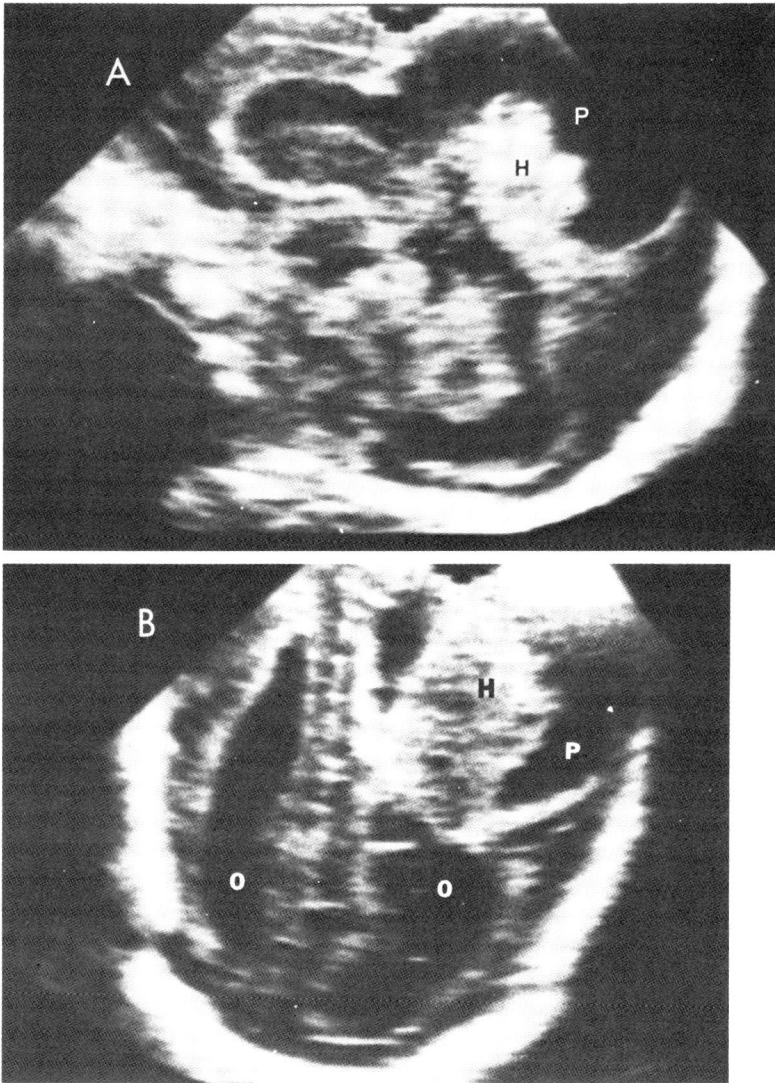

Fig 6–12.—Intraparenchymal hemorrhage in occipital lobe on sagittal **(A)** and coronal **(B)** ultrasound scans. Parieto-occipital hematoma *(H)* is in clot retraction stage. Porencephaly *(P)* is developing as the clot resolves. Occipital horns *(O)* are not dilated. (From Johnson et al.[13] Reproduced with permission.)

2–3 months necrosis and phagocytosis are complete, leaving an area of porencephaly which is completely anechoic.[36]

Long term follow-up of these lesions over several years has shown some decrease in the size of porencephaly, which is apparently due to new growth of brain during the first 2 years of life. Porencephalic areas do not always decrease in size with shunting, probably because they represent areas of brain necrosis and the brain cannot simply reexpand.

Unusual Locations of Parenchymal Hemorrhage

Less frequently, the subependymal hematomas may extend or occur further posteriorly. Extension of SEH into the temporal horn is relatively rare. The occipital lobe may be affected by extension of a massive intraparenchymal hemorrhage that originated in the parietal lobe (Fig 6–12). The occipital lobe is not primarily affected with this type of hemorrhage, probably because of the lack of the germinal matrix in this location.

Thalamic hemorrhage has followed massive hemorrhage in three of our cases (Fig 6–13).[25] Each case has been associated with acute hydrocephalus, marked midline shift, and a fatal outcome. This constellation of findings has occurred in extremely preterm infants.

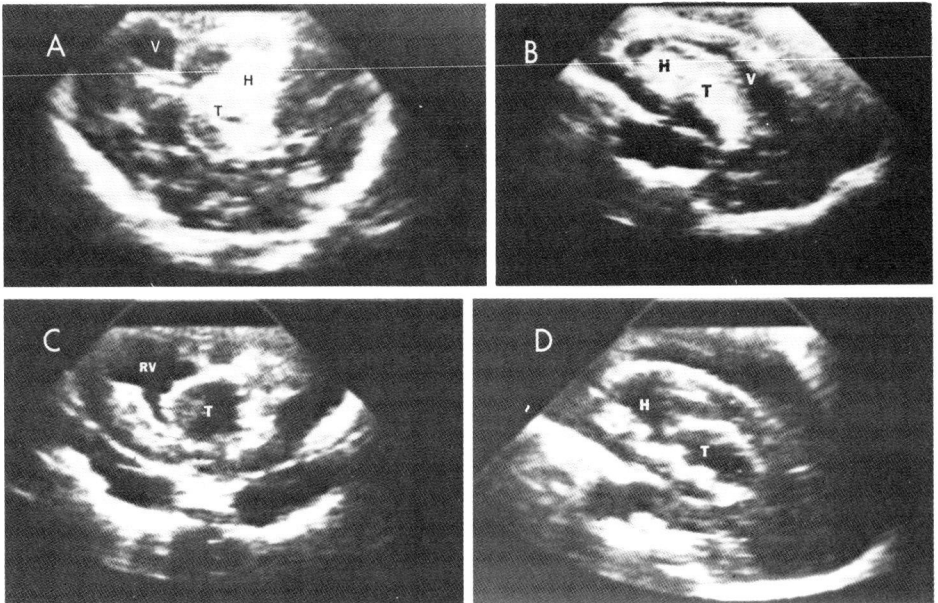

Fig 6–13.—Thalamic hemorrhage in a neonate born at 26 weeks' gestation. Coronal **(A)** and sagittal **(B)** ultrasound scans show a large acute parenchymal hematoma *(H)* extending into the thalamus *(T)*. The ventricles *(V)* are dilated, and there is a marked left-to-right shift. Coronal **(C)** and sagittal **(D)** ultrasound scans 5 days later show central sonolucency in the thalamic hematoma. The right lateral ventricle *(RV)* is dilated and shifted toward the right. (From Rumack and Johnson.[127] Reproduced with permission.)

TABLE 6–3.—GRADING SYSTEM FOR ICH*

Grade I	Isolated SEH
Grade II	SEH or CPH with IVH, no ventricular dilatation
Grade III	SEH or CPH with IVH, with ventricular dilatation
Grade IV	SEH or CPH with IVH and IPH

*Modified from Papile et al.[4]

GRADES OF INTRACRANIAL HEMORRHAGE

Because the severity of ICH plays a role in the prognosis and outcome of the infant, a number of grading systems have been devised to define the location and extent of ICH (Table 6–3). Most grading systems for ICH in current use are modifications of the one developed by Burstein et al.[5, 37–39] In this classification, grade 0 ICH is "normal," that is, there is no evidence of hemorrhage (Fig 6–14). Grade I ICH is ICH confined to the subependymal germinal matrix (Fig 6–15). A grade II hemorrhage is diagnosed when the germinal matrix hematoma ruptures into the ventricles without ventricular dilatation (Fig 6–16). With the development of ventricular dilatation following IVH, a grade III hemorrhage is diagnosed (Fig 6–17). Grade IV ICH implies IVH with parenchymal hemorrhage due to rupture of the initial SEH into the parenchyma (Fig 6–18).

Present grading systems have a number of significant problems that must be resolved before such systems can be used reliably to evaluate the prognosis and outcome of infants after ICH. These systems do not provide for certain less common hemorrhages. Occasionally an IVH will appear to arise from the choroid without a preexisting SEH. This could reasonably be classified as grade II or III hem-

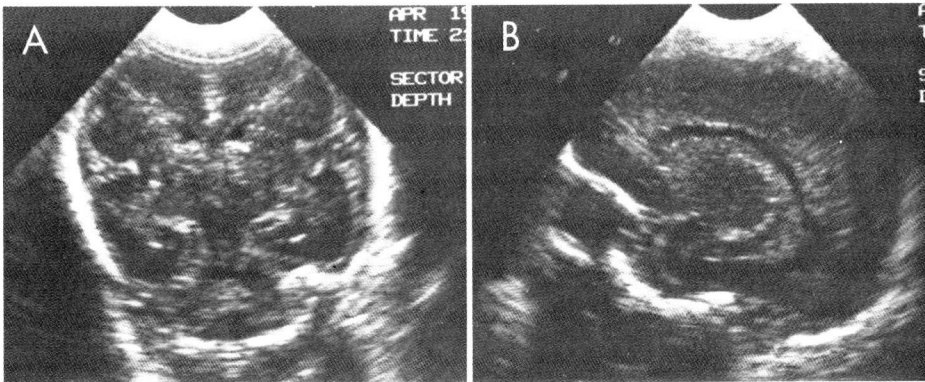

GRADE 0 – NORMAL

Fig 6–14.—Normal ultrasound scans at 28 weeks' gestation. **A,** Coronal scan at the level of the bodies of the lateral ventricles. **B,** Sagittal scan demonstrating normal lateral ventricle. No sulci are evident.

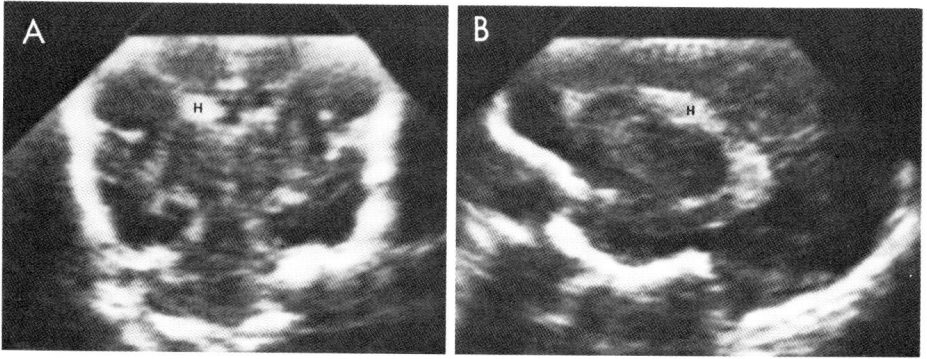

GRADE I – SEH

Fig 6–15.—Grade I—Subependymal hemorrhage *(H)* on coronal **(A)** and sagittal **(B)** scans.

orrhage, depending on the ventricular size. Some parenchymal hemorrhages exist as isolated lesions that do not extend from areas of SEH. These could be classified as grade IV ICH, or perhaps grade V would be more appropriate.

Another difficulty has been the confusing terminology used by workers in this field. The term "intraventricular hemorrhage" has been extensively misused in the literature as a substitute for "intracranial hemorrhage." The term "intraventricular hemorrhage" should only be used when actual intraventricular hemorrhage is present.

There is a lack of consistency between the grading systems presently proposed and the fact that these grading systems do not define the time at which the infant

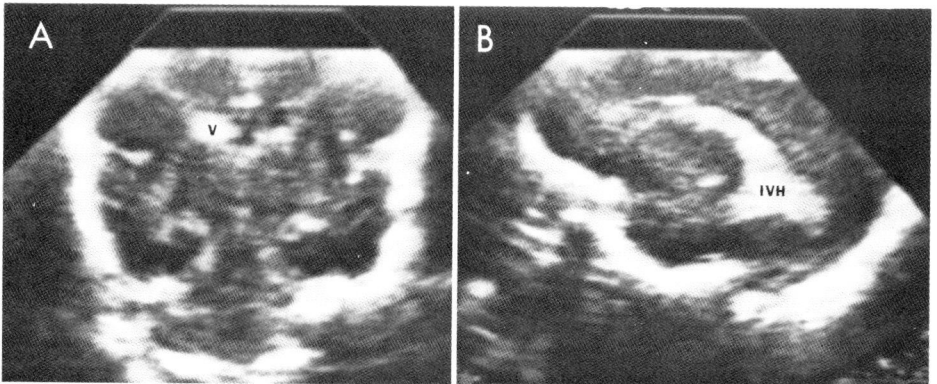

GRADE II – IVH

Fig 6–16.—Grade II—Subependymal and intraventricular hemorrhage *(IVH)* on coronal **(A)** and sagittal **(B)** ultrasound scans.

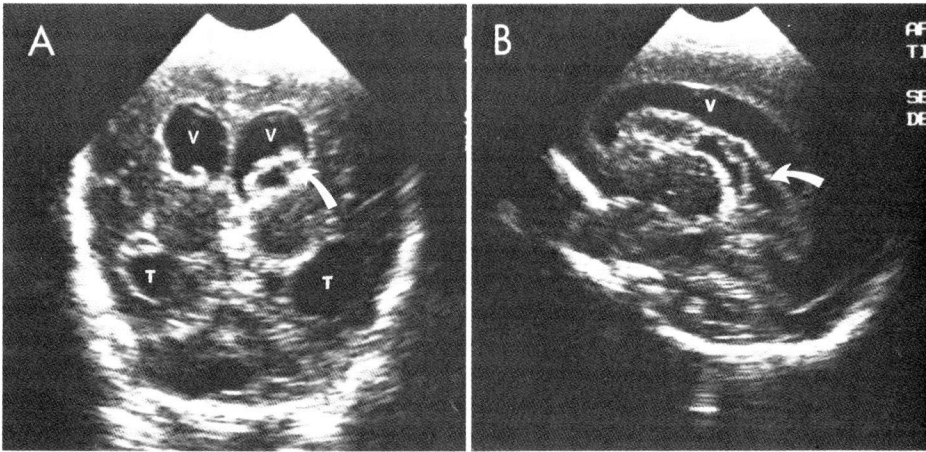

GRADE III – IVH & HC

Fig 6–17.—Grade III—Subependymal and intraventricular hemorrhage with hydrocephalus on coronal **(A)** and sagittal **(B)** ultrasound scans. Note dilated ventricles *(V)* and temporal horns *(T)*. *Arrow* points to intraventricular hemorrhage.

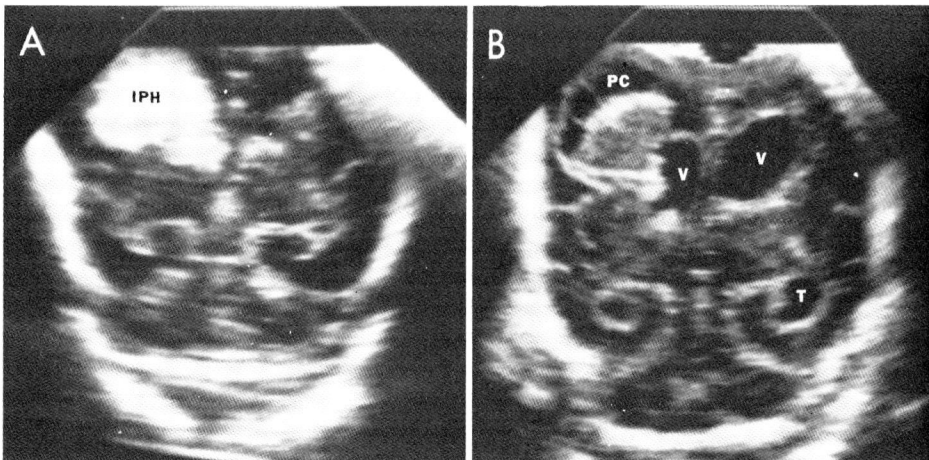

GRADE IV – IPH

Fig 6–18.—Grade IV—Subependymal, intraventricular, and intraparenchymal *(IPH)* hemorrhage. Acute **(A)** and resolving **(B)** hematoma develops porencephaly *(PC)* and ventricular enlargement *(V)*, including the temporal horns *(T)*.

is examined. Obviously a grade I hemorrhage can progress to a grade IV hemorrhage, and clearly the hemorrhage must be graded at its worst to determine the prognosis.

The extent and location of the hemorrhage are also of major importance: a 2-mm hematoma is most likely less damaging than a 2-cm hematoma. Because permanent neurologic deficits would be expected to correlate with location and volume of brain parenchymal destruction, prognosis based on the most extensive hemorrhage and the greatest severity of hydrocephalus should be of greatest clinical value.

TIMING OF ICH

Subependymal germinal matrix hemorrhage typically occurs during the first week of life[35, 40, 41] but has been reported even later associated with episodes of acute respiratory distress[41, 42] and disseminated intravascular coagulation (DIC) (unpublished data). SEH and IVH occur only rarely in utero and can be diagnosed on obstetric ultrasound.[43] Serial ultrasound examinations in preterm infants have demonstrated that most ICH occurs in the first 3 days of life (Fig 6–19).

Clinical predictions of the occurrence of hemorrhage have been quite unreliable, with many false positives and false negatives reported, so ultrasound screening of all infants born before 32 weeks' gestation should be routine.[44]

Most of the severe hemorrhages associated with hydrocephalus or parenchymal hemorrhage (grades III or IV) have their onset in the first 24 hours of life. Because ICH may progress, ultrasound screening is recommended at 5–7 days to document the most severe extent of hemorrhage. In our experience, 91% of neonates with ICH will have hemorrhaged by 5 days. A clinical problem may indicate the necessity for a scan on day 1. However, a normal study on day 1 does not exclude the

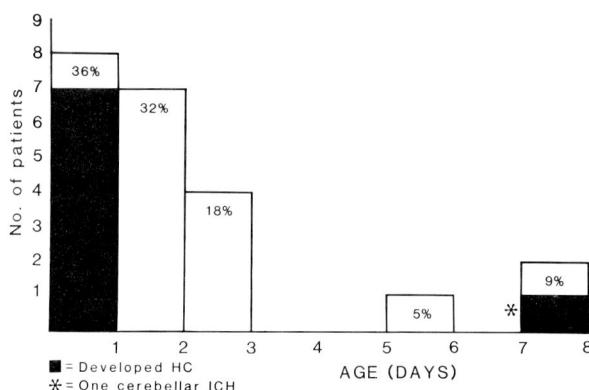

Fig 6–19.—Timing of initial intracranial hemorrhage (ICH). Eighty-six percent of patients had developed hemorrhage by 3 days. However, only 36% hemorrhaged on day 1. If screening for hemorrhage is done between 5 and 7 days, 91% of these patients will be identified. Most of the severe cases of hemorrhage began on day 1, although evidence of hydrocephalus frequently took 1–2 weeks to appear. One patient developed cerebellar hemorrhage and hydrocephalus at 7 days. (From Rumack et al.[35] Reproduced with permission.)

later development of ICH. It is extremely important to repeat the study when clinical findings emerge that are known to be related to the pathogenesis of intracranial hemorrhage.

POSSIBLE ETIOLOGIC FACTORS IN SUBEPENDYMAL GERMINAL MATRIX HEMORRHAGE

Germinal matrix hemorrhage in the neonate is strongly correlated with extreme prematurity (<32 weeks' gestation) and birth asphyxia[46–47] (Table 6–4). The incidence of germinal matrix hemorrhage is 67% at 28 weeks' gestation and decreases progressively to less than 5% at term (Fig 6–20).[48] The observed *gestational dependence* of ICH relates to developmental changes in the germinal matrix.[30] The germinal matrix is a loosely organized sheet of primitive neural cells that is richly supplied with a poorly supported capillary bed and thin-walled veins. It becomes relatively avascular by 36 weeks when the cortical migration of neural cells is completed and the germinal matrix has involuted. The germinal matrix lies above the entire caudate nucleus in the subependymal region of the lateral ventricle. It is also present in the roof of the third and fourth ventricles.

Birth weight does not correlate as well with the incidence of hemorrhage, probably because it is not as good an indicator of brain maturity as gestational age (Fig 6–21).[48–52] However, most investigators have found a high incidence in low birth

TABLE 6–4.—POSSIBLE ETIOLOGIC FACTORS
IN SUBEPENDYMAL HEMORRHAGE

Gestational age[4, 45–50]	28–34 wk
Low birth weight[4, 48–52]	<1,500 gm
Sex[4, 47]	Male 2:1
Multiple gestation[4]	
Trauma at delivery[56, 57]	Vaginal or C section
	Forceps?
Longer labor[56, 57]	
Anoxic episode[47, 52, 57, 59, 64, 65]	Low Apgar at 5 minutes
	Respiratory distress syndrome
Hyperosomolarity	Hypernatremia[64, 70, 71]
Hypocoagulation[76,77]	Maternal ingestion of aspirin[79]
	Hypocoagulation on day 1[78]
Direct intracranial pressure	Superior sagittal sinus
	compression[56]
	Occipital osteodiastasis[57]
Lack of autoregulation[68]	Increased cerebral blood flow[73]
(transmits systemic blood	Pneumothorax[62, 63]
pressure to brain)	Patent ductus arteriosus with/
	without ligation[64]
	Hypertension[125, 126]
	Intravascular bolus[48, 69, 73]
	Hypercapnia[64]
	Decreased cerebral blood flow[73]
	Hypotension[67]
	Hyperventilation
	Patent ductus arteriosis
	Hyperviscosity
	Hydrocephalus

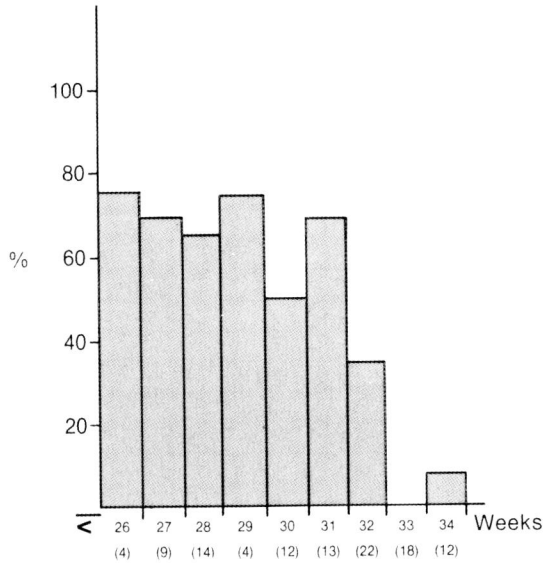

Fig 6–20.—Incidence of germinal matrix hemorrhage by gestational age.

weight infants (<1,500 gm) with germinal matrix hemorrhage.[4, 50–52]

The fragile vessels in the germinal matrix are thought to be related to the cause of hemorrhage, and on this basis, three theories have been proposed to explain the mechanism of germinal matrix hemorrhage: arterial rupture, capillary rupture,[53] and venous thrombosis.[54] The most accepted theory is capillary rupture.

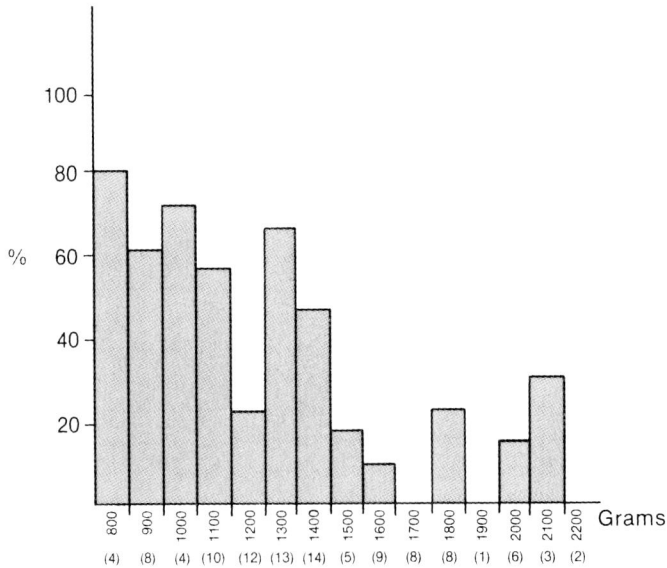

Fig 6–21.—Incidence of germinal matrix hemorrhage by birth weight.

An etiologic model has been developed by Wigglesworth and Pape[55] which integrates the various mechanisms that have been associated with hemorrhage at a pathophysiologic level. The basic final pathways appear to be controlled by cerebral blood flow in this model (Fig 6–22).

The role of extrinsic *trauma* appears to be less important in premature infants than in term infants,[56, 57] although some studies have reported a correlation between ICH and vaginal breech delivery.[58, 59] Most current etiologic theories are based on abnormalities of blood flow in an anatomical region which has a poor ability to limit hemorrhage.[60]

Most studies of ICH have shown a correlation with hypoxia. Many indices of *hypoxia* have been evaluated, including fetal heart tones, Apgar scores,[59] ventilator dependence,[4, 61] episodes of pneumothorax,[62, 63] and arterial blood gases.[64] While the relative contribution of individual indicators has varied among reports, hypoxia and respiratory distress[65–67] have been consistently observed. The mechanism by which hypoxia promotes ICH has been more difficult to document.

The systemic blood pressures commonly recorded in hypoxic neonates are at the lower end of the autoregulation curve by which the brain is able to control cerebral blood flow by changes in arterial vascular tone. An apparent *loss of autoregulation* in neonates makes the premature particularly at risk to sudden changes in systemic blood pressures, which would not be transmitted to the brain in the adult.[68]

Hypoxia and hypercarbia both cause *increased cerebral blood flow*, which could distend fragile arterioles and capillaries in a pressure-passive system. In an experimental model, ICH has been induced in the beagle puppy by pharmacologic agents[64] or by volume pushes into hypovolemic puppies[69]; in both settings, the insult was associated with carotid arterial hypertension. An intravascular bolus has been associated with a higher incidence of germinal matrix hemorrhage in some clinical series,[47, 70–72] particularly with sodium bicarbonate infusions. Changes in

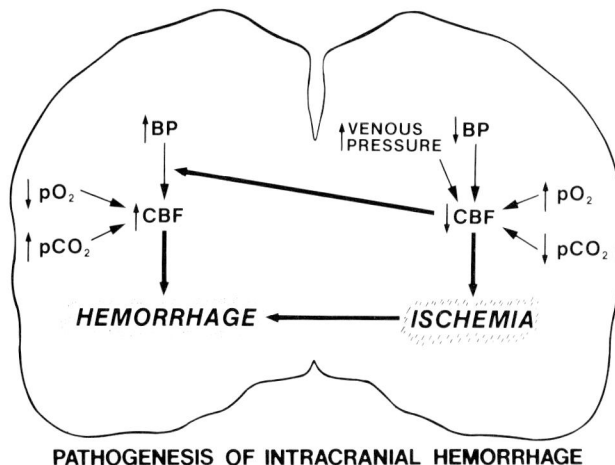

PATHOGENESIS OF INTRACRANIAL HEMORRHAGE

Fig 6–22.—Pathogenesis of intracranial hemorrhage. *BP,* blood pressure; *CBF,* cerebral blood flow; Pco_2, carbon dioxide level; Po_2, oxygen level. (Modified from Wigglesworth and Pape.[55])

cerebral blood flow have been estimated from blood velocity measurements made with Doppler ultrasound. Sudden increases in *brain blood velocity* have been documented in human preterm infants following ligation of the patent ductus arteriosus, pneumothorax, seizures, and endotracheal suctioning.[73]

The Doppler studies have reported *decreased brain blood velocity* in association with hyperventilation, patent ductus arteriosus, and hyperviscosity.[73] The images generated by position emission tomography scanners corroborate this finding and document hypoperfusion in an area of intraparenchymal hemorrhage, generally extending well beyond the area of actual hemorrhage and sometimes occupying the entire cerebral hemisphere.[74, 75] This vascular insufficiency is most commonly in the distribution of the middle cerebral artery.

An alternative hypothesis of primary venous thrombosis with subsequent hemorrhagic infarction has been proposed on the basis of pathologic material found in a rabbit model of ICH. Evidence of DIC and fibrin generation has been reported in association with ICH in a number of studies.[76,77] In its more fulminant forms, this clotting activation has progressed to a depletion of clotting factors, resulting in a hypocoaguable state which can aggravate the extent of the intracranial bleed. The exact role of coagulopathy is controversial, but there clearly appears to be an association. A recent Colorado study demonstrated that *hypocoagulation on day 1* was associated with ICH in the first 72 hours.[78] This correlates well with an earlier CT study that showed an association between *maternal ingestion of aspirin* in the week prior to delivery and neonatal ICH.[79] Clearly, the care of the critically ill premature infant is fraught with many pitfalls in therapeutic measures.

The etiology of neonatal ICH is most likely multifactorial and probably includes all of the mechanisms outlined above. Further advances in imaging techniques, including nuclear magnetic resonance imaging, positron emission tomography, and Doppler ultrasound, will permit clinical assessment of the entire extent and degree of impairment of normal brain physiology.[74, 75, 80] Currently neither CT nor ultrasound is sensitive in detecting ischemic or infarcted tissue unless hemorrhage is present. Investigation of blood flow will help to identify infants at high risk for ICH and monitor the effects of supportive care.

ANATOMICAL SEQUELAE OF ICH

The major anatomical sequelae of ICH are subependymal cysts, porencephaly, and hydrocephalus. The majority of subependymal hematomas will resolve completely on ultrasound and CT without obvious clinical findings. A minority will result in a small subependymal cyst of unknown clinical significance. Almost all parenchymal hemorrhages result in porencephaly; the pattern of resolution was discussed earlier under intraparenchymal hemorrhage.[29, 30]

Posthemorrhagic Hydrocephalus

Posthemorrhagic hydrocephalus develops in approximately 70% of infants with IVH.[28] An abnormally increasing head size can be noted at age 2–3 weeks. However, ventriculomegaly can be reliably diagnosed by ultrasound, and screening for hydrocephalus should be done weekly after IVH.[81, 82] The optimum time for the diagnosis of hydrocephalus is age 2 weeks.[83] Screening is essential because there is frequently severe hydrocephalus by the time the clinical diagnosis is made.[84, 85] The extreme compressibility of the premature brain allows mantle thinning first, with no significant change in head size. Since the sutures and fontanelles are open, increased intracranial pressure is a late finding.

IVH may cause diffuse dilatation of part or all of the ventricles acutely, simply by expansion by large amounts of blood. Several days to 2 weeks after IVH, there is frequently aqueductal obstruction, presumably by clot in the aqueduct or from ventriculitis.[30]

Extraventricular obstructive hydrocephalus may develop secondary to subarachnoid hemorrhage, which occurs when the intraventricular blood exits the ventricular system, and may block the foramina of Magendie and Luschka,[31] or the arachnoid granulations.

Posthemorrhagic hydrocephalus is the only type of hydrocephalus that may return to normal ventricular size a few weeks after the ventricular dilatation. In other types of hydrocephalus, the ventricular size may arrest with only mild dilatation not requiring a shunt, but it would be unlikely to return to normal.

Approximately two thirds of newborns with IVH develop ventricular dilatation.[28] In our experience with the survivors, one third return to normal ventricular size, one third have arrested hydrocephalus with mild to moderate ventricular dilatation and only one third of the survivors required a shunt.

In follow-up studies, CT may reveal asymmetric cortical growth, asymmetric ventricular dilatation, a subdural fluid collection, or encephalomalacia of the affected hemisphere. Therefore, while neither CT nor ultrasound performed in the first week of life can reasonably predict the long-term outcome of ICH, CT or US repeated between ages 2 and 6 months may give useful information, especially in the infant with asymmetric neurologic findings, when the hemorrhage seems small compared to the neurologic deficit. Evidence for more extensive damage later is caused by necrosis of the adjacent infarcted areas which were not hemorrhagic.

NEUROLOGIC SEQUELAE AND PROGNOSIS OF ICH

While the pattern of anatomical resolution of ICH has been well delineated by CT and ultrasound, the long-term clinical significance remains controversial.[86, 87] Newborns with SEH or IVH without hydrocephalus (grade I or II ICH) have a good potential for normal development in the first few years of life. Their short-term outcome does not appear to be different from that of infants without ICH who survive equivalent degrees of prematurity and respiratory distress.

In the majority of cases, IVH results in some ventricular dilatation but it may be transient and revert to normal ventricular size without intervention[28] or become stable mild hydrocephalus. Currently, the remaining infants (one third of the survivors of IVH) with progressively expanding ventricles and clinical signs of hydrocephalus are treated with various drugs,[88–90] serial lumbar punctures,[91–94] or ventriculoperitoneal shunting procedures.

There is increasing mortality with increasing grade or severity of hemorrhage. However, the hemorrhage was not always the cause of death and probably should be used as an indicator of the severity of the initial insult. Intraparenchymal, intraventricular (grade IV) hemorrhage has the worse prognosis (54% mortality) and has been more frequently fatal when specific types of parenchymal extension were seen—for example, extension into the thalamus and posterior fossa—or when midline shift occurred.[28]

The neurologic and developmental prognosis also worsens with increasing grades of ICH, and most infants with persistent hydrocephalus suffer residual deficits. Koons et al.[95] reported normal Bayley developmental scores in 40 of 66 (87%) infants with no evidence of hemorrhage or with isolated SEH (grade 0 or I) and abnormal tone in only one of these infants. However, the 15 infants with IVH and hydrocephalus (grade III) included some with retardation and four with significant motor problems. All four of these infants with grade IV ICH were severely impaired. This was corroborated in the study of Stewart et al.,[96] in which 7 of 81 (9%) infants with no hemorrhage or with SEH alone (grade 0–I) were reported to have sequelae, compared to 15 of 28 (54%) infants with intraventricular and/or intraparenchymal hemorrhage and hydrocephalus (grades II–IV) IVH.

Papile et al. conducted 12-to-24-month follow-up on 117 infants after germinal matrix hemorrhage.[97] Their data agreed that low birth weight infants with only subependymal (grade I) or subependymal and intraventricular hemorrhage (grade II) had the same risk for major handicaps (12%–18%) as low birth weight infants without hemorrhage. However, infants with grade III or IV hemorrhage (intraventricular and intraparenchymal hemorrhage) had a markedly increased incidence of major handicaps.

There is no published follow-up study of ICH beyond age 4 years. Some anatomical sites including the frontal lobe, affect brain functions requiring abstract reasoning, and are not amenable to evaluation in the preschool years. The study of Ment and colleagues[98] revealed a trend toward decreasing developmental scores with time in the survivors of IVH, which is worrisome.

At present, there is insufficient long-term follow-up to fully define the natural history of ICH. Collation of reports from many centers yields a consensus of experience regarding the following points. The survivors of IVH, hydrocephalus, and/or parenchymal hemorrhage (grade III–IV ICH) are at high risk for motor disabilities and at considerable risk for cognitive deficits. Intraparenchymal hemorrhage often results in contralateral hemiparesis. Cerebral atrophy is frequently accompanied by a profound developmental delay. However, while it is apparent that there are trends of worsened outcome with increasingly severe grades of ICH, there is tremendous variability within each grade, and caution should be exercised in predicting the outcome for an individual child.

PERIVENTRICULAR HEMORRHAGIC INFARCTION

Diffuse cerebral hemorrhage in the white matter is probably caused by infarction. It has been termed periventricular leukomalacia.[99] Typically it originates lateral to the frontal horn and bodies of the lateral ventricles rather than in the floor of the ventricles, as in typical SEH. In this position, ventricular rupture is less common and periventricular infarction usually presents as a diffuse cortical hemorrhage without IVH (Fig 6–23).[100–106] If there is only infarction without hemorrhage it may still be highly echogenic.[25] Usually diffuse cerebral infarction is somewhat less echogenic than acute hemorrhage. A comparison CT scan may be necessary to differentiate between hemorrhage and infarction without hemorrhage.

CORTICAL HEMORRHAGE

Isolated cortical hemorrhage is unusual and a specific etiology should be sought. Possible causes include major coagulation disorders (Fig 6–24), arteriovenous mal-

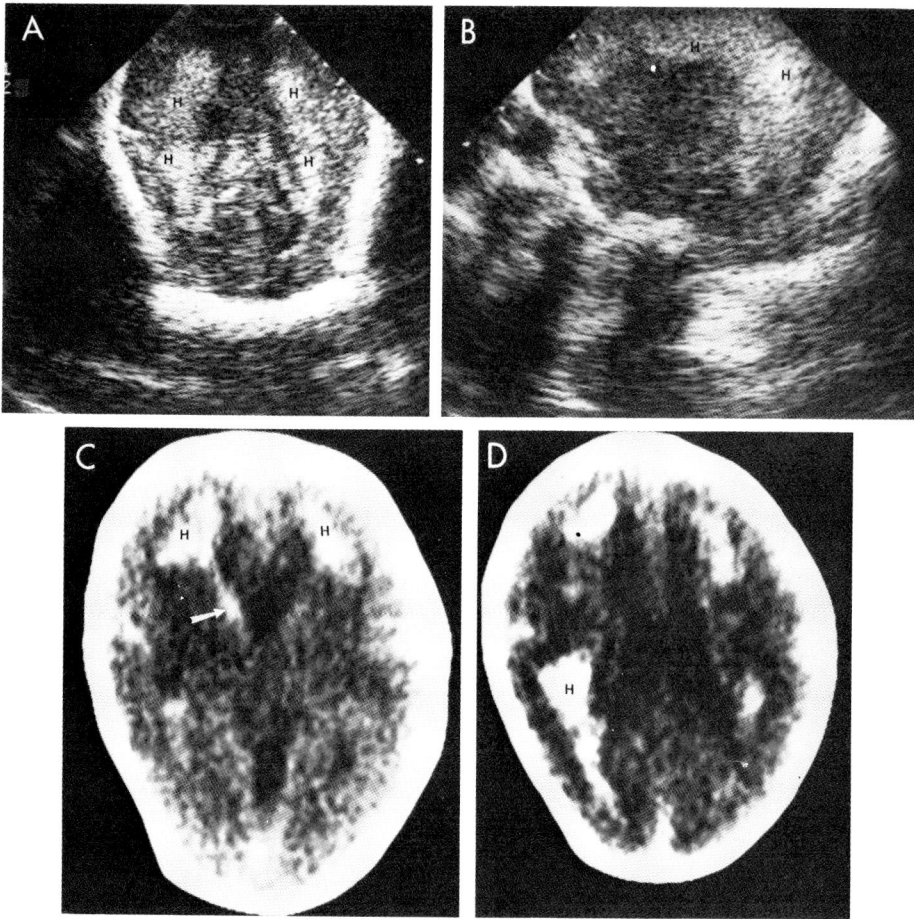

Fig 6–23.—Periventricular hemorrhagic infarction. Coronal **(A)** and sagittal **(B)** ultrasound scans demonstrate extensive parenchymal hemorrhage *(H)* lateral to the ventricles. **C** and **D,** axial CT scans in the same patient show periventricular hemorrhage *(H)* with only a small amount in the region of the caudate *(arrow,* **C**).

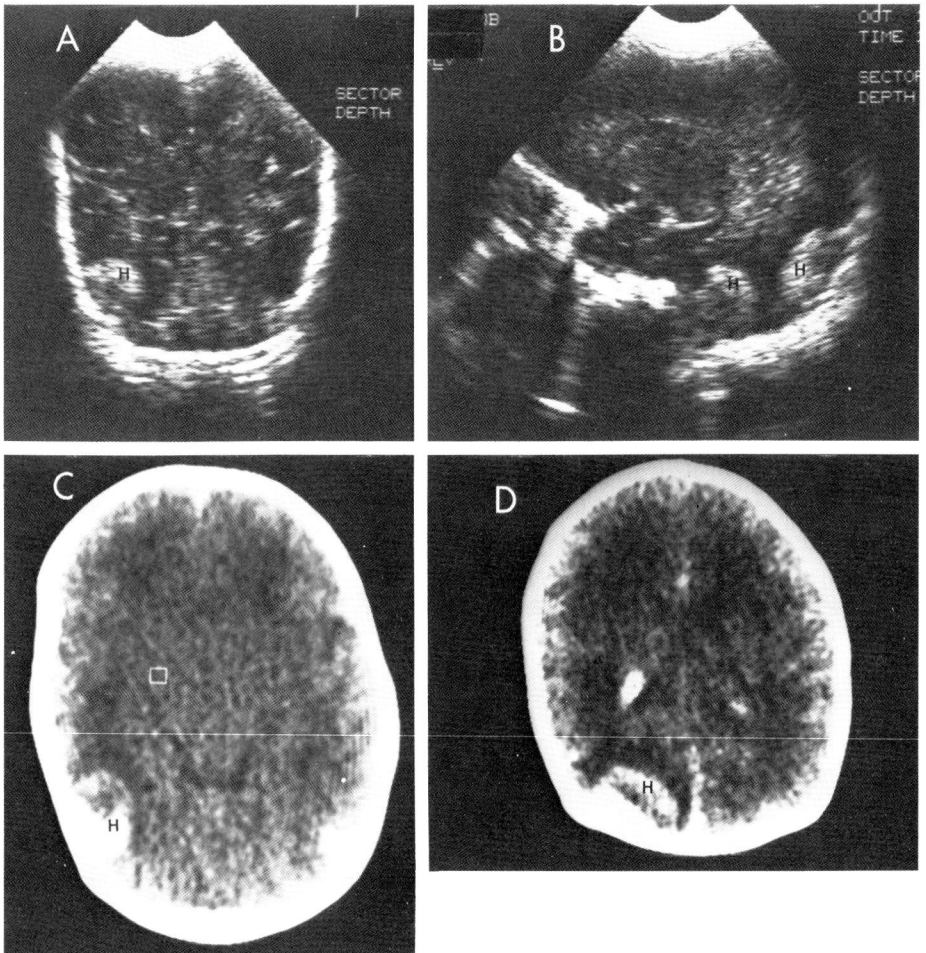

Fig 6–24.—Cortical hematomas in occipital cortex in 6-day-old infant with thrombocytopenia and seizures. Coronal **(A)** and sagittal **(B)** ultrasound scans demonstrate two discrete occipital hematomas *(H)*. Axial CT scans **(C** and **D)** demonstrate hematomas *(H)*, but it is more difficult to separate the two lesions.

formations, tumor, abscess, and trauma. Spontaneous hemorrhage is very uncommon in the parenchyma at any distance from the subependymal region.

DIFFUSE HEMORRHAGE

Occasionally an infectious process may be associated with intracerebral hemorrhage. The initial insult is presumably a vasculitis causing cerebral necrosis and secondary hemorrhage.[6]

CEREBELLAR HEMORRHAGE

Intracerebellar hemorrhage is relatively uncommon but frequently fatal. It has been reported at autopsy in 15%–25% of infants less than 32 weeks' gestation.[107] Various etiologies have been suggested, but the cause of death is probably related to the confined space of the posterior fossa and the resulting pressure on the brain stem. In premature infants, it is frequently associated with germinal matrix hemorrhage and may be caused by the same mechanism from the germinal matrix in the fourth ventricle.[6] In term infants, it appears to be caused by direct intracranial pressure in breech deliveries and by the mask ventilation previously used in nurseries. Occipital compression results in overlapping of the sutures of the occipital bone and, if severe, may cause subdural tears and cerebellar hemorrhage. Superior sagittal sinus compression from head compression with mask ventilation has been demonstrated. In addition to direct pressure on the sinus, venous flow was compressed diffusely and intracranial pressure dropped, causing ischemia and infarction. Discontinuing mask ventilation resulted in a sudden drop in ICH—in this case, primarily intracerebellar hemorrhage.[56, 57] Intracerebellar hemorrhage has also been reported with defects in coagulation.

The diagnosis of cerebellar hemorrhage requires careful attention to the posterior fossa.[108, 109] On CT it may be obscured by posterior fossa artifact from bone and thus requires windows which are good for subdural hematomas. On ultrasound, the cerebellum is quite echogenic, and careful attention to the cerebellar architecture, particularly for symmetry, is necessary to diagnose cerebellar hemorrhage (Fig 6–25).[110–112]

If the infant survives the acute hemorrhage, pressure from the posterior fossa hemorrhage may cause hydrocephalus particularly if it is in the subdural space. Extensive hemorrhage with acute hydrocephalus may require immediate surgery for decompression.[113, 114] Although decompression of the hematoma is possible, no patient that survived cerebellar hemorrhage in the study by Scotti et al.[108] was normal.

Conservative therapy has been successful in some patients with only minor symptoms. In fact, Fishman et al.[113] describe a patient with opisthotonic posturing, nystagmus, and bradycardia who began to recover. Surgery was not undertaken and the hydrocephalus resolved spontaneously.

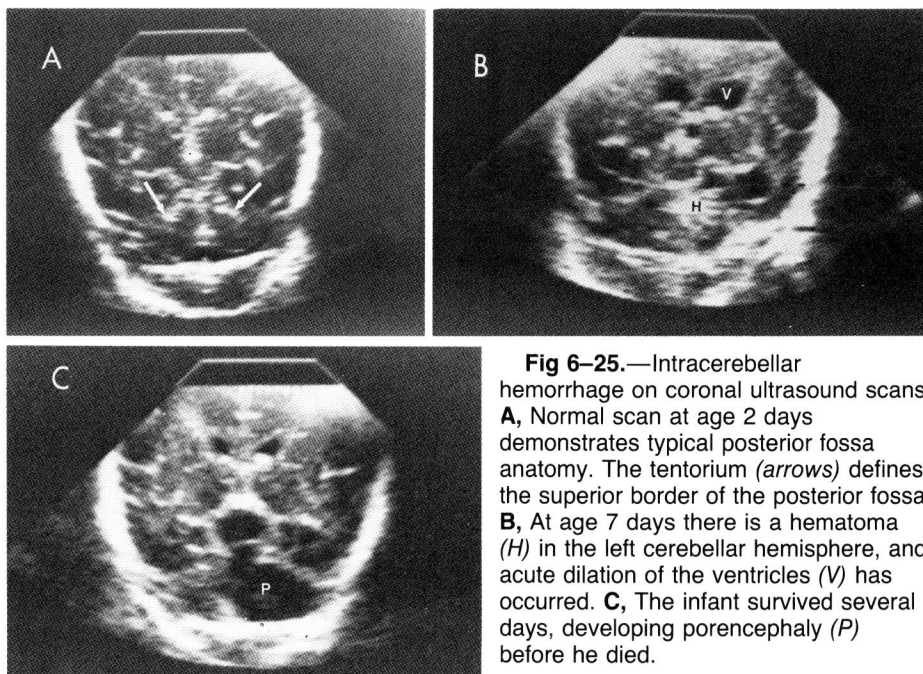

Fig 6–25.—Intracerebellar hemorrhage on coronal ultrasound scans. **A,** Normal scan at age 2 days demonstrates typical posterior fossa anatomy. The tentorium *(arrows)* defines the superior border of the posterior fossa. **B,** At age 7 days there is a hematoma *(H)* in the left cerebellar hemisphere, and acute dilation of the ventricles *(V)* has occurred. **C,** The infant survived several days, developing porencephaly *(P)* before he died.

BRAIN STEM HEMORRHAGE

Brain stem hemorrhage is very difficult to diagnose with any of the presently available modalities. From autopsy studies, it appears to be rare in infancy, usually microscopic, and probably represents a terminal event.[6]

EXTRACEREBRAL HEMORRHAGE

Epidural and Subdural Hemorrhage

Subdural hemorrhage may result from tearing of the unsupported veins bridging the brain and dural sinuses, tearing of the dural fold where it extends into the venous sinuses, or laceration of a sinus by the margin of a fractured or separated skull. The hematoma will collect between the dura and the skull if it is epidural; in this case, the origin is typically arterial and under high pressure. The hematoma will collect between the dura and the brain if there is a venous laceration, giving rise to the term "subdural." A subdural hematoma may extend into the bases of the tentorium or the falx, depending on the site of origin.

Epidural hematomas are rare and typically occur in term infants.[6] Subdural hematomas occur after trauma secondary to torn bridging veins in either the supratentorial or infratentorial regions. With improvements in obstetric care, posttraumatic lesions have greatly decreased in incidence and subdurals are uncommon and are rarely fatal.

Peripheral hemorrhage such as posttraumatic subdural or epidural hematomas are best evaluated with CT. The site and extent of peripheral hemorrhage may be identified on CT particularly well with changes from bone to brain windows (Fig 6–26).[115]

If peripheral hematomas or effusions are large enough (at least 1 cm thick) they can be diagnosed with ultrasound (Fig 6–27).[116] The near field artifact may obscure smaller lesions if they are directly under the transducer. Effusions lateral to the anterior fontanelle and directly beneath bone are even more difficult to detect with ultrasound. If one sees the surface of the brain, there is most likely an extracerebral fluid collection, and CT is recommended.

Posterior Fossa Subdural Hematomas

Infratentorial subdural hematomas are rare and usually fatal. They have occasionally caused obstructive hydrocephalus[117, 118] Autopsy findings have demonstrated that a retrocerebellar subdural hematoma may result from tears in the tentorium or rupture of bridging veins to the cerebellum or vein of Galen.[118] A massive hematoma may cause compression of the cerebellum and brain stem with obliteration of the prepontine cisterns and cisterna ambiens and obstruction of the caudal end of the fourth ventricle.

CT can be valuable, but several pitfalls are noted. Care should be taken not to misinterpret the normal, slightly higher density of the cerebellum relative to the cerebral cortex.[119] Coronal views may be extremely helpful in determining that the

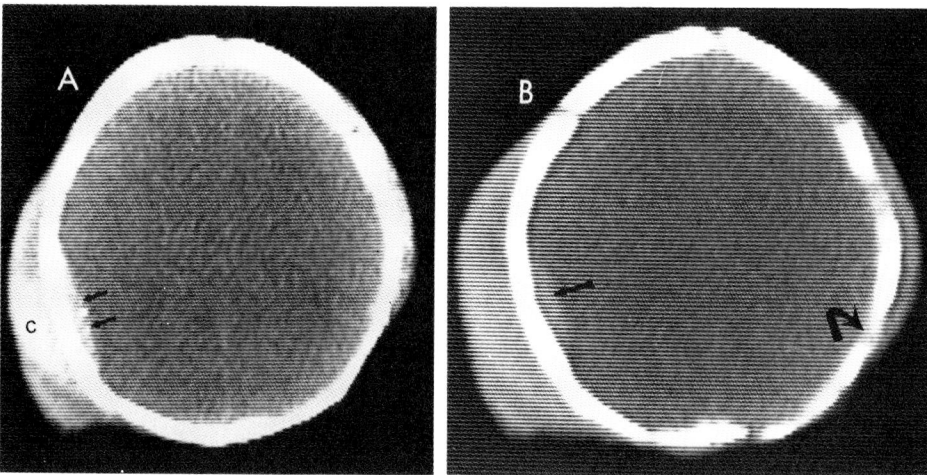

Fig 6–26.—Acute epidural hematoma on axial CT scans. **A,** Soft tissue swelling from birth trauma is cephalohematomas *(C)*. A small epidural hematoma lies just under the skull. **B,** Bone window settings demonstrate the hematoma *(arrow)* and skull fractures *(curved arrow)*.

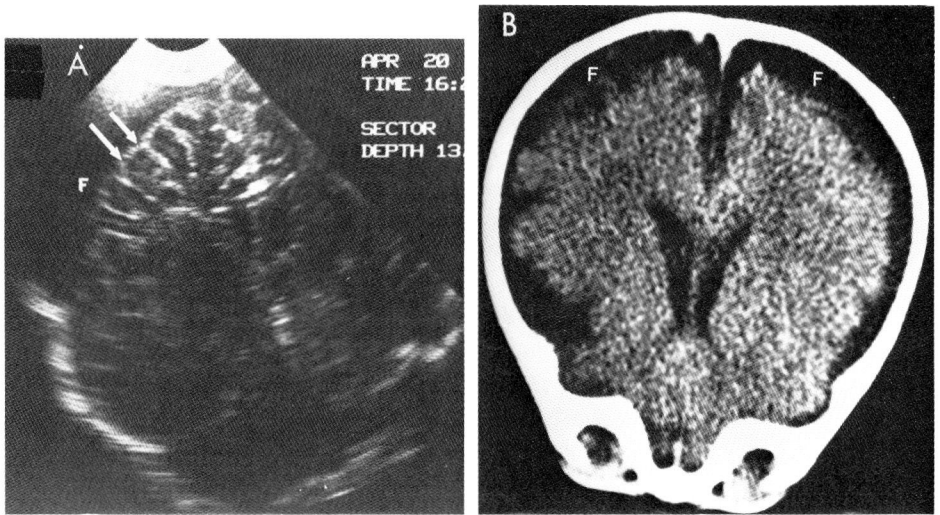

Fig 6–27.—Subdural hematoma. **A,** Coronal ultrasound scan demonstrates fluid *(F)* in the chronic subdural hematoma. *Arrows* point to surface of brain—this surface is not normally seen. From Rumack and Johnson.[127] Reproduced with permission. **B,** Coronal CT scan demonstrates fluid *(F)* in chronic subdural hematomas.

hemorrhage is clearly within or below the tentorium.[108] In fact, hemorrhage may extend both supratentorially and infratentorially.

Ultrasound diagnosis may be difficult. The cerebellum is normally quite echogenic, and asymmetric echogenicity may indicate hemorrhage. Subdural tentorial hemorrhage has not been described on ultrasound but should be most apparent on coronal scans as thickening of the tentorium.

Decompression of the subdural hematomas may be necessary in the posterior fossa if there is hydrocephalus.[108, 116, 117] The prognoses of the survivors appear to be better than for intracerebellar hemorrhage, although only 3 of 9 infants studied by Scotti et al.[108] were normal.

Subdural Collections in Infancy

Studies of the battered child syndrome have made it clear that subdural hematomas rarely occur spontaneously in infancy.[120] Although the outcome of a small subdural hematoma may be benign,[121–124] the risk to the unprotected infant is death. The diagnosis of a subdural hematoma may be made with CT or ultrasound. Small collections may be missed with ultrasound, so CT is essential to document the injury if there is any possibility of this diagnosis.

Subdural effusions may be a complication of meningitis, particularly after *Hemophilus influenzae* infection, and frequently resolve spontaneously.

Subarachnoid Hemorrhage

Isolated subarachnoid hemorrhage usually occurs in term infants who have been asphyxiated and occasionally after trauma or DIC. It may occur secondary to IVH with flow of CSF through the foramina of Magendie and Luschka.

Subarachnoid hemorrhage has been a difficult diagnostic problem. When there is clearly blood in the suprasellar cistern and sylvian fissures on CT, it is easy to diagnose. However, in neonates lying supine, subarachnoid hemorrhage tends to collect in the posterior fossa and along the falx, where it may be confused with normal vascular structures on CT (Fig 6–28,A).

On ultrasound examination, subarachnoid hemorrhage may thicken the appearance of normal vessels in the sylvian fissure, but a large amount of blood must be present to be certain that these structures are abnormally thick (Fig 6–28,B). In our experience, small amounts of subarachnoid hemorrhage have not been detected on ultrasound. Because subarachnoid hemorrhage can be missed on ultrasound, CT should be the primary method of investigation of subarachnoid hemorrhage.

RECOMMENDATIONS

The most likely site of hemorrhage should influence the choice of modality. CT is the method of choice in term neonates and infants who infrequently experience intracranial hemorrhage which is more commonly peripheral. Ultrasound is the method of choice for screening all neonates born at less than 32 weeks' gestation, in whom hemorrhage is more commonly central. Any infant that has neurologic or hematologic signs suggesting hemorrhage but a normal ultrasound examination should undergo a CT examination.

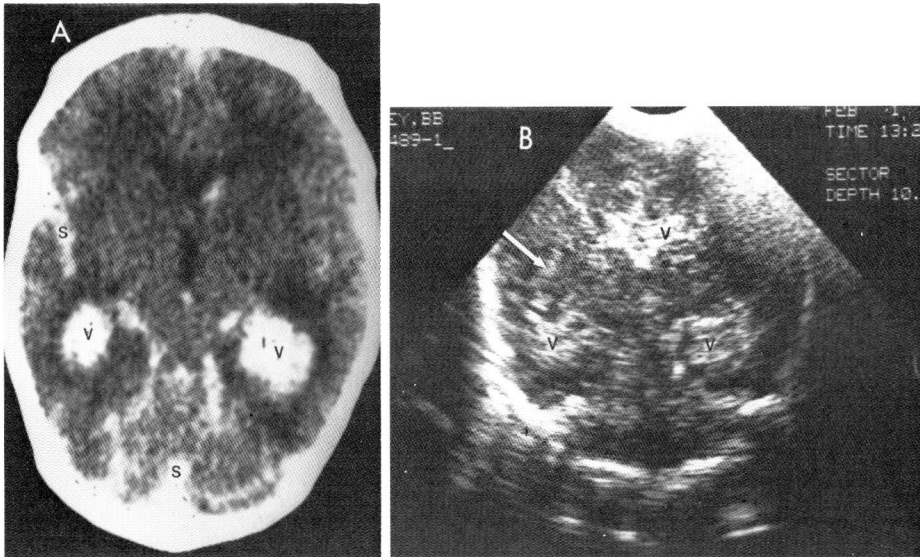

Fig 6–28.—Subarachnoid hemorrhage. **A,** Axial CT scan shows subarachnoid hemorrhage (S) in right sylvian fissure and layered over the tentorium and vascular structures posteriorly. Intraventricular hemorrhage (V) is also present. **B,** Coronal ultrasound scan in same patient. Subarachnoid hemorrhage (arrow) thickens the normal vessels in the sylvian fissure. Intraventricular hemorrhage (V) is also present.

REFERENCES

1. Leech R.W., Kohnen P.: Subependymal and intraventricular hemorrhages in the newborn. *Am. J. Pathol.* 77:465–475, 1974.
2. Valdes-Dapena M.A., Arey J.B.: The causes of neonatal mortality: An analysis of 501 autopsies on newborn infants. *J. Pediatr.* 77:366–375, 1970.
3. Volpe J.J.: *Neurology of the Newborn.* Philadelphia, W.B. Saunders Co., 1981.
4. Papile L., Burstein J., Burstein R., et al.: Incidence and evolution of subependymal and intraventricular hemorrhage: A study of infants with birthweight less than 1500 grams. *J. Pediatr.* 92:529–534, 1978.
5. Burstein J., Papile L., Burstein R.: Intraventricular hemorrhage and hydrocephalus in premature newborns: A prospective study with CT. *AJR* 132:631–635, 1979.
6. Pape K.E., Wigglesworth J.S.: *Hemorrhage, Ischemia and the Perinatal Brain.* London, England, Lavenham Press, Ltd., 1979.
7. Pevsner P.H., Garcia-Bunuel R., et al.: Subependymal and intraventricular hemorrhage in neonates. *Radiology* 119:111–114, 1976.
8. Butzer J.F., Cancilla P.A., Cornell S.H.: Computerized axial tomography of intracerebral hematoma. *Arch. Neurol.* 33:206–214, 1976.
9. Rumack C.M., McDonald M.M., O'Meara O.P., et al.; CT detection and course of intracerebral hemorrhage in premature infants. *AJR* 131:493–497, 1978.
10. Scott W.R., New P.F.J., Davis K.R., et al.: Computerized axial tomography of intracerebral hematoma and IVH. *Radiology* 112:73–80, 1974.
11. Burstein J., Papile L., Burstein R.: Subependymal germinal matrix and intraventricular hemorrhage in premature infants: Diagnosis by CT. *AJR* 128:971–976, 1977.
12. Flodmark O., Becker L.E., Harwood-Nash D.C., et al.: Correlation between CT and autopsy in premature and full-term neonates that have suffered asphyxia. *Radiology* 137:93–103, 1980.
13. Johnson M.L., Rumack C.M., Mannes E.J., et al.: Detection of neonatal intracranial hemorrhage utilizing static and real-time ultrasound. *JCU* 9:427–433, 1981.
14. Babcock D.S., Han B.K.: The accuracy of high resolution, real-time ultrasonography of the head in infancy. *Radiology* 139:665–676, 1981.
15. Mack L.A., Wright K., Hirsch J., et al.: Intracranial hemorrhage in premature infants: Accuracy of sonographic evaluation. *AJR* 137:245–250, 1981.
16. Babcock D.S., Bove K.E., Han B.K.: Intracranial hemorrhage in premature infants: Sonographic-pathologic correlation. *AJNR* 3:309–317, 1982.
17. Grant E.G., Borts F.T., Schellinger D., et al.: Real-time ultrasonography of neonatal intraventricular hemorrhage and comparison with CT. *Radiology* 139:687–691, 1981.
18. Saurbrei E.E., Digney M., Harrison P.B., et al.: Ultrasonic evaluation of neonatal intracranial hemorrhage and its complications. *Radiology* 139:677–685, 1981.
19. Silverboard G., Horder M.H., Ahmann P.A., et al.: Reliability of ultrasound in the diagnosis of intracerebral hemorrhage and posthemorrhagic hydrocephalus: Comparison with CT. *Pediatrics* 66:507–514, 1980.
20. London P.A., Carroll B.A., Enzmann D.R.: Sonography of ventricular size and germinal matrix hemorrhage in premature infants. *AJNR* 1:295–300, 1980.
21. New P.F.J., Aronow S.: Attenuation measurements of whole blood and blood fractions in computed tomography. *Radiology* 121:635–640, 1976.
22. Birnholz J.C., Farrell E.E.: Physical foundations of ultrasound imaging of intraventricular hemorrhage. Presented at the First Perinatal Intracranial Hemorrhage Conference, Washington, D.C., Dec. 5, 1980.
23. Dolinskas C.A., Bilaniuk L.T., Zimmerman R.A., et al.: Computed tomography of intracerebral hematomas: Transmission CT observations on hematoma resolution. *AJR* 129:681–688, 1977.
24. Rumack C.M., Johnson M.L.: Ultrasonic evaluation of intracranial hemorrhage in premature infants. *Semin. Ultrasound* 3:209–215, 1982.

25. Rumack C.M., Johnson M.L.: Role of CT and US in neonatal brain imaging. *J. Comput. Tomogr.* 7:17–29, 1983.
26. Edwards M.K., Brown D.L., Chua G.T.: Complicated infantile meningitis: Evaluation by real-time sonography. *AJNR* 3:431–434, 1982.
27. Albright L., Fellows R.: Sequential CT scanning after neonatal intracerebral hemorrhage. *AJR* 136:949–953, 1981.
28. Lee B.C.P., Grassi A.E., Schechner S., et al.: Neonatal intraventricular hemorrhage: A serial computed tomography study. *J. Comput. Assist. Tomog.* 3:483–490, 1979.
29. Rumack C.M., Johnson M.L., Johnson J.A., et al.: Patterns of intracranial hemorrhage resolution: Correlation with clinical prognosis. Presented at annual meeting of the American Roentgen Ray Society, Atlanta, April 1983.
30. Grant E.G., Kerner M., Schellinger D., et al.: Evolution of porencephalic cysts from intraparenchymal hemorrhage in neonates: Sonographic evidence. *AJR* 138:467–470, 1982.
31. Norman M.G.: Perinatal brain damage. *Perspect. Pediatr. Pathol.* 4:41–92, 1978.
32. Larroche J.C.: Post-haemorrhagic hydrocephalus in infancy: An anatomic study. *Biol. Neonate* 20:287–299, 1972.
33. Larroche J.C.: Subependymal pseudocysts in the newborn. *Biol. Neonate* 21:170, 1972.
34. Sherwood A., Hopp A., Smith J.F.: Cellular reactions to subependymal plate haemorrhage in the human neonate. *Neuropathol. Appl. Neurobiol.* 4:245–261, 1978.
35. Rumack C.M., Johnson M.L., McDonald M.M.: Timing of intracranial hemorrhage. Presented at the Society for Pediatric Radiology, San Francisco, March 1981.
36. Mack L.A., Rumack C.M., Johnson M.L.: Ultrasound evaluation of cystic lesions. *Radiology* 137:451–455, 1980.
37. Krishnamoorthy K.S., Fernandez R.A., et al.: Evaluation of neonatal intracranial hemorrhage by computerized tomography. *Pediatrics* 59:165–172, 1977.
38. Shankaran S., Slovis T.L., Bedard M.P., et al.: Sonographic classification of intracranial hemorrhage: A prognostic indicator of mortality, morbidity and short term neurologic outcome. *J. Pediatr.* 100:469, 1982.
39. Ahmann P.A., Lazarra A., Dykes F.D., et al.: Intraventricular hemorrhage in the high risk preterm infant: Incidence and outcome. *Ann. Neurol.* 7:118, 1980.
40. Tsiantos A., Victorin L., Relier J.P., et al.: Intracranial hemorrhage in the prematurely born infant: Timing of clots and evaluation of clinical signs and symptoms. *J. Pediatr.* 85:854–859, 1974.
41. Emerson P., Fujimara M., Howat P., et al.: Timing of intraventricular hemorrhage. *Arch. Dis. Child.* 52:183–187, 1977.
42. Hect E.T., Filly R.A., Callen T.W., et al.: Late onset of intracranial hemorrhage in the preterm newborn. *Radiology*, to be published.
43. Kim M.S., Elyaderani M.K.: Sonographic diagnosis of cerebroventricular hemorrhage in utero. *Radiology* 142:479–480, 1982.
44. Lazzaru A., Ahmann P., Dykes F., et al.: Clinical predictability of intraventricular hemorrhage in preterm infants. *Pediatrics* 65:30–34, 1980.
45. Thorburn R.J., Lipscomb A.P., Stewart A.L.: Timing and antecedents of hemorrhage into the geminal layer and ventricles and of cerebral atrophy in very preterm infants. *J. Reprod. Biol.*, to be published.
46. Dykes F., Lazzara A., Ahmann P., et al.: Intraventricular hemorrhage: A prospective evaluation of etiopathogenesis. *Pediatrics* 66:42, 1980.
47. Cole V.A., Durbin G.M., Olaffson A., et al.: Pathogenesis of intraventricular hemorrhage in newborn infants. *Arch. Dis. Child.* 49:722–728, 1974.
48. Guggenheim M.A., Rumack C.M., Langendoerfer S., et al.: Clinical factors associated with neonatal germinal matrix hemorrhage, unpublished manuscript.
49. Fedrich J., Butler N.R.: Certain causes of neonatal death: II. Intraventricular haemorrhage. *Biol. Neonate* 15:257–290, 1970.
50. Wigglesworth J.S., Davies P.A., Keith I.H., et al.: Intraventricular hemorrhage in the

preterm infant without hyaline membrane disease. *Arch. Dis. Child.* 52:447–451, 1977.

51. Harcke H.T., Jr., et al.: Perinatal cerebral intraventricular hemorrhage. *J. Pediatr.* 80:37–42, 1972.

52. Spears R.L., et al.: Relationship between hyaline membrane disease and intraventricular hemorrhage as cause of death in low birth weight infants. *Am. J. Obstet. Gynecol.* 105:1028–1031, 1969.

53. Hambleton G., Wigglesworth J.S.: Origin of intraventricular haemorrhage in the preterm infant. *Arch. Dis. Child.* 51:651–659, 1976.

54. Towbin A.: Central nervous system damage in the human fetus and newborn infant. *Am. J. Dis. Child.* 119:529–542, 1970.

55. Wigglesworth J.S., et al.: An integrated model for haemorrhagic and ischemic lesions in the newborn brain. *Early Hum. Dev.* 2:179–199, 1978.

56. Newton T.H., Gooding C.A.: Compression of superior sagittal sinus by neonatal calvarial moulding. *Radiology* 115:635–639, 1975.

57. Wigglesworth J.S., Husemeyer R.P.: Intracranial birth trauma in vaginal breech delivery: The continued importance of injury to the occipital bone. *Br. J. Obstet. Gynaecol.* 84:684–691, 1977.

58. Bejar R., Coen R.W., Gluch L.: Hypoxic-ischemic and hemorrhagic brain injury in the newborn. *Perinatol. Neonatol.* 6:69, 1982.

59. McDonald M.M., Johnson M.L., Rumack C.M., et al.: Etiology and timing of intracranial hemorrhage, unpublished manuscript.

60. Gilles F.H., Price R.A., et al.: Fibrinolytic activity in the ganglionic eminence of the premature human brain. *Biol. Neonate* 18:426–432, 1971.

61. Thomas D.B., Burnard E.D.: Prevention of intraventricular hemorrhage in babies receiving artificial ventilation. *Med. J. Aust.* 1:933–936, 1973.

62. Lipscomb A.P., Reynolds E.O.R., Blackwell R.J., et al.: Pneumothorax and cerebral haemorrhage in preterm infants. *Lancet* 1:414–416, 1981.

63. Hill A., Perlman J.M., Volpe J.J.: Relationship of pneumothorax to occurrence of IVH in the premature newborn. *Pediatrics* 69:144–149, 1982.

64. Goddard J., Lewis R.M., Alcala H., et al.: Intraventricular hemorrhage: An animal model. Child Neurology Society Abstracts. Presented at the Child Neurology Society, Key Stone, Colorado, 1978.

65. Wigglesworth J.S., Keith I.H., Girling D.J., et al.: Hyaline membrane disease, alkali and intraventricular hemorrhage. *Arch. Dis. Child.* 51:755, 1976.

66. Marshall R.E., et al.: Intracranial haemorrhage and hyaline membrane disease. *Lancet* 1:880, 1974.

67. Fujimura M., et al.: Clinical events relating to intraventricular hemorrhage in the newborn. *Arch. Dis. Child.* 54:409–414, 1979.

68. Lou H.C., Lassen N.A., et al.: Impaired autoregulation of cerebral blood flow in the distressed newborn infant. *J. Pediatr.* 94:118–121, 1979.

69. Goddard-Finegold J., Armstrong D., Zeller R.S.: Intraventricular hemorrhage following volume expansion after hypovolemic hypotension in the newborn beagle. *J. Pediatr.* 100:796–799, 1982.

70. Papile L., Burstein J., et al.: Relationship of I.V. sodium bicarbonate infusions and cerebral intraventricular hemorrhage. *J. Pediatr.* 93:834–836, 1978.

71. Finberg L.: The relationship of intravenous infusions and intracranial hemorrhage: A commentary. *J. Pediatr.* 91:777, 1977.

72. Simmons M.A., Adcock E.W., Bard H., et al.: Hypernatremia and intracranial hemorrhage in neonates. *N. Engl. J. Med.* 291:6–10, 1974.

73. Volpe J.J.: Cerebral blood flow in the newborn infant: Relation to hypoxic-ischemic brain injury and periventricular hemorrhage. *J. Pediatr.* 94:170–173, 1979.

74. Volpe J.J., Perlman J., et al.: Positron emission tomography (PET) in the assessment of regional cerebral blood flow in the newborn. *Ann. Neurol.* 12:225(A), 1982.

75. Deuel R.K.: Pathophysiology, live. *Pediatrics* 70:312, 1982.

76. Hathaway W.E., Mull M.M., Pechet G.S.: Disseminated intravascular coagulation in the newborn. *Pediatrics* 43:233–240, 1969.

77. Chessells J.M., Wigglesworth J.S.: Coagulation studies in preterm infants with respiratory distress and intracranial hemorrhage. *Arch. Dis. Child.* 47:564–570, 1970.

78. McDonald M.M., Johnson M.L., Rumack C.M., et al.: The role of coagulopathy in newborn intracranial hemorrhage, unpublished manuscript.

79. Rumack C.M., Guggenheim M.A., Rumack B.H., et al.: Neonatal intracranial hemorrhage and maternal use of aspirin. *Obstet. Gynecol.* 58:52s–56s, 1981.

80. Delpy D.T., Gordon R.E., Hope P.L., et al.: Noninvasive investigation of cerebral ischemia by phosphorus. *Pediatrics* 70:310, 1982.

81. Horbar J.D., Walters C.L., Phillip A.G.S., et al.: Ultrasound detection of changing ventricular size in post-hemorrhagic hydrocephalus. *Pediatrics* 66:674–678, 1980.

82. Bowerman R.A., Donn S.M., Silver T.M., et al.: Sonographic evolution and natural history of neonatal periventricular/intraventricular hemorrhage. Presented at the Second Special Conference on Perinatal Intracranial Hemorrhage, Washington, D.C., 1982.

83. Partridge J.C., Babcock D.S., Steichen J.J., et al.: Optimal timing for diagnostic cranial ultrasound in low birth weight infants: Detection of intracranial hemorrhage and ventricular dilatation. Presented at the Second Special Conference on Perinatal Intracranial Hemorrhage, Washington, D.C., Dec. 1982.

84. Volpe J.J., Pasternak J.F., Allan W.C.: Ventricular dilatation preceding rapid head growth following neonatal intracranial hemorrhage. *Am. J. Dis. Child.* 131:1212, 1977.

85. Korobkin R.: The relationship between head circumference and the development of communicating hydrocephalus in infants following intraventricular hemorrhage. *Pediatrics* 56:74–77, 1975.

86. Hack M., Fanaroff A.A., Merkatz I.R.: The low birth weight infant: Evolution of a changing outlook. *N. Engl. J. Med.* 22:1162, 1979.

87. Fitzhardinge P.M., Kalman E., Ashby S., et al.: Present status of the infant of very low birth weight treated in a referral neonatal intensive care unit in 1974. *Ciba Found. Symp.* 59:139, 1978.

88. Donn S.M., Roloff D.W., Goldstein G.W.: Prevention of intraventricular hemorrhage in preterm infants by phenobarbitone: A controlled trial. *Lancet* 2:215–217, 1981.

89. Bergman E., Epstein M., Freeman J.: Medical management of hydrocephalus with acetazolamine and furosemide, abstracted. *Ann. Neurol.* 4:189, 1978.

90. Morgan M.E., Benson J.W., Cooke R.W.: Ethamsylate reduces the incidence of periventricular haemorrhage in very low birth weight infants. *Lancet* 2:830–831, 1981.

91. Papile L.A., Burstein J., Burstein R., et al.: Post hemorrhagic hydrocephalus in low birth-weight infants: Treatment by serial lumbar puncture. *J. Pediatr.* 97:273–277, 1980.

92. Mantovani J.F., Pasternak J.E., Mathew O.P., et al.: Failure of daily lumbar punctures to prevent development of hydrocephalus after IVH. *J. Pediatr.* 97:278–281, 1980.

93. Behrman R.E.: Serial lumbar punctures and intraventricular hemorrhage, editorial. *J. Pediatr.* 97:250, 1980.

94. Hill A., Taylor D.A., Volpe J.J.: Treatment of posthemorrhagic hydrocephalus by serial lumbar puncture: Factors that account for success or failure. *Ann. Neurol.* 10:284–285, 1981.

95. Koons A., Sun S., Kamtorn V., et al.: Neurodevelopmental outcome related to intraventricular hemorrhage and perinatal events. Read before the Second Special Conference on Perinatal Intracranial Hemorrhage, Washington, D.C., 1982.

96. Stewart A.L., Thorburn R.J., Hope P.L., et al.: Relationship between ultrasound appearance of the brain in very preterm infants and neurodevelopmental outcome at 18

months of age. Read before the Second Special Conference on Perinatal Intracranial Hemorrhage, Washington, D.C., Dec. 1982.

97. Papile, L.A., Munsick-Bruno G., Schaefer A.: The relationship of cerebral intraventricular hemorrhage and early childhood handicaps. Read before the Second Special Conference on Perinatal Intracranial Hemorrhage, Washington, D.C., Dec. 1982.

98. Ment L.R., Scott D.T., Ehrenkranz R.A., et al.: Neonates of ≤ 1250 grams birth weight: Prospective neurodevelopmental evaluation during the first year post term. *Pediatrics* 70:292–295, 1982.

99. Armstrong D., Norman M.G.: Periventricular leukomalacia in infants: Complications and sequelae. *Arch. Dis. Child.* 49:367–375, 1974.

100. Grant E., Schellinger D., Richardson J., et al.: Periventricular halo: Normal sonographic finding or neonatal cerebral hemorrhage. *AJR* 140:793–796, 1983.

101. DiPietro M.A., Brody B.A., Teele R.L.: The periventricular echogenic "blush" on cranial ultrasound. Read before the annual meeting of the Society for Pediatric Radiology, Atlanta, April 1983.

102. Hill A., Melson G.L., Clark B., et al.: Hemorrhagic periventricular leukomalacia: Diagnosis by real-time ultrasound and correlation with autopsy findings. *Pediatrics* 69:282–284, 1982.

103. Levene M.J., Wigglesworth J.S., Dubowitz V.: Hemorrhagic periventricular leukomalacia in the neonate: A real-time ultrasound study. *Pediatrics* 71:794–797, 1983.

104. Babcock D.S., Ball W.: Ultrasound diagnosis and short term prognosis of post-asphyxiation encephalopathy in term infants. Presented at the Second Special Conference on Perinatal Intracranial Hemorrhage, Washington, D.C., Dec. 1982.

105. Manger M.N., Feldman R.C., Brown W.J., et al.: Intracranial ultrasound diagnosis of neonatal periventricular leukomalacia: A case report and review of the literature. Presented at the Second Special Conference on Perinatal Intracranial Hemorrhage, Washington, D.C., Dec. 1982.

106. Slovis T.L., Shankaran S., Bederd, M., et al.: Unusual intracranial complications of hemorrhage in the hypoxic-ischemic infant. Presented at the Second Special Conference on Perinatal Intracranial Hemorrhage, Washington, D.C., Dec. 1982.

107. Martin R., Roessmann U., Fanaroff A.: Massive intracerebellar hemorrhage in low birthweight infants. *J. Pediatr.* 89:290–293, 1976.

108. Scotti G., Floodmark O., Harwood-Nash D.C.: Posterior fossa hemorrhages in the newborn *J. Comput. Assist. Tomogr.* 5:68–72, 1981.

109. Rom S., Serfontein G.L., Humphreys R.P.: Intracerebellar hematoma in the neonate. *J. Pediatr.* 93:486–488, 1978.

110. Perlman J.M., Nelson J.S., McAllister W.H., et al.: Intracerebellar hemorrhage in a premature newborn: Diagnosis by real-time ultrasound. *Pediatrics* 71:159–162, 1983.

111. Roeder J.D., Setzer E.S., Kaude J.V.: Ultrasonic detection of perinatal intracerebellar hemorrhage. *Pediatrics* 70:385–386, 1982.

112. Grant E., Schellinger D., Richardson J.: Real time ultrasonography of the posterior fossa. *J. Ultrasound Med.* 2:73–87, 1983.

113. Fishman M.A., Percy A.K., Cheek W.R., et al.: Successful conservative management of cerebellar hematomas in term neonates. *J. Pediatr.* 98:466–468, 1981.

114. Ravenel S.D.: Posterior fossa hemorrhage in the term newborn: Report of two cases. *Pediatrics* 64:39–42, 1979.

115. Koshu K., Hirashima Y., Endo S., et al.: CT findings in a case of neonatal acute subdural hematoma. *Neuroradiology* 21:223–225, 1981.

116. Slovis T.L., Kelly J.K., Eisenbrey A.B., et al.: Detection of extracerebral fluid collections by real-time sector scanning through the anterior fontanelle. *J. Ultrasound Med.* 1:41–44, 1982.

117. Gilles F.H., Shillito J.: Infantile hydrocephalus: Retrocerebellar subdural hematoma. *J. Pediatr.* 76:529–537, 1970.

118. Blank N.K., Strand R., Gilles F.H., et al.: Posterior fossa subdural hematomas in neonates. *Arch. Neurol.* 35:108–111, 1978.

119. Potter E.L., Craig J.M.: *Pathology of the Fetus and Infant*. Chicago, Year Book Medical Publishers, Inc., 1975.

120. French B., Dublin A.B.: Infantile chronic subdural hematoma of the posterior fossa diagnosed by CT. *J. Neurosurg.* 47:949–952, 1977.

121. Kempe C.H., Silverman F.N., Steele B.F., et al.: The battered-child syndrome. *JAMA* 181:17–24, 1962.

122. Robertson W.C., Chun R.W.M., Orrison W.W., et al.: Benign subdural collections of infancy. *J. Pediatr.* 94:382–385, 1979.

123. Mori K., Handa H., Itoh M., et al.: Benign subdural effusion in infants. *J. Comput. Assist. Tomogr.* 4:466–471, 1980.

124. Briner S., Bodensteiner J.: Benign subdural collections of infancy. *Pediatrics* 67:802–804, 1981.

125. Goddard J., Lewis R.M., Armstrong D.L., et al.: Moderate rapidly induced hypertension as a cause of intraventricular hemorrhage in the newborn beagle model. *J. Pediatr.* 96:1057–1060, 1980.

126. Lou H.C., Lassen N.A., Friis-Hansen B.: Is arterial hypertension crucial for the development of cerebral hemorrhage in premature infants? *Lancet* 1:1215, 1979.

127. Rumack C.M., Johnson M.L.: Real time evaluation of the neonatal brain. *Clinics in U.S.* 10:179–198, 1983.

CHAPTER 7

Hydrocephalus

Carol M. Rumack, M.D.
Michael L. Johnson, M.D.

HYDROCEPHALUS is dilation of the ventricular system when there is an anatomical obstruction to the normal movement of cerebrospinal fluid (CSF) and an associated elevation of CSF pressure.[1] Hydrocephalus implies that there is an obstructive process. Ventricular enlargement or ventriculomegaly, as opposed to hydrocephalus, should be diagnosed when there are large ventricles that are the result of a previous destructive process or congenital malformation but the ventricles are not under pressure. The differential diagnosis in the infant is most often made by correlation with head size. If there are large ventricles and a large or normal-sized head, the condition is most likely hydrocephalus. If there are large ventricles and a small head, the condition has most likely been caused by destruction or malformation.

NORMAL CEREBROSPINAL FLUID CIRCULATION

Cerebrospinal fluid (CSF) surrounds the brain and spinal cord. It acts as a protective fluid, as an excretory system from brain to blood, as a controlled chemical environment, and possibly as a nutritive transport system.[1] It has been called the third circulation[2, 3] to emphasize that CSF flow represents a major shift of fluid.

Approximately 40% of CSF production is in the choroid plexus of the lateral, third, and fourth ventricles.[4] The other 60% is produced by extracellular fluid movement from blood through the brain and into the ventricles.[5]

CSF flow moves from the lateral ventricles, through the foramina of Monro, third ventricle, aqueduct of Sylvius, and fourth ventricle, and exits through the foramen of Magendie in the midline or laterally through the foramina of Luschka into the basal subarachnoid cisterns (Fig 7–1). Anterior flow continues upward through the chiasmatic cisterns, sylvian fissure, and the pericallosal cisterns up over the hemispheres, where it is reabsorbed by the arachnoid granulations in the sagittal sinus. Posteriorly, CSF flow continues around the cerebellum, through the tentorial incisura, the quadrigeminal cistern, the posterior callosal cistern, and up over the hemispheres. A small amount flows down the spinal subarachnoid space.

CSF flow results from continued production of CSF by the choroid plexus and from respiratory and vascular pulsations, ciliary action in the ependyma, and a

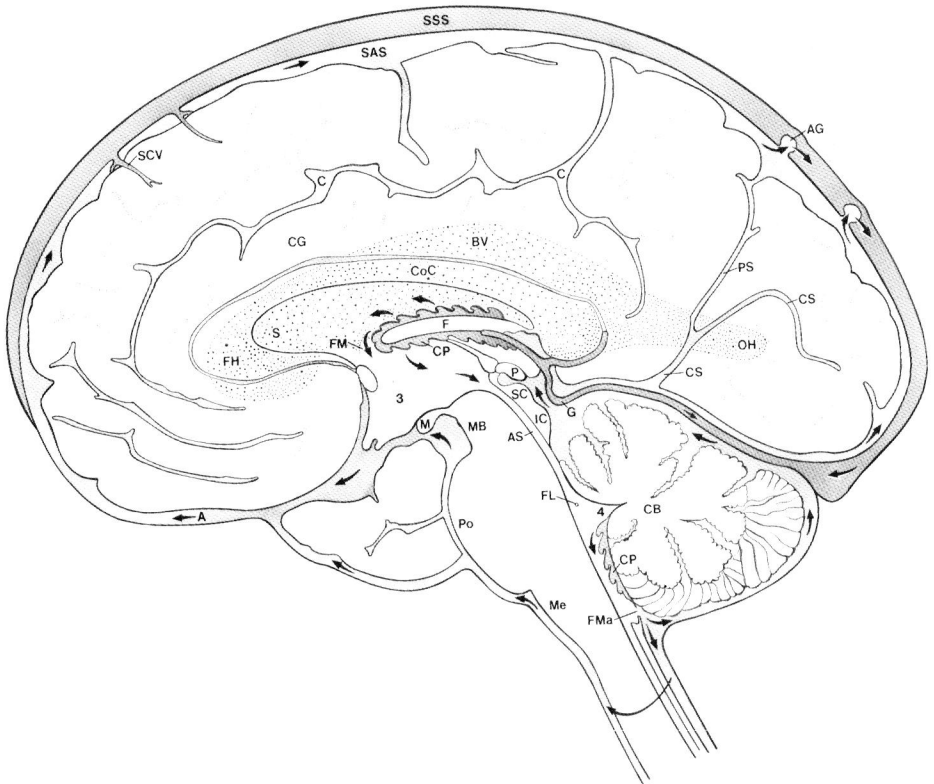

Fig 7–1.—Normal cerebrospinal fluid circulation. *A,* arachnoid; *AG,* arachnoidal granulation; *AS,* aqueduct of sylvius; *BV,* body of ventricle; *C,* cingulate gyrus; *CoC,* corpus callosum; *CP,* choroid plexus; *CS,* calcarine sulcus; *F,* fornix; *FH,* frontal horn of lateral ventricle; *FL,* foramen of luschka; *FM,* foramen of Monro (intraventricular foramen); *FMa,* foramen of Magendie; *G,* great cerebral vein; *IC,* inferior colliculi; *M,* mammary body; *MB,* midbrain; *Me,* medulla; *OH,* occipital horn of lateral ventricle; *P,* pineal body; *PS,* parieto-occipital sulcus; *Po,* pons; *S,* septum pellucidum; *SAS,* subarachnoid space; *SC,* superior colliculi; *SCV,* superior cerebral vein; *SSS,* superior sagittal sinus; *3,* third ventricle; *4,* fourth ventricle.

downhill pressure gradient between the subarachnoid spaces and the venous sinuses.[6, 7]

CSF production does not change significantly with increased intracranial pressure below a rather large range of pressures.[8] CSF production in children varies between 532 and 576 ml/day. Since a normal volume of CSF is approximately 50–150 ml, this means that CSF normally turns over several times a day unless there is significantly increased intracranial pressure.[1]

VENTRICULAR SIZE

Normal Ventricular Measurements

Measurements of the ventricular system can be very detailed. In our experience, ventricular size can be most easily evaluated by direct observation. It has been our policy to measure the lateral ventricular ratio (LVR)[9–11] (Fig 7–2) on ultrasound as a guide since the comparable previous scan may have been done at a sector depth that makes the ventricles look smaller. A normal LVR in premature neonates ranges from 24% to 34%, with a mean of 31%.[10] In normal term infants the LVR is slightly lower. The mean width of the lateral ventricle in the premature neonate is 1.0 cm, with a mean hemispheric width of 3.1 cm. Several investigators have shown that ultrasound can evaluate ventricular size as reliably as computerized tomography (CT).[10–19]

On CT examination, a consistent attempt is made to use the same circle size so that comparisons are valid. If necessary, a bifrontal ratio can be measured (Fig 7–3).[20, 21] Progressive ventricular dilatation should be determined over a series of scans to demonstrate hydrocephalus conclusively. However, equilibrium may have already been reached at maximum ventricular size at the time of the first scan.

Parts of the neonatal ventricular system appear relatively large compared to the adult system. The fourth ventricule is normally larger at birth, which may be related to the small size of the cerebellum and its growth over the first year of life.[22] The third ventricle is slitlike. Small temporal horns are visualized frequently.[4] The occipital horns of the lateral ventricles are slightly larger than the frontal horns.

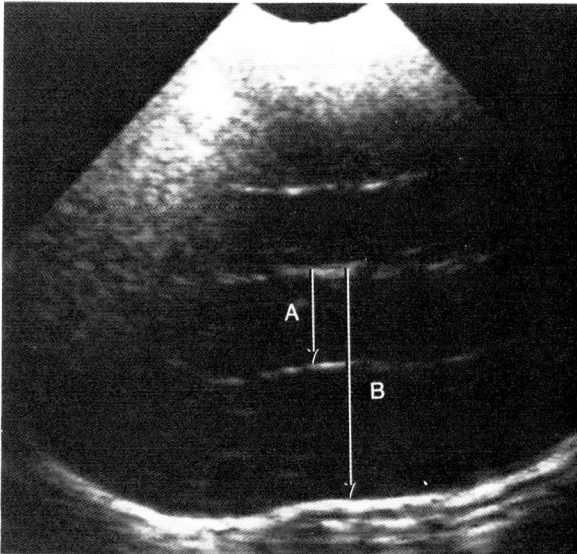

Fig 7–2.—Lateral ventricular ratio = lateral ventricular width/hemispheric width = A/B. Measurements are made from the middle of the midline echo to the inner surface of lateral ventricle (A) or the inner table of the skull (B) on axial ultrasound scans.

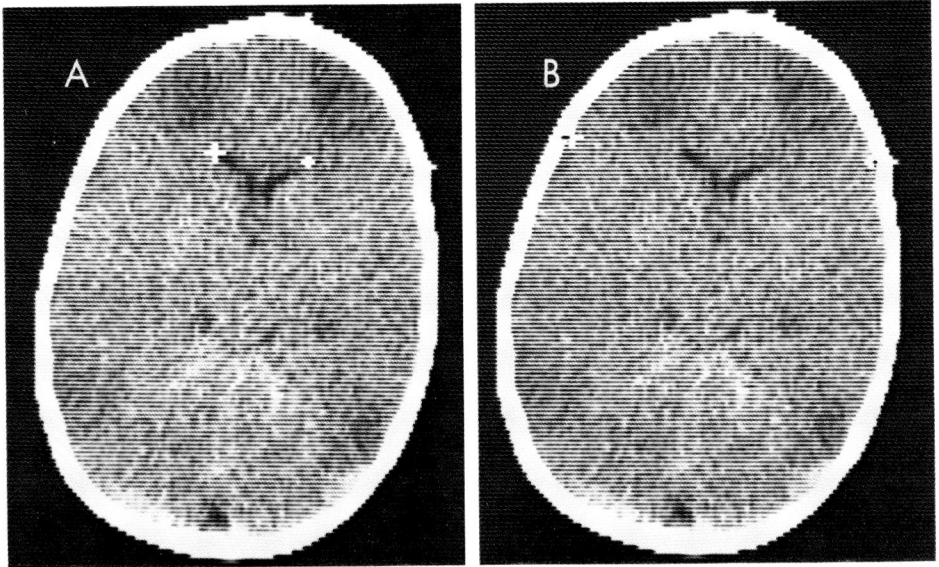

Fig 7–3.—Bifrontal ventricular ratio = lateral edge of both frontal horns/inner diameter of skull. Measurements are made at the same level in the brain. Calipers mark the edges of ventricles **(A)** and brain **(B)**. This is an axial CT scan in a 6-month-old infant.

More detailed measurements have been described by many investigators.[23–25] However, visualization of the entire ventricular system is more reliable than any one measurement (Fig 7–4).

Brain Mantle

When there is severe hydrocephalus, the brain mantle may decrease to less than 1 cm, and several investigators have reported a poorer prognosis in neonates with this condition.[26–28] Previous studies were based on pneumoencephalography.

Brain mantle is the distance from the edge of the ventricle to the inner table of the skull. It has been measured from the top of the skull to the ventricle on pneumoencephalography.[29] On CT, it is measured from the ventricle to the inner surface of the skull and may vary in different parts of the brain.[30] On ultrasound, brain mantle can best be measured in the axial view from the lateral wall of the lateral ventricle to the inner surface of the skull.[10]

Diagnosis of Hydrocephalus

With the development of hydrocephalus, the occipital horns will enlarge first and the frontal horns enlarge last.[31] However, if frontal horn shape is considered, ballooning of the frontal horns will be appreciated before an actual change in size occurs in hydrocephalus; this situation would be unlikely in atrophic ventricular enlargement (Fig 7–5).[21, 32] Enlargement of the ventricular system can be thought

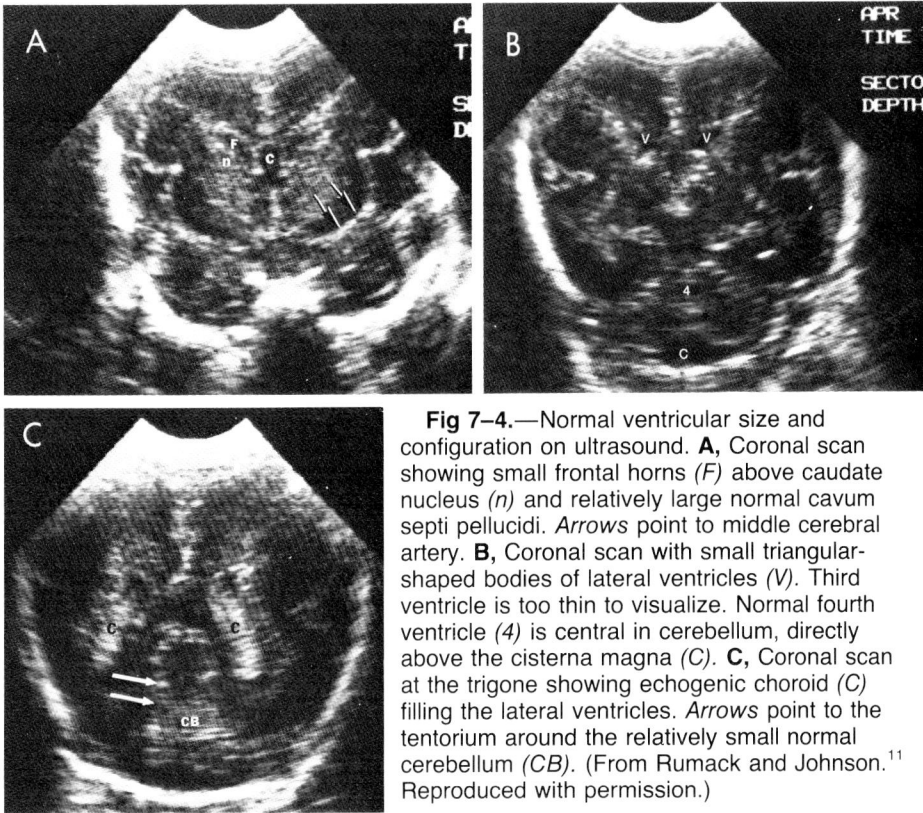

Fig 7–4.—Normal ventricular size and configuration on ultrasound. **A,** Coronal scan showing small frontal horns *(F)* above caudate nucleus *(n)* and relatively large normal cavum septi pellucidi. *Arrows* point to middle cerebral artery. **B,** Coronal scan with small triangular-shaped bodies of lateral ventricles *(V)*. Third ventricle is too thin to visualize. Normal fourth ventricle *(4)* is central in cerebellum, directly above the cisterna magna *(C)*. **C,** Coronal scan at the trigone showing echogenic choroid *(C)* filling the lateral ventricles. *Arrows* point to the tentorium around the relatively small normal cerebellum *(CB)*. (From Rumack and Johnson.[11] Reproduced with permission.)

of as similar to distention of a balloon, with the larger segments enlarging more easily than the smaller ones because they require less pressure for distention.

The LVR should be no more than 34%. If a previous scan is available, enlargement may already be evident. An LVR of 35%–40% is mild ventricular dilation, an LVR of 41%–50% is moderate ventricular dilation, and an LVR over 50% is severe dilation.

Cortical mantle thinning in children with hydrocephalus is often asymmetric, being more severe posteriorly. Apparently the occipital cortex is less resistant to pressure than the basal ganglia adjacent to the frontal horns.

INDICATIONS FOR VENTRICULAR EVALUATION

Intrauterine screening for hydrocephalus should be performed if there is a history of a previous abnormality such as an X-linked aqueductal stenosis,[33] a myelomeningocele, or other neural tube defect. If there are maternal signs such as polyhydramnios to suggest the possibility of a brain malformation that would interfere with the normal swallowing of amniotic fluid, the brain should be evaluated. Normal ventricular size varies tremendously during brain development in utero, so

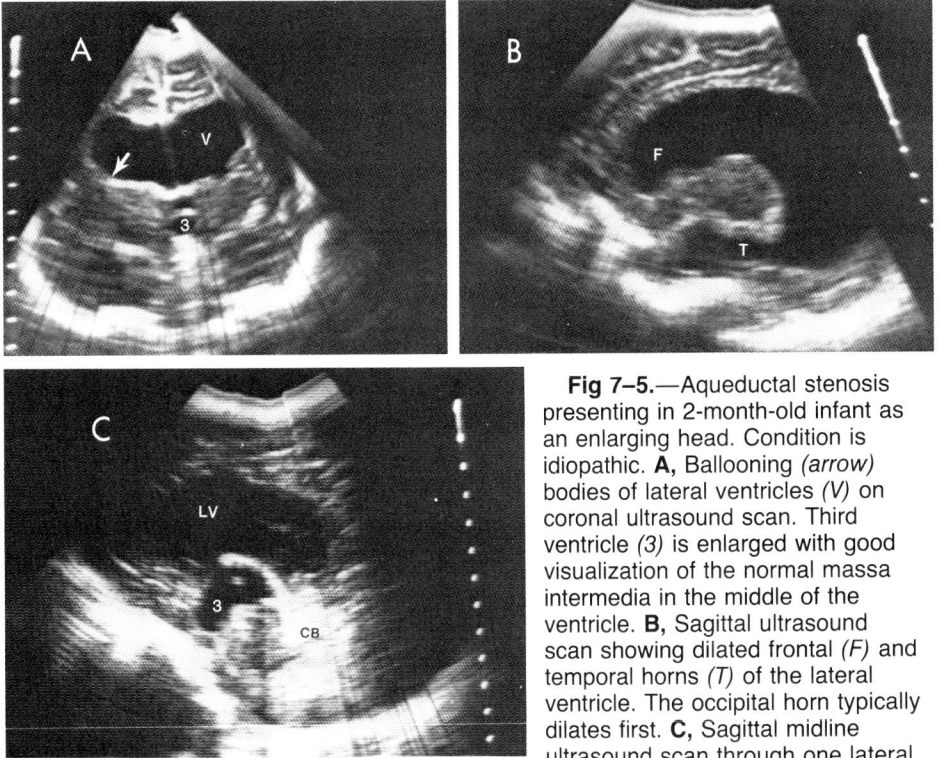

Fig 7–5.—Aqueductal stenosis presenting in 2-month-old infant as an enlarging head. Condition is idiopathic. **A,** Ballooning *(arrow)* bodies of lateral ventricles *(V)* on coronal ultrasound scan. Third ventricle *(3)* is enlarged with good visualization of the normal massa intermedia in the middle of the ventricle. **B,** Sagittal ultrasound scan showing dilated frontal *(F)* and temporal horns *(T)* of the lateral ventricle. The occipital horn typically dilates first. **C,** Sagittal midline ultrasound scan through one lateral ventricle *(LV)* showing the foramen of Monro and the third ventricle *(3)*. The fourth ventricle is barely visible as a notch in the normal echogenic cerebellum *(CB)*.

the size of the ventricles must be correlated with the normal range for that gestational age.[34]

Neonatal screening for hydrocephalus should be performed routinely on premature infants less than 32 weeks' gestational age because intracranial hemorrhage occurs frequently in this age group. Infants with intraventricular hemorrhage should be followed at least to 1 month of age for the development of ventricular dilation. The optimum time to screen for the onset of hydrocephalus is at age 2 weeks.[35] Weekly follow-up examinations should be sufficient to determine whether the ventricles return to normal size, arrest with mild hydrocephalus, or continue to progressively enlarge. Screening is necessary because ventricular enlargement precedes head growth and significant brain damage may occur before hydrocephalus is clinically evident.[36, 37]

After the neonatal period hydrocephalus should be sought if there is evidence of increasing head circumference. A CT examination should be performed for an enlarging head if hydrocephalus is not diagnosed by ultrasound, because unsuspected subdural hematomas may be the cause of the head enlargement.

INTRAVENTRICULAR OBSTRUCTIVE HYDROCEPHALUS

The terms communicating and noncommunicating hydrocephalus are confusing, as all obstructive hydrocephalus is noncommunicating. Therefore this discussion will be based on the level of obstruction, which is determined by the transition from an enlarged ventricle to a normal ventricle (Table 7–1).

FORAMEN OF MONRO OBSTRUCTION.—If the foramen of Monro is obstructed on one side, there may be unilateral ventricular enlargement. In our experience, asymmetric hydrocephalus is more often caused by destruction rather than by obstruction of the side of the larger ventricle (Fig 7–6). A midline shift toward the smaller side would be expected if there were unilateral obstruction (Fig 7–7).[4] However, if only one ventricle responds to shunting, positive contrast evaluation of the ventricles may be necessary to prove the failure of communication. Bilateral foramen of Monro obstruction is quite rare in infants and typically occurs in older children secondary to a large suprasellar cyst or a tumor adjacent to the foramen of Monro.

AQUEDUCTAL OBSTRUCTION.—Aqueductal obstruction is a common obstruction resulting from hemorrhage in premature newborns. Aqueductal stenosis has been reported as an X-linked dominant trait (see Fig 7–5).[33] The third and lateral ventricles are enlarged and the fourth ventricle is normal or small. Unusual causes of aqueductal obstruction may be a vein of Galen aneurysm,[38] or, very rarely, a tumor or cyst.

Aqueductal stenosis is frequently present in the Chiari II malformation, in addition to obstruction of the basilar cisterns. The fourth ventricle is tiny if visualized at all. Aqueductal stenosis is described in detail along with its many associated findings in the chapter on congenital malformations.

TABLE 7–1.—HYDROCEPHALUS

Intraventricular obstructive
  Posthemorrhagic
    Aqueductal obstruction
    Fourth ventricle obstruction
    Posterior fossa subdural
  Chiari malformation
  Dandy-Walker syndrome
  Aqueductal stenosis
  Postinfectious: intraventricular septations
  Vein of Galen aneurysm
  Tumor or cyst
Extraventricular obstructive
  Posthemorrhagic
  Postinfectious
  Achondroplasia
  Absence of arachnoid granulations
Nonobstructive
  Choroid plexus papilloma
  Superior vena cava obstruction
  Vein of Galen aneurysm

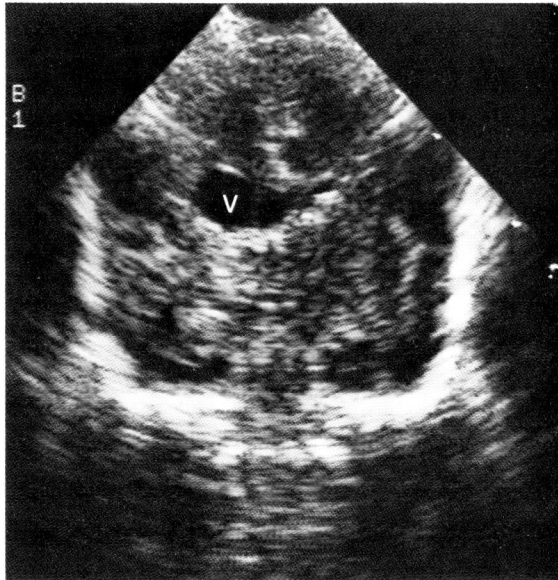

Fig 7–6.—Asymmetric ventricular dilation toward a region of infarction in a coronal ultrasound scan through the bodies of the lateral ventricles shows right lateral ventricle *(V)* much larger than the left. Right hemisphere is echogenic from infarction.

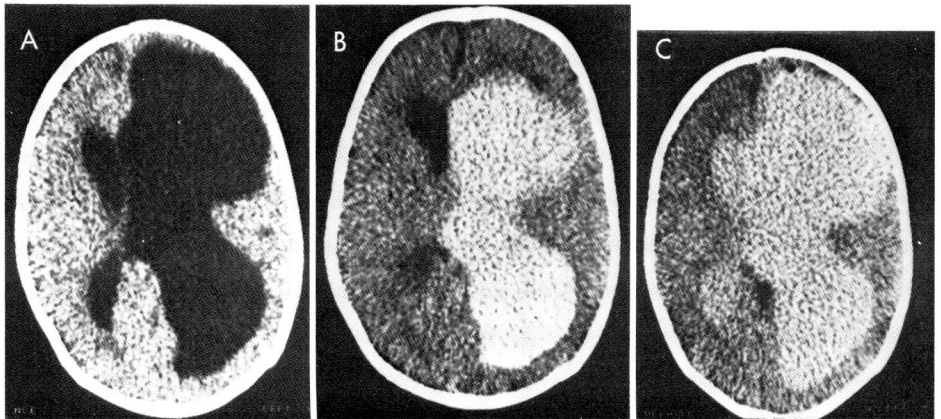

Fig 7–7.—Unilateral foramen of Monro obstruction, partial. **A,** Axial CT scan in a 5½-month-old infant who was a premature twin, breech delivery, with an enlarging head, probably from posthemorrhagic hydrocephalus. Left lateral ventricle is much larger than the right. **B,** Metrizamide injected into the left lateral ventricle showed no contrast in the right lateral ventricle. **C,** Metrizamide has moved into the right lateral ventricle on this delayed scan, indicating a partial obstruction.

OBSTRUCTION OF THE FORAMINA OF MAGENDIE AND LUSCHKA.—Obstruction of these foramina (the outlets of the fourth ventricle) is part of the mechanism causing the Dandy-Walker syndrome.[1] Most investigators report that the lesion develops so early (first month of gestation) that vermian dysgenesis, cerebellar hypoplasia, and failure of neural tube closure lead to small, separate cerebellar hemispheres with a large cyst, in addition to the obstructive elements. Typically, shunting for hydrocephalus in a patient with a Dandy-Walker cyst decompresses the lateral ventricles much more effectively than the cyst. A second shunt may be necessary to decompress the cyst (Fig 7–8).

Posterior fossa hemorrhage may compress the fourth ventricle directly or it may

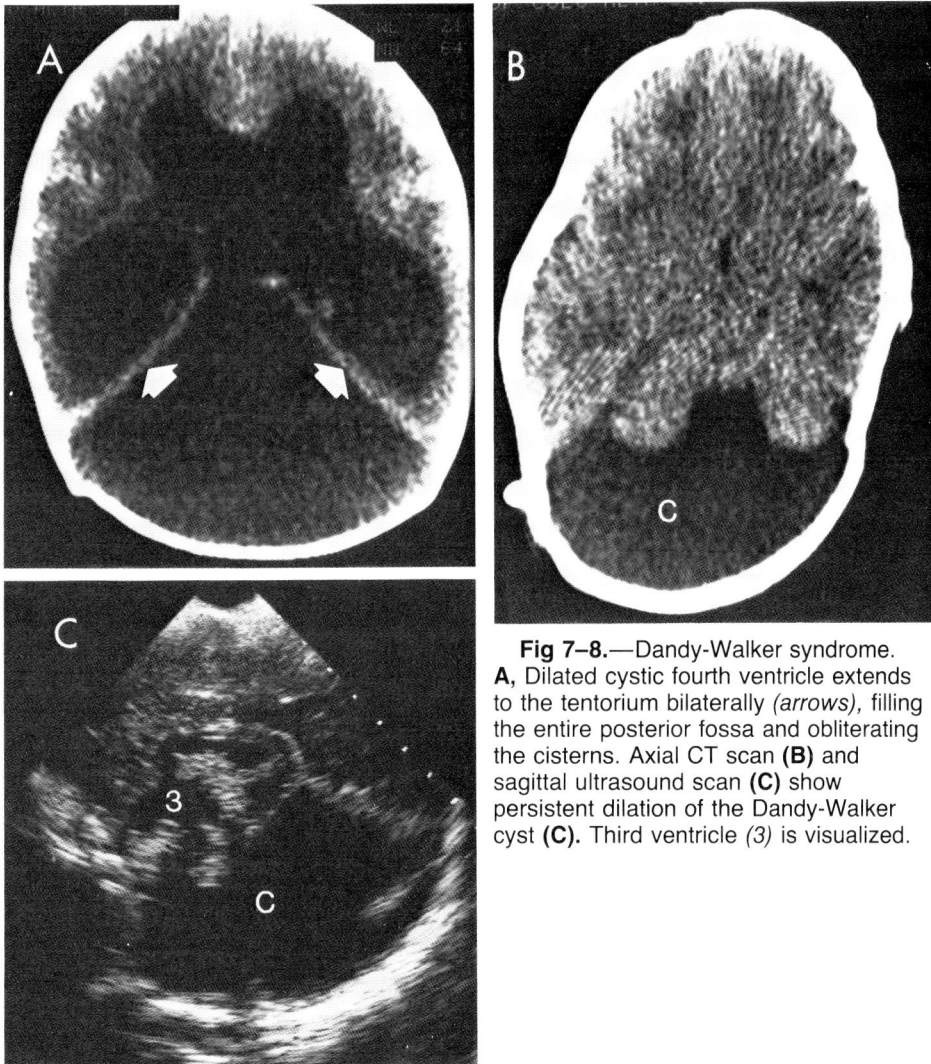

Fig 7–8.—Dandy-Walker syndrome. **A,** Dilated cystic fourth ventricle extends to the tentorium bilaterally *(arrows)*, filling the entire posterior fossa and obliterating the cisterns. Axial CT scan **(B)** and sagittal ultrasound scan **(C)** show persistent dilation of the Dandy-Walker cyst **(C).** Third ventricle *(3)* is visualized.

extend within the tentorium and compress the aqueduct or third ventricle.[39, 40]

Intraventricular septations may develop at any level within the ventricular system.[41] These are typically postinfectious and can loculate a portion of the ventricles.[42]

Intracranial tumors are very rare in utero or in the first 2 years of life but they can cause extrinsic compression of a portion of the ventricular system.

EXTRAVENTRICULAR OBSTRUCTIVE HYDROCEPHALUS

An extraventricular block can occur anywhere peripheral to the foramina of Magendie and Luschka (see Table 7–1). After meningitis, the basilar cisterns may become obstructed or the arachnoid granulations may develop arachnoiditis (Fig 7–9,A). The scarring that occurs with healing can block the normal flow of CSF.

Posthemorrhagic hydrocephalus may cause an extraventricular block with adhesions from hemorrhage (Fig 7–9,B and C).

A rare cause of extraventricular obstructive hydrocephalus is thought to be due to a constricted foramen magnum in achondroplasia.[43] An even rarer cause is absence of the arachnoid granulations such that resorption of the CSF cannot occur.[44]

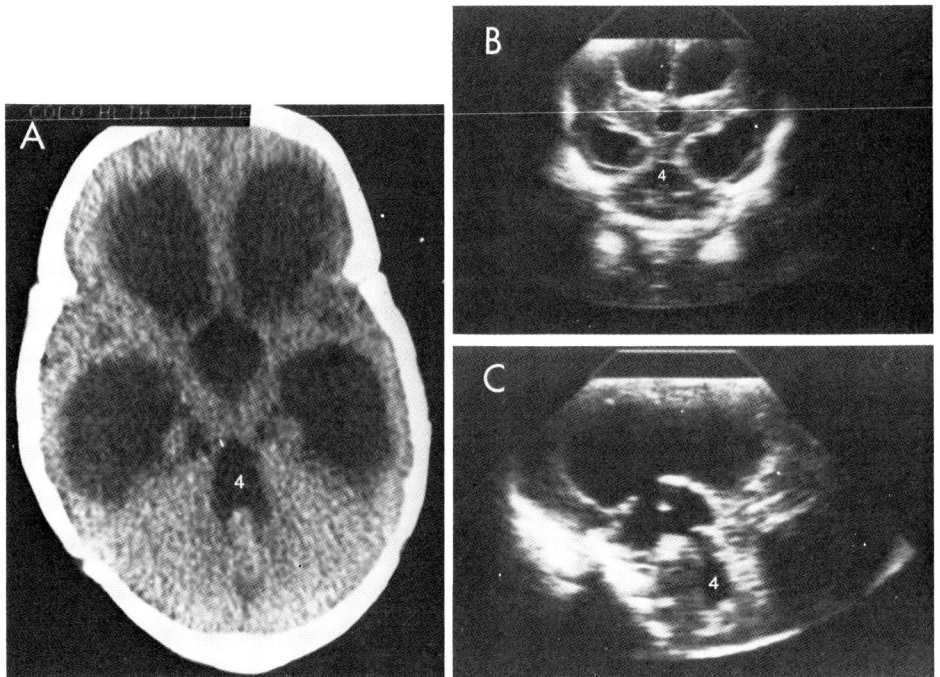

Fig 7–9.—Extraventricular obstructive hydrocephalus in two infants. (A), Axial CT scan in 9-month-old female who developed enlarging head several months after pneumococcal meningitis. All of the ventricular system is enlarged, including the fourth ventricle (4). Coronal (B) and Sagittal (C) ultrasound scans of a 3-month-old infant with posthemorrhagic hydrocephalus show involvement of all the ventricles, including the fourth (4).

NONOBSTRUCTIVE HYDROCEPHALUS

Several causes have been postulated for nonobstructive hydrocephalus, including increased CSF production from a choroid plexus papilloma.[45] Increased venous pressure interfering with the CSF gradient of flow from subarachnoid fluid into the dural sinuses may occur from obstruction of the superior vena cava in infancy,[46] as a complication of total parenteral nutrition,[47] or from other causes of increased intracranial venous pressure.[48] Hydrocephalus may also occur from an increase in intraventricular CSF pulse pressure.[49] Although the vein of Galen aneurysm appears to cause hydrocephalus by obstruction of the aqueduct, it may transmit an increased pulse pressure, thus leading to hydrocephalus, as reported in adults with aneurysms of the basilar artery.[50–52]

Normal pressure hydrocephalus does not appear to occur in children.[53]

BRAIN DAMAGE FROM HYDROCEPHALUS

Damage to the brain caused by hydrocephalus appears to occur in a series of steps beginning with periventricular edema, which presently is best evaluated with CT (Fig 7–10).[54, 55] On the basis of animal studies by McLone et al.[56] and Rubin et

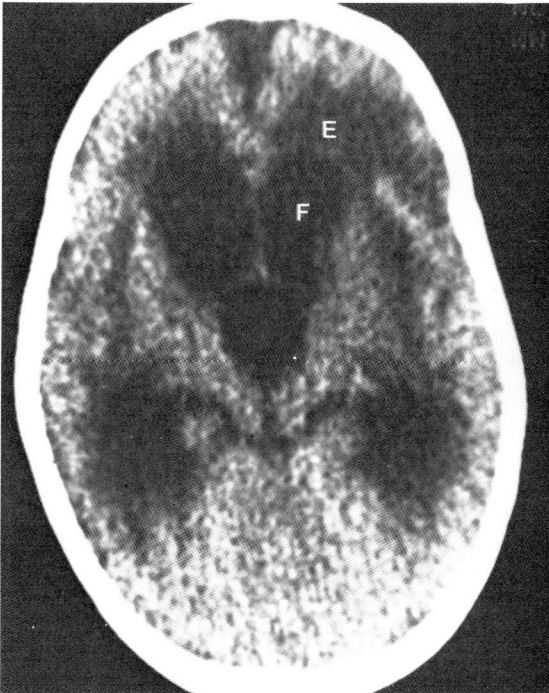

Fig 7–10.—Periventricular edema from increased intracranial pressure. Axial CT scan demonstrates enlarged third and lateral ventricles with edema *(E)* anterior to the frontal horn *(F)* in a 3-month-old infant postmeningitis.

al.,[57] ependymal disruption leading to white matter edema is the first result of increased intracranial pressure. Periventricular white matter edema is reversible, but the subsequent events of axonal degeneration, myelin disruption, and astrocytic scarring are not reversible.[58] Initially with hydrocephalus, brain function is not affected. With continued pressure, more axons are stretched and broken, leading eventually to ataxia and spasticity. White matter is destroyed first, as it collects large amounts of extracellular edema.[59, 60]

These same investigators showed that cerebral arteries are stretched by hydrocephalus and thus compromise cerebral blood flow, leading to further brain loss.

After the intracranial pressure increases to a certain point, transventricular absorption of CSF occurs and an equilibrium is reached.[61, 62] The open sutures and fontanelle allow the ventricles to enlarge more than they would after the sutures are closed.[63] Thus a larger ventricular size develops in infants at equilibrium than would develop after the sutures close in the older child and adult.

VENTRICULAR PERITONEAL SHUNTING FOR HYDROCEPHALUS

Two important unanswered questions are how large the ventricles must become before shunting is required, and, particularly in posthemorrhagic hydrocephalus, how long to wait to determine if hydrocephalus will resolve or arrest spontaneously. Several investigators have tried drug therapy such as phenobarbitone, or repeated lumbar taps to remove CSF as temporizing measures.[64–66] Posthemorrhagic hydrocephalus may resolve spontaneously, and thus the complications of a ventriculoperitoneal shunt may be avoided.[67] Ventricular taps may be necessary if there is intraventricular obstructive hydrocephalus, because removal of spinal CSF will not remove CSF from the ventricles. The different responses to taps in posthemorrhagic hydrocephalus reported in the literature may be related to the origin of the hydrocephalus.[66, 68, 69] In extraventricular obstructive hydrocephalus, a spinal tap may result in decompression, whereas if the patient has predominantly intraventricular obstructive hydrocephalus, improvement is unlikely.

Hydrocephalus which occurs as a result of meningitis or on the basis of an obstructive malformation may arrest with only mild ventricular dilation, but once there is severe ventricular dilation there is no reason to expect the ventricles to return to normal size.

Intraoperative Placement of Shunt Catheters

Ultrasound allows real-time intraoperative assessment and positioning of the intraventricular catheter (Fig 7–11).[70, 71] Previously used methods, such as plain radiography or intraoperative fluoroscopy, could not reveal ventricular or parenchymal detail.

The ventriculoperitoneal shunt catheter must be placed accurately so that it is in the ventricular system and the most anterior end of the shunt is placed into the frontal horn anterior to the foramen of Monro. The choroid plexus does not extend into the frontal horns, so the choroid cannot enter the shunt tip or side holes, where it might obstruct the flow of CSF.

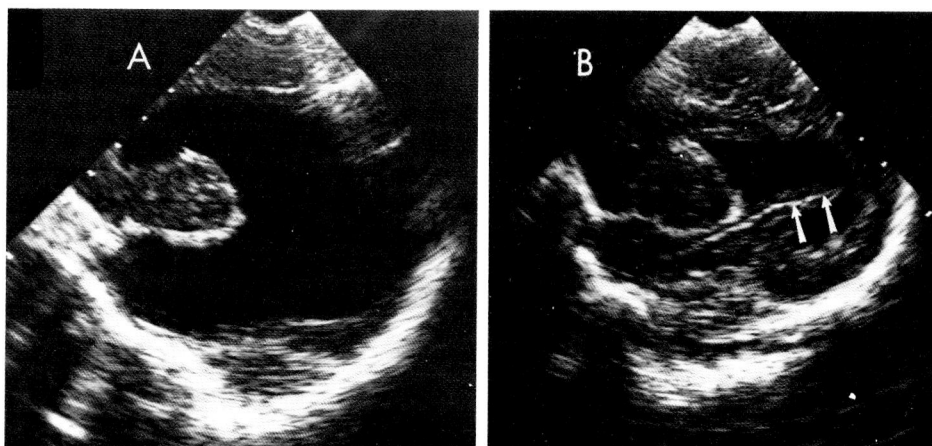

Fig 7–11.—Aqueductal stenosis from intraventricular hemorrhage in 3-month-old premature infant. **A,** dilated lateral ventricle. **B,** sagittal ultrasound scan showing shunt in lateral ventricle *(arrows)* which is now decompressed.

Anatomical Changes After Shunting

Chiari Malformation

Decompression of a Chiari malformation can be done effectively with just a lateral ventricular shunt, but some changes occur after shunting that can be confusing.[4, 72] The posterior fossa cisterns are large and the tentorial opening is wide. After decompression, the cerebellum may rise up or "tower" into the tentorial opening, causing a pseudomass that is best seen in an axial projection on CT. On ultrasound, the coronal projection makes it immediately obvious that this is just the cerebellum surrounded by the larger cisterns and the large tentorial incisura. The interhemispheric fissure may be enlarged (Fig 7–12). The falx is frequently deficient, resulting in interdigitation of the sulci across the midline.

Postshunt Changes

Inversion of the cortical mantle and more marked decompression of one ventricle than the other may occur after any type of ventricular decompression.[73, 74]

SHUNT COMPLICATIONS

TRAPPED FOURTH VENTRICLE.—After ventriculoperitoneal shunt placement, the fourth ventricle may remain enlarged or become enlarged due to obstruction at the aqueduct and its outlet.[75, 76] The typical patient has a Chiari malformation with an associated aqueductal block. Some patients have repeated infections leading to entrapment. A Dandy-Walker cyst may not respond to shunting of the lateral ventricles, requiring a second shunt for fourth ventricular decompression.

RECURRENT OBSTRUCTION.—Shunt malfunction may be due to distal tubing ob-

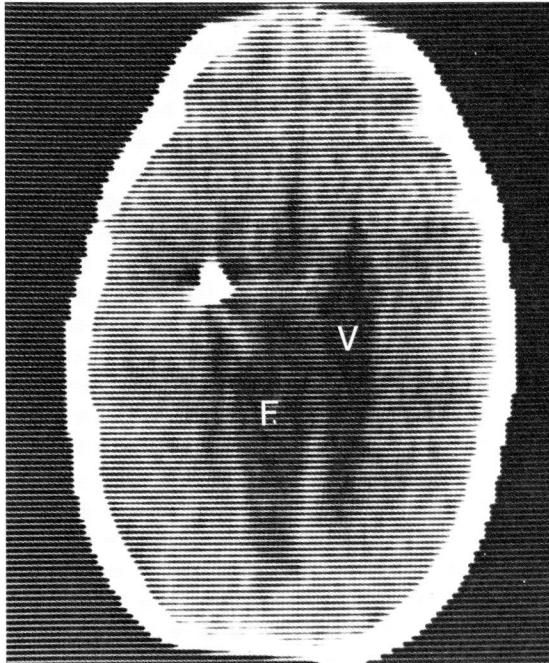

Fig 7–12.—Chiari II malformation on axial CT scan in infant with shunt. Left lateral ventricle *(V)* is larger than the slitlike right lateral ventricle, which contains the shunt. A large interhemispheric fissure *(F)* is typically present.

struction or proximal intraventricular obstruction. The shunt may pull out of the abdomen as the child grows or may break off in the soft tissues; the peritoneal absorption may become blocked with accumulation of fluid in the abdomen, pleural space, or anywhere the shunt is placed. A CSF pseudocyst may develop in the abdomen, causing shunt obstruction.[77–79]

Within the ventricular system, infection may occasionally develop, leading to loculation and obstruction of the ventricular system by ventricular septations.[42] Multiple tubes may be necessary to decompress the loculations.

The intracranial end of the catheter may be placed into the brain, which could cause obstruction, but usually the side holes will still function. However, porencephaly may develop at the end of the shunt tip. The choroid plexus may be caught in the side holes of the ventricular catheter and cause intraventricular obstruction of the catheter.

HEMORRHAGE.—Intraventricular hemorrhage has been seen with placement of a ventricular catheter both in utero (our experience) and neonatally.[70] However, the small amount of hemorrhage that results has not led to clotting of the catheter and resolves without incident.

SUBDURAL HEMATOMAS.—Subdural hematomas may occur if there is a sudden, marked change in ventricular size. Subdural hematomas are very rare in newborn infants and become a problem only as the skull becomes less compliant (Fig 7–13).

INFECTION.—Shunt revision may be necessary if meningitis or encephalitis develops. Particular concern is required because meningitis may extend into encephalitis and ventriculitis in newborn infants, who typically have difficulty confining an intracranial infection to one area of the brain.

RECOMMENDATIONS FOR POSTSHUNT EVALUATION

As long as the anterior fontanelle is open, evaluation for hydrocephalus can easily be done with ultrasound very quickly and inexpensively.[80] If there is concern about postoperative subdural hematomas, a CT examination may be necessary (see Fig 7–13).[81] At times positive contrast ventriculography or radionuclide evaluation may be necessary.[82–85]

A baseline examination for ventricular size should be done approximately 1 week after surgery. At least 24 hours should be allowed for decompression after shunting

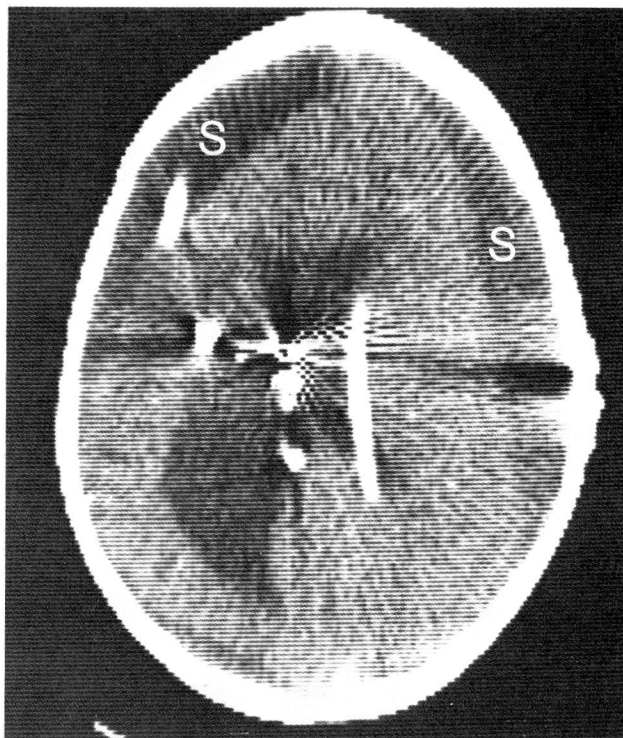

Fig 7–13.—Two-year-old child with shunt, after removal of a choroid plexus papilloma from the right lateral ventricle. Bilateral subdural hematomas (S) required shunts for drainage. Axial CT scan.

before reevaluation of ventricular size and complications, although in neonates decompression may occur within hours.

SHUNT DEPENDENCE.—Ventricular shunting may be so effective that slitlike ventricles develop (see Fig 7–12).[53] In some infants this is of no concern, but in others this is an indication of shunt dependency. Presumably adhesions or perhaps fibrosis develop, which do not allow the ventricles to expand. Then, when shunt malfunction occurs, the ventricles do not appear enlarged, even if the patient is symptomatic.

PROGNOSIS OF HYDROCEPHALUS

The prognosis of hydrocephalus depends on multiple factors. In those infants presenting with clinically apparent hydrocephalus at birth, McCullough and Balzer-Martin reported that those with very severe hydrocephalus (98th percentile) usually died in the first few months of life if their condition remained untreated.[56] Of patients who were treated, 86% survived 1–16 years. About two-thirds of the treated patients had a normal intellect (53% probability of an IQ over 80) or borderline intellect (19% probability of an IQ between 65 and 80).

The type of hydrocephalus does affect the eventual outcome. Dandy-Walker syndrome was associated with a high mortality and a very low IQ in survivors. Congenital aqueductal atresia was associated with intellectual impairment. According to Dennis et al.,[87] an aqueductal obstruction has a worse prognosis, probably because of the selective thinning of the vertex and occipital lobe. Deformities of the aqueduct and tectal plate may interfere with visual development and thus may particularly affect nonverbal intelligence.

Children with meningomyelocele perform poorly on tasks of persistent motor control and eye-hand coordination.[87] Meningomyelocele patients with hydrocephalus scored lower on perceptual motor functioning.[88] The higher the level of myelomeningocele, and thus the higher the sensory loss, the lower the IQ. However, the higher the level of the sac, the more often was hydrocephalus present, which probably affects the IQ.

One factor not clearly delineated by these studies is the age at onset of hydrocephalus. The children studied by McCullough and Balzer-Martin all had an enlarged head at birth, but this was not a criterion of the other studies. With the knowledge of intrauterine anatomy gained from ultrasound, we have seen that patients with a Dandy-Walker cyst already have hydrocephalus early in utero. Patients with meningomyelocele often have only slight dilation of the ventricles at birth, which becomes worse after the repair of the meningomyelocele. The poorer outcome in patients with the Dandy-Walker cyst may be related in part to the earlier onset of hydrocephalus.

Multiple factors have been used to predict the outcome of the hydrocephalus. According to the reports of Young et al.[26] and Shurtleff et al.,[28] a cerebral mantle of 2.8 cm correlates with good intelligence in treated patients. These reports support the concept that hydrocephalus must be recognized and treated early. Persons treated before age 2 months tend to have a high IQ.[89] In the children studied by

McCullough and Balzer-Martin, there was a clear relationship between a large head circumference at birth, small ventricular-skull distance (mantle), and limited brain mass[28] with impaired intelligence. A ventricular-skull distance of 1 cm or less predicted an IQ of less than 80.

REFERENCES

1. Harwood-Nash D.C., Fitz C.R.: *Neuroradiology in Infants and Children.* St. Louis, C.V. Mosby Co., 1976.
2. Milhorat T.H.: The third circulation revisited. *J. Neurosurg.* 42:628–645, 1975.
3. Cushing H.: *Studies in Intracranial Physiology and Surgery: The Third Circulation; The Hypophysis: The Gliomas.* London, Oxford University Press, 1926.
4. Naidich T.P., Schott L.H., Baron R.L.: Computed tomography in evaluation of hydrocephalus. *Radiol. Clin. North Am.* 20:143–168, 1982.
5. Cserr H.F. Ostrach L.H.: Bulk flow in interstitial fluid after intracranial injection of blue Dextran 2000. *Exp. Neurol.* 45:50–60, 1974.
6. Drayer B.P., Rosenbaum A.E.: Studies of the third circulation with Amipaque CT cisternography and ventriculography. *J. Neurosurg.* 48:946–956, 1978.
7. Shulman K., Yarnell P., Ransohoff J.: Dural sinus pressure in normal and hydrocephalic dogs. *Arch Neurol.* 10:575–580, 1964.
8. Lorenzo A.V., Page L.K., Watters G.V.: Relationship between cerebrospinal fluid formation absorption and pressure in human hydrocephalus. *Brain* 93:679–693, 1970.
9. Johnson M.L., Rumack C.M.: Ultrasonic evaluation of the neonatal brain. *Radiol. Clin. North Am.* 18:117–132, 1980.
10. Johnson M.L., Rumack C.M.: B-mode echoencephalography in the normal and high risk infant. *AJR* 133:375–381, 1979.
11. Rumack C.M., Johnson M.L.: Real-time ultrasound evaluation of the neonatal brain. *Clin. Ultrasound* 10:179–202, 1982.
12. Garrett W.J., Kossoff G., Jones R.F.C.: Ultrasonic cross-sectional visualization of hydrocephalus in infants. *Neuroradiology* 8:279–288, 1975.
13. Edwards M.K., Brown D.L., Muller J.: Cribside neurosonography: Real-time sonography for intracranial investigation of the neonate. *AJR* 136:271–276. 1981.
14. Grant E.G., Schellinger D., Borts F.T., et al.; Real-time sonography of the neonatal and infant head. *AJR* 136:265–270, 1981.
15. Slovis T., Kuhns L.R.: Real-time sonography of the brain through the anterior fontanelle. *AJR* 136:277–286, 1981.
16. Babcock D.S., Han B.K., LeQuesne G.W.: B-mode gray-scale ultrasonography of the head in the newborn and young infant. *AJR* 134:457–468, 1980.
17. Garrett W.J., Kossoff G., Warren P.: Cerebral ventricular size in children: A two dimensional ultrasonic study. *Radiology* 136:711–715, 1980.
18. Skolnick M.L., Rosenbaum A.E., Matzuk T., et al.: Detection of dilated cerebral ventricles in infants: A correlative study between US and CT. *Radiology* 131:447–451, 1979.
19. Morgan C.L., Trought W.S., Rothman S.J., et al.: Comparison of gray-scale ultrasonography and CT in the evaluation of macrocrania in infants. *Radiology* 132:119–123, 1979.
20. Hahn F.J.Y., Rim K.: Frontal ventricular dimensions on normal computed tomography. *AJR* 126:593–596, 1976.
21. LeMay M., Hochberg F.H.: Ventricular differences between hydrostatic hydrocephalus and hydrocephalus ex vacuo by computed tomography. *Neuroradiology* 17:191–195, 1979.
22. Birnholz J.C.: Newborn cerebellar size. *Pediatrics* 70:284–287, 1982.
23. Evans W.A. Jr.: An encephalographic ratio for estimating the size of the cerebral ventricles: Further experience with serial observations. *Am. J. Dis. Child.* 64:820–830, 1942.
24. Pedersen H., Gyldensted M., Gyldensted C.: Measurement of the normal ventricular

system and supratentorial subarachnoid space in children with computed tomography. *Neuroradiology* 17:231–237, 1979.

25. Haug G.: Age and sex dependence of the size of normal ventricles on computed tomography. *Neuroradiology* 14:201–204, 1977.

26. Young H.F., Nielsen F.E., Weiss M.H., et al.: The relationship of intelligence and cerebral mantle in treated infantile hydrocephalus. *Pediatrics* 52:38–44, 1973.

27. Villani R., Giani S.M., Giovanelli M., et al.: Skull changes and intellectual status in hydrocephalic children following CSF shunting. *Dev. Med. Child. Neurol.* 18:78–81, 1976.

28. Shurtleff D.B., Foltz E.L., Loeser J.D.: Hydrocephalus: A definition of its progression and relationship to intellectual function, diagnosis and complications. *Am. J. Dis. Child.* 125:688–693, 1973.

29. Shurtleff D.B., Foltz E.L., Chapmen J.T.: Ventriculo-skull distance: Its reliability as an estimate of "cerebral mantle" in the normocephalic child. *Am. J. Dis. Child.* 111:262–266, 1966.

30. Huckman M.S., Fox J., Topel J.: The validity of criteria for the evaluation of cerebral atrophy by computed tomography. *Radiology* 116:85–92, 1975.

31. Epstein F., Naidich T., Kricheff I., et al.: Role of computerized axial tomography in diagnosis, treatment and follow-up of hydrocephalus. *Child's Brain* 3:91–100, 1977.

32. Heinz E.R., Ward A., Drayer B.P., et al.: Distinction between obstructive and atrophic dilatation of ventricles in children. *J. Comput. Assist. Tomogr.* 4:320–325, 1980.

33. Edwards, J.H.: The syndrome of sex-linked hydrocephalus. *Arch. Dis. Child.* 36:486–493, 1961.

34. Rumack C.M., Johnson M.L., Zunkel D.: Antenatal diagnosis. *Clin. Ultrasound* 8:210–230, 1981.

35. Partridge J.C., Babcock D.S., Steichen J.J., et al.: Optimal timing for diagnostic cranial ultrasound in low birthweight infants: Detection of intracranial hemorrhage and ventricular dilatation. Presented at the Second Perinatal Intracranial Conference, Washington, D.C., Dec. 1982.

36. Volpe J.J., Pasternak J.F., Allan W.C.: Ventricular dilatation preceding rapid head growth following neonatal intracranial hemorrhage. *Am. J. Dis. Child.* 131:1212, 1977.

37. Korobkin R.: The relationship between head circumference and the development of communicating hydrocephalus following intraventricular hemorrhage. *Pediatrics* 56:74, 1975.

38. Cubberly D.A., Jaffe R.B., Nixon G.W.: Sonographic demonstration of Galenic arteriovenous malformations in the neonate. *AJNR* 3:435–439, 1982.

39. Scotti G., Floodmark D., Harwood-Nash D.C.: Posterior fossa hemorrhages in the newborn. *J. Comput. Assist. Tomogr.* 5:68–72, 1981.

40. Perlman J.M., Nelson J.S., McAllister W.J., et al.: Intracerebellar hemorrhage in a premature newborn: Diagnosis by real-time US and correlation with autopsy findings. *Pediatrics* 71:159–162, 1983.

41. Savolaine E.R., Gerber A.M.: Computerized tomography studies of congenital and acquired cerebral intraventricular membranes. *J. Neurosurg.* 54:388–391, 1981.

42. Schultz P., Leeds N.E.: Intraventricular septations complicating neonatal meningitis. *J. Neurosurg.* 38:320–326, 1973.

43. Friedman W.A., Mickle J.P.: Hydrocephalus in achondroplasia: A possible mechanism. *Neurosurgery* 7:150-153, 1980.

44. Gilles F.H., Davidson R.I.: Communicating hydrocephalus associated with deficient dysplastic parasagittal arachnoidal granulations. *J. Neurosurg.* 35:421–426, 1971.

45. Eisenberg H.M., McComb J.G., Lorenzo A.V.: Cerebrospinal fluid overproduction and hydrocephalus associated with choroid plexus papilloma. *J. Neurosurg.* 40:381–385, 1974.

46. Hooper R.: Hydrocephalus and obstruction of the superior vena cava in infancy: Clinical study of the relationship between cerebrospinal fluid pressure and venous pressure. *Pediatrics* 28:792–799, 1961.

47. Steward D.R., Johnson D.G., Myers G.G.: Hydrocephalus as a complication of jugular catheterization during total parenteral nutrition. *J. Pediatr. Surg.* 10:771–777, 1975.

48. Rosman N.P., Shands K.N.: Hydrocephalus caused by increased intracranial venous pressure: A clinicopathological study. *Ann. Neurol.* 3:445–460, 1978.

49. Pettorossi V.E., DiRocco C., Mancinelli R., et al.: Communicating hydrocephalus induced by mechanically increased amplitude of the intraventricular cerebrospinal fluid pulse pressure: Rationale and method. *Exp. Neurol.* 59:30–39, 1978.

50. Ekbom K., Greitz T., Kugelberg E.: Hydrocephalus due to ectasia of the basilar artery. *J. Neurol. Sci.* 8:465–477, 1969.

51. Ekbom K., Greitz T.: Syndrome of hydrocephalus caused by saccular aneurysm of the basilar artery. *Acta Neurochir.* 24:71–77, 1971.

52. Greitz T., Ekbom K., Kugelberg E., et al.: Occult hydrocephalus due to ectasia of the basilar artery. *Acta Radiol. (Diagn.)* 9:310–316, 1969.

53. Fitz C.R.: The ventricles and subarachnoid spaces in children, in: Lee S.H., Rao K.C.V.G.: *Cranial Computed Tomography.* New York, McGraw-Hill Book Co., 1983.

54. Pasquini U., Bronzini M., Gozzoli E., et al.: Periventricular hypodensity in hydrocephalus: A clinico-radiological and mathematical analysis using computed tomography. *J. Comput. Assist. Tomogr.* 1:443–448, 1977.

55. Mori K., Murata T., Nakano Y., et al.: Periventricular lucency in hydrocephalus on computerized tomography. *J. Comput. Assist. Tomogr.* 4:204–209, 1980.

56. McLone D.G., Bondareff W., Raimondi A.J.: Hydrocephalus-3, a murine mutant: II. Changes in the brain extracellular space. *Surg. Neurol.* 1:233–242, 1973.

57. Rubin R.C., Hochwald G.M., Tiell M., et al.: Hydrocephalus: I. Histological and ultrastructural changes in the pre-shunted cortical mantle. *Surg. Neurol.* 5:109–114, 1976.

58. Lux W.E., Jr. Hochwald G.M., Sahar A., et al.: Periventricular water content: Effect of pressure in experimental chronic hydrocephalus. *Arch. Neurol.* 23:475–479, 1970.

59. Rowlatt U.: The microscopic effects of ventricular dilatation with increase in head size. *J. Neurosurg.* 48:957–961, 1978.

60. Drapkin A.J., Sahar A.: Experiment hydrocephalus: Cerebrospinal fluid dynamics and ventricular distensibility during early stages. *Child's Brain* 4:278–288, 1978.

61. Sahar A., Hochwald G.M., Sadik A.R., et al.: Cerebrospinal fluid absorption in animals with experimental obstructive hydrocephalus. *Arch. Neurol.* 21:638–644, 1969.

62. Hopkins L.N., Bakay L., Kinkel W.R., et al.: Demonstration of transventricular CSF absorption by computed tomography. *Acta Neurochir.* 39:151–157, 1977.

63. Hochwald G.M., Epstein F., Malhan C., et al.: The role of the skull and the dura in experimental feline hydrocephalus. *Dev. Med. Child. Neurol. (Suppl.)* 27:65–69, 1972.

64. Morgan M.E., Benson J.W., Cooke R.W.: Ethamsylate reduces the incidence of periventricular hemorrhage in very low birthweight infants. *Lancet* 2:830–831, 1981.

65. Bergman E., Epstein M., Freeman T.: Medical management of hydrocephalus with acetazolamide and furosemide. *Ann. Neurol.* 4:189, 1978.

66. Papile L.A., Burstein J., Burstein R., et al.: Post-hemorrhagic hydrocephalus in low birthweight infants: Treatment by serial punctures. *J. Pediatr.* 97:273–277, 1980.

67. Rumack C.M., Johnson M.L., McDonald M., et al.: Patterns of intracranial hemorrhage resolution. Presented at the Society for Pediatric Radiolgy, New Orleans, May 1982.

68. Mantovani J.F., Pasternak J.F., Mathew O.P., et al.: Failure of daily lumbar punctures to prevent development of hydrocephalus after IVH. *J. Pediatr.* 97:278–281, 1980.

69. Behrman R.E.: Serial lumbar punctures and IVH, editorial. *J. Pediatr.* 97:250, 1980.

70. Shkolnik A., McLone D.G.: Intraoperative real-time ultrasonic guidance of ventricular shunt placement in infants. *Radiology* 141:515–517, 1981.

71. Knake J.E., Chandler W.F., McGillicuddy J.E., et al.: Intraoperative sonography for brain tumor localization and ventricular shunt placement. *AJR* 139:733–738, 1982.

72. Emery J.L.: Intracranial effects of long standing decompression of the brain in children with hydrocephalus and myelomeningocoele. *Dev. Med. Child. Neurol.* 7:302–309, 1965.

73. Kaufman B., Weiss M.H., Young H.F., et al.: Effects of prolonged cerebrospinal fluid shunting on the skull and brain. *J. Neurosurg.* 38:288–297, 1973.

74. Steinbok P., Berry K.: Inversion of the cortical mantle as a complication of ventricular shunting. *J. Neurosurg.* 49:129–131, 1978.

75. Zimmerman R.A., Bilaniuk L.T., Gallo E.: Computed tomography of the trapped fourth ventricle. *AJR.* 130:503–506, 1978.

76. Scotti G., Musgrave M.A., Fitz C.R., et al.: The isolated fourth ventricle in children: CT and clinical review of 16 cases. *AJR* 135:1233–1238, 1980.

77. Lee T.G., Parsons P.M.: Ultrasound diagnosis of cerebrospinal fluid abdominal cyst. *Radiology* 127:220, 1978.

78. Goldfine S.L., Turetz F., Beck R., et al.: Cerebrospinal fluid intraperitoneal cyst: An unusual abdominal mass. *AJR* 130:568–569, 1978.

79. Price H.I., Rosenthal S.I., Batkitzky S., et al.: Abdominal pseudocysts as a complication of ventriculoperitoneal shunt. *Neuroradiology* 21:273–276, 1981.

80. Fried A.M., Adams W.E., Ellis G.T., et al.: Ventriculoperitoneal shunt function: Evaluation by sonography. *AJR* 134:967–970, 1980.

81. Murtagh F.R., Quencer R.M., Poole C.A.: Cerebrospinal fluid shunt function and hydrocephalus in the pediatric age group. *Radiology* 132:385–388, 1979.

82. Faria M.A., O'Brien M.S., Tindall C.T.: A technique for evaluation of ventricular shunts using Amipaque and CT. *J. Neurosurg.* 53:92–96, 1980.

83. Gilday D.L., Kellam J.: In-DTPA evaluation of CSF diversionary shunts in children. *J. Nucl. Med.* 14:920–923, 1973.

84. Sty J.R., D'Souza B.J., Daniels D.: Nuclear anatomy of diversionary central nervous system shunts in children. *Clin. Nucl. Med.* 3:271, 1978.

85. Hayden P.W., Rudd T.G., Shurtleff D.B.: Combined pressure-radionuclide evaluation of suspected cerebrospinal fluid shunt function: A seven year clinical experience. *Pediatrics* 66:679–684, 1980.

86. McCullough D.C., Balzer-Martin L.A.: Current prognosis in overt neonatal hydrocephalus. *J. Neurosurg.* 5:378–383, 1982.

87. Dennis M., Fitz C.R., Netley C.T., et al.: The intelligence of hydrocephalic children. *Arch. Neurol.* 38:607–615, 1981.

88. Soare P.L., Raimondi A.J.: Intellectual and perceptual-motor characteristics of treated myelomeningocele children. *Am. J. Dis. Child.* 131:199–204, 1977.

89. Lorber J.: The results of early treatment of extreme hydrocephalus. *Dev. Med. Child. Neurol. (Suppl.)* 16:21–29, 1968.

CHAPTER 8

Intracranial Neoplasms, Cysts, and Vascular Malformations

Carol M. Rumack, M.D.
Michael L. Johnson, M.D.

INTRACRANIAL NEOPLASMS

INTRACRANIAL TUMORS are very uncommon in children less than 2 years of age and very rare in the newborn period. Wei and Norman reported 75 tumors in the first two years of life, with 26 seen in the first year of life and 49 during the second year (Table 8–1). These cases were collected over 52 years. Forty percent had a supratentorial lesion and 60% infratentorial. Listed in order of frequency were: astrocytoma (supratentorial); ependymoma, medulloblastoma, and astrocytoma (infratentorial). The most common sites of childhood neoplasms are indicated schematically in Figure 8–1. Several authors have reported large series of tumors, including data on older children.[1, 2] This chapter discusses congenital tumors (those presenting in the first 60 days of life) and tumors presenting in infancy before the fontanelle closes.[3]

Knowledge of normal anatomical structural detail is essential to the recognition

TABLE 8–1.—NEOPLASMS PRESENTING IN FIRST TWO YEARS OF LIFE, TORONTO HOSPITAL FOR SICK CHILDREN*

NEOPLASM	SUPRATENTORIAL	INFRATENTORIAL	TOTAL 1920–1972
Astrocytoma	25	10	35
Ependyoma	2	16	18
Medulloblastoma	0	13	13
Undifferentiated neuroectodermal (e.g., choroid plexus papilloma)	2	3	5
Medulloepithelioma	1	0	1
Nonbiopsied brain stem glioma	0	3	3
			75

*Modified from Harwood-Nash and Fitz.[1]

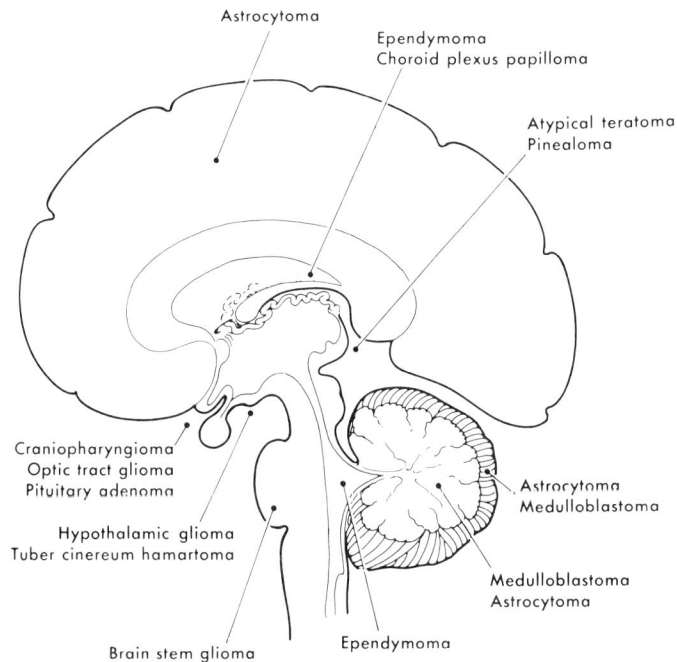

Fig 8–1.—Sites of childhood neoplasms. (From Harwood-Nash and Fitz.[1] Reproduced with permission.)

of the brain tumors, which are quite rare and may be quite subtle. Careful review of the chapter on normal anatomy may be valuable, and there are excellent discussions of the normal anatomical appearance of the posterior fossa on computerized tomography (CT) by Segall et al.[4] and on ultrasound by Grant et al.[5] The posterior fossa is particularly important in children because approximately half of the tumors are infratentorial.[1]

Most infants with intracranial neoplasms present with signs of hydrocephalus and occasionally with hemorrhage or failure to thrive.

Ultrasound Findings

Several small series of intracranial tumors have been evaluated with ultrasound. In most cases the patients presented with hydrocephalus in the first year of life (Fig 8–2).[6–10] The tumors described to date also include intraoperative experience in children and adults.[10–13] All intracranial tumors reported since high-resolution gray scale equipment became available have displayed a change in brain architecture that caused primarily echogenic masses, with cystic areas in some tumors (Table 8–2). Previously reported tumors were studied with A-mode or bistable ultrasound equipment and detail was very poor.[14–16] Calcification within a tumor is recognized as very highly echogenic material with acoustic shadowing. The parenchymal changes may be subtle, and with more experience we may find a tumor which is

TABLE 8–2.—Brain Tumors Studied by Gray Scale Ultrasound*

STUDY, YEAR	AGE	SYMPTOMS	NO	DIAGNOSIS	CALCIUM	↑ ECHOES	CYSTIC
CHILDREN							
Shkolnik,[6] 1975	3 mo.	HC	1	Craniopharyngioma	+	+	+
Slovis, 1981[7]	3 mo.	HC	1	Neuroectodermal tumor	−	+	Rim of fluid
Slovis and Kuhns,[8] 1981	3 mo.	HC	1	Racemose hemangioma	−	+	−
Han et al.,[9] 1983	2 yr	HC	7	Optic glioma (3)	+	+	+ (1)
				Teratoma	+	+	+
				Thalamic glioma		+	
				Ependymoma		+	+†
				Medulloblastoma		+	
Shkolnik and McLone,[10] 1983	6 mo.–14 yr	HC/OR	8	Choroid plexus papilloma	−	+	−
				Cerebellar astrocytoma	−	+	+
				Ependymoma	−	+	+
				Primary neuroectodermal	−	+	+
ADULT							
Gooding et al.[11] 1983	Adults		3	Astrocytoma or glioma		+	+ (1)
				Metastases		+	+ (1)
				Angioma		+	
Rubin et al.[12] 1982	41–83 yr	OR	4	Meningioma	+	+	−
Knake et al.[13] 1982	11–47 yr	OR	15	Ganglioma (1)	+	+	+ (2)
				Astrocytoma (5)		+	+
				Endodermal sinus tumor (1)		+	
				Metastatic from lung, breast, kidney (6)		+	(Breast)
				Metastatic small cell tumor (1)		+	
				Unclassified glioma (1)		+	

*Most studies reported on 5-MHz or 7.5 MHz sector scanners. HC, hydrocephalus; OR, operating room.
†Cystic with nodule in wall.

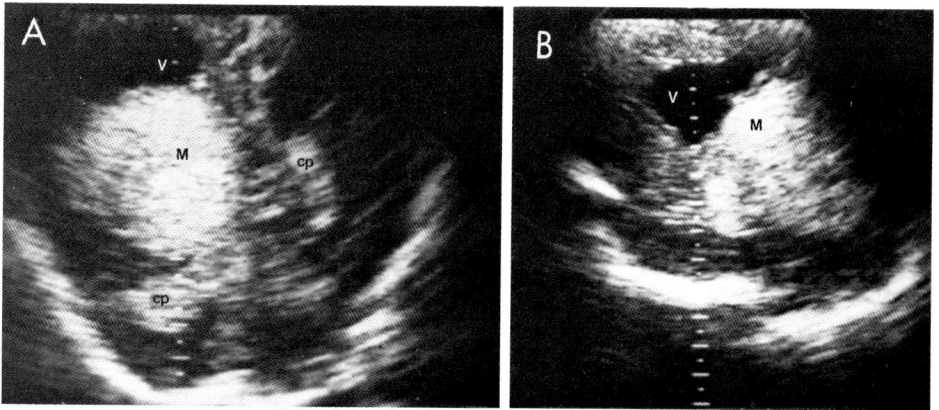

Fig 8–2.—Choroid plexus papilloma in 6-month-old infant with a rapidly growing head. Coronal **(A)** and sagittal **(B)** ultrasound scans demonstrate a large echogenic (solid) mass *(M)* within the dilated lateral ventricle *(V)*. The echogenicity of the mass was similar to that of the normal choroid plexus *(cp)*. Dilation of left lateral ventricle was less prominent. (From Shkolnik and McLone.[10] Reproduced with permission.)

isoechogenic with brain. However, since ultrasound depends not only on density but also on architectural differences, the tumor would have to have an architecture very similar to that of normal brain.

Differentiation of a neoplasm from hemorrhage, infarction, or infection may not always be possible. An acute hematoma may be very echogenic, but if there is any question, it should be followed, since it will become sonolucent centrally about 1–2 weeks after hemorrhage. Infarction may be quite echogenic from 1 to 2 weeks after onset and then become sonolucent. Neonatal CNS infections may simulate a tumor but often are not as well localized as tumors, since the infant brain has difficulty in walling off infection. If a tumor is detected in utero, no symptoms may present in the mother and it may be particularly difficult to distinguish among infection, tumor, and infarction.

On the basis of reported correlative CT scans and intraoperative scanning, ultrasound appears to show the site and extent of the tumor quite well.[6–13] Larger series will be necessary to understand brain tumors in detail and the limitations of ultrasound.

Intraoperative localization of tumors by ultrasound undoubtedly will become more frequent, and thus will lead to a more definitive understanding of brain tumors. A 3.5 or 5.0-MHz, transducer with portable real-time equipment is necessary for good detail. A 7-MHz transducer may be required for near field evaluation, and the smallest possible transducer is desirable for visualization through a craniotomy site or burr hole.[11] Intraoperative ultrasound can identify and localize tumors for accurate biopsy and complete removal.[10-13, 17, 18] Brain exploration can be minimized, thus decreasing the damage to normal brain. Stereotaxic CT for surgery can be eliminated, which in turn will reduce the complexity, time, and cost of the procedure.

CT Findings

CT scanning can fairly accurately suggest a specific diagnosis based on the characteristic CT findings (Table 8–3). Segall et al.[4] report that CT is not only the procedure of choice, replacing angiography and nuclear medicine scanning, but it is extremely accurate even in nonenhancing brain stem gliomas, if thin slices are made routinely down through the foramen magnum. Contrast enhancement with metrizamide has been necessary in only a few cases in their experience.

Hydrocephalus

Obstructive hydrocephalus is the most common presentation of *medulloblastomas* in infants.[19] Medulloblastomas are embryonic tumors composed of poorly differentiated primitive cells arising in the neuroepithelial roof of the ventricle. These cells migrate upward and laterally to form the external granular layer within the cerebellar hemisphere. They most frequently arise in the midline of the vermis, behind and above the fourth ventricle, but they may extend to the cerebellopontine angle cisterns.[20] Tumor density is typically slightly greater than that of normal brain, with homogeneous enhancement and sharply marginated borders. Occasionally small cystic areas or calcification[20] may be seen within the tumor. Seeding has been reported along CSF pathways of the ventricles and cisterns.

Obstructive hydrocephalus is a common presentation of all of these rare tumors. A *teratoma* that was diagnosed in utero (Figure 8–3) arose behind the third ventricle, compressing it and causing obstructive hydrocephalus. The tumor had typical areas of calcification and cyst formation, and markedly distorted and even replaced large areas of the brain.[21] These tumors usually occur in the pineal region[1] but can occur in the suprasellar region or even in the ventricles and cause hydrocephalus.

Overproduction of CSF is thought to be the cause of hydrocephalus from a *choroid plexus papilloma*. Typically it presents in the first 2 years of life.[22] It appears to cause hydrocephalus without obstruction in some patients (Fig 8–4). CT findings may include a frondlike mass with calcification which is slightly hyperdense and enhances well with contrast. If the papilloma becomes malignant, it will usually extend beyond the ependymal lining into the brain parenchyma.

TABLE 8–3.—CHARACTERISTIC CT FINDINGS

LESION	NON-CONTRAST-ENHANCED	CALCIUM	CONTRAST-ENHANCED
Astrocytoma	Cyst fluid density >CSF Well marginated	Occasionally	Enhancing (nodule in cyst)
Medulloblastoma	↑ Density Centrally located		Homogeneous enhancement Sharp borders
Ependymoma	↑, ↓ or isodense rim of CSF from ventricular extension	50%	Patchy or homogeneous

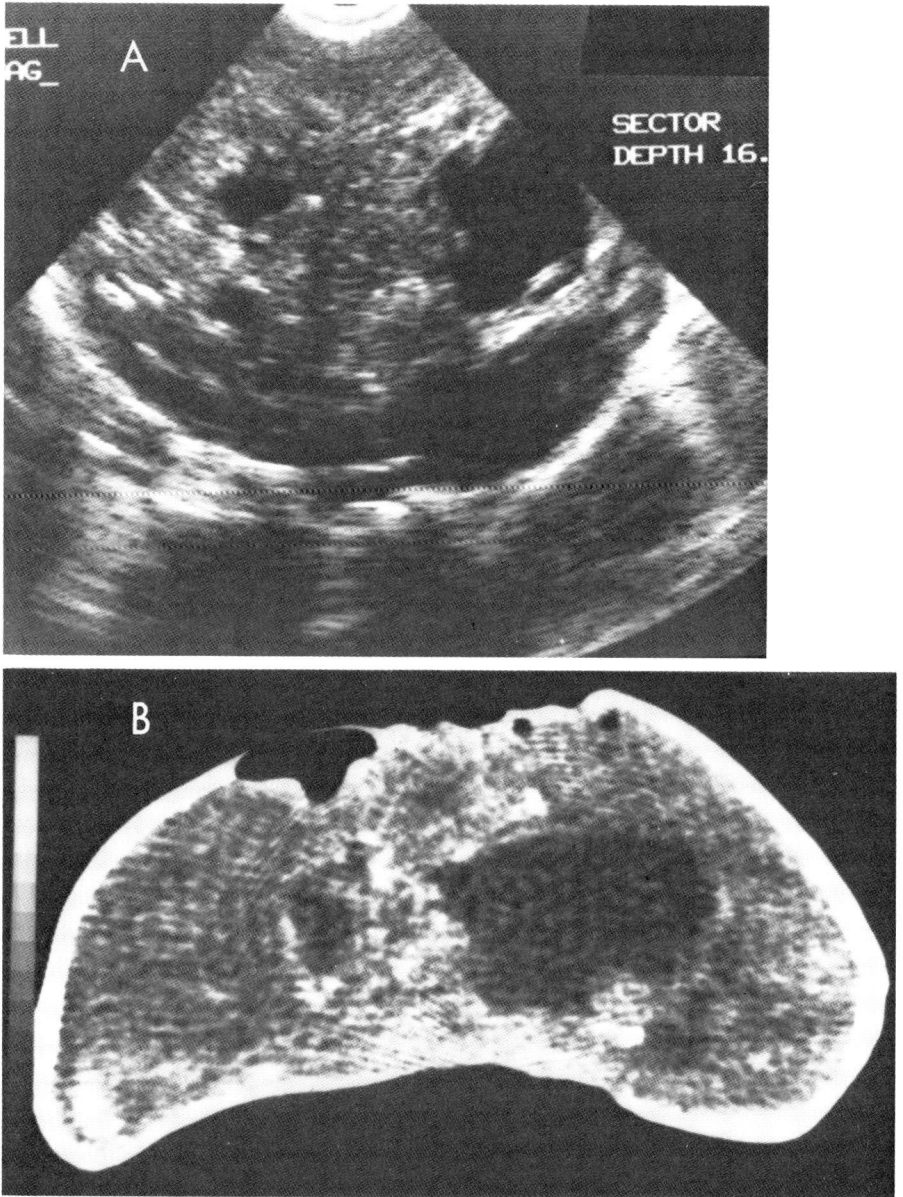

Fig 8–3.—Teratoma. **A,** Sagittal ultrasound scan in newborn shows complete distortion of the normal brain architecture by the cystic and solid mass of the teratoma. Cartilage formation with calcium and dysplastic ventricle formation were present in the tumor. **B,** Axial CT scan shows similar cystic, solid, and calcific changes in the teratoma. The tumor originated behind the third ventricle, caused massive obstructive hydrocephalus of the lateral ventricles, and compressed the rest of the brain into the posterior fossa. The brain is distorted due to ventricular decompression prior to delivery.

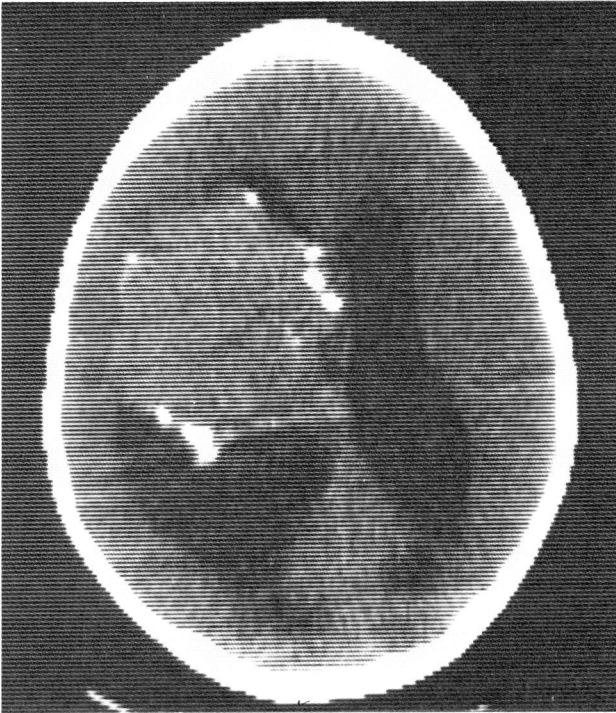

Fig 8–4.—Choroid plexus papilloma. This is a non-contrast-enhanced CT scan in 10-month-old infant with an intraventricular mass with calcium at the edges of the tumor. Hydrocephalus is thought to be due to overproduction of CSF, although this tumor may be compressing the foramina of Monro.

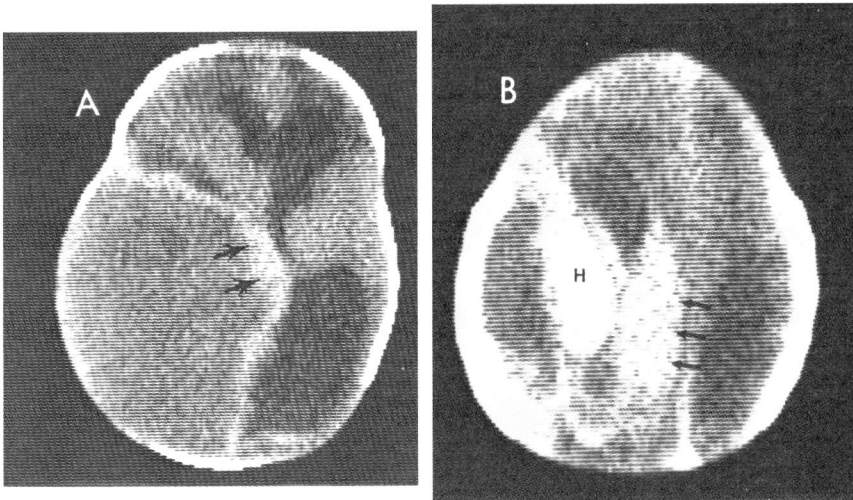

Fig 8–5.—Astrocytoma (grade IV). **A,** CT scan without contrast on first day of life in a term infant diagnosed as hydrocephalus and hemorrhage *(arrows).* **B,** Marked mass effect led to contrast CT scan demonstrating enhancing region near the vertex *(arrows)* adjacent to an area of hemorrhage *(H).*

Hemorrhage

Unusual types of hemorrhage should be evaluated with contrast CT to rule out a neoplasm or vascular malformation. In our experience, two patients who were initially diagnosed as having hemorrhage actually had tumors. One newborn presented with a large head, hydrocephalus, and a midline shift. Intraventricular and parenchymal hemorrhage did not account for the marked shift. The cyst fluid was of higher density than the CSF, as has been reported by other authors.[4, 23] With contrast, an enhancing area on the superior edge of the cyst was identified. An *astrocytoma* (grade IV) was diagnosed; this is the commonest neonatal tumor in most series (Fig 8–5).[1, 24]

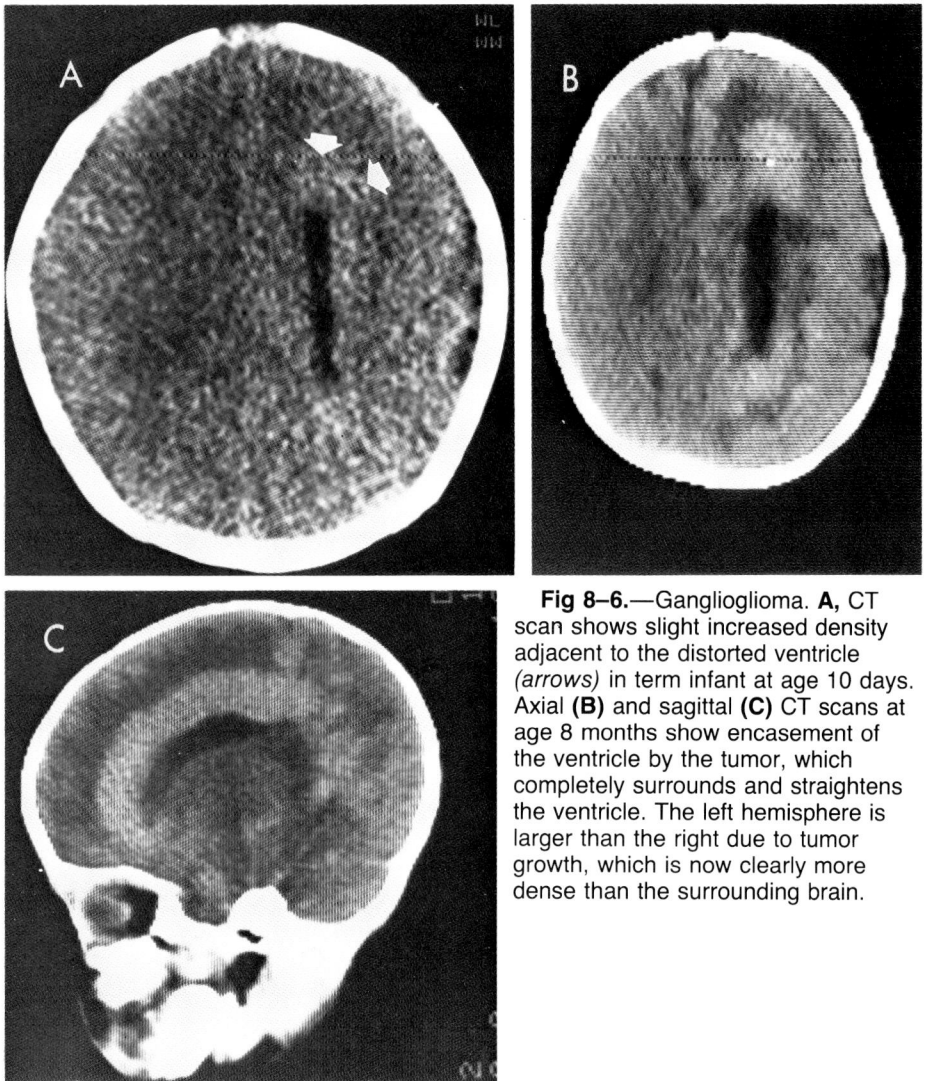

Fig 8–6.—Ganglioglioma. **A,** CT scan shows slight increased density adjacent to the distorted ventricle *(arrows)* in term infant at age 10 days. Axial **(B)** and sagittal **(C)** CT scans at age 8 months show encasement of the ventricle by the tumor, which completely surrounds and straightens the ventricle. The left hemisphere is larger than the right due to tumor growth, which is now clearly more dense than the surrounding brain.

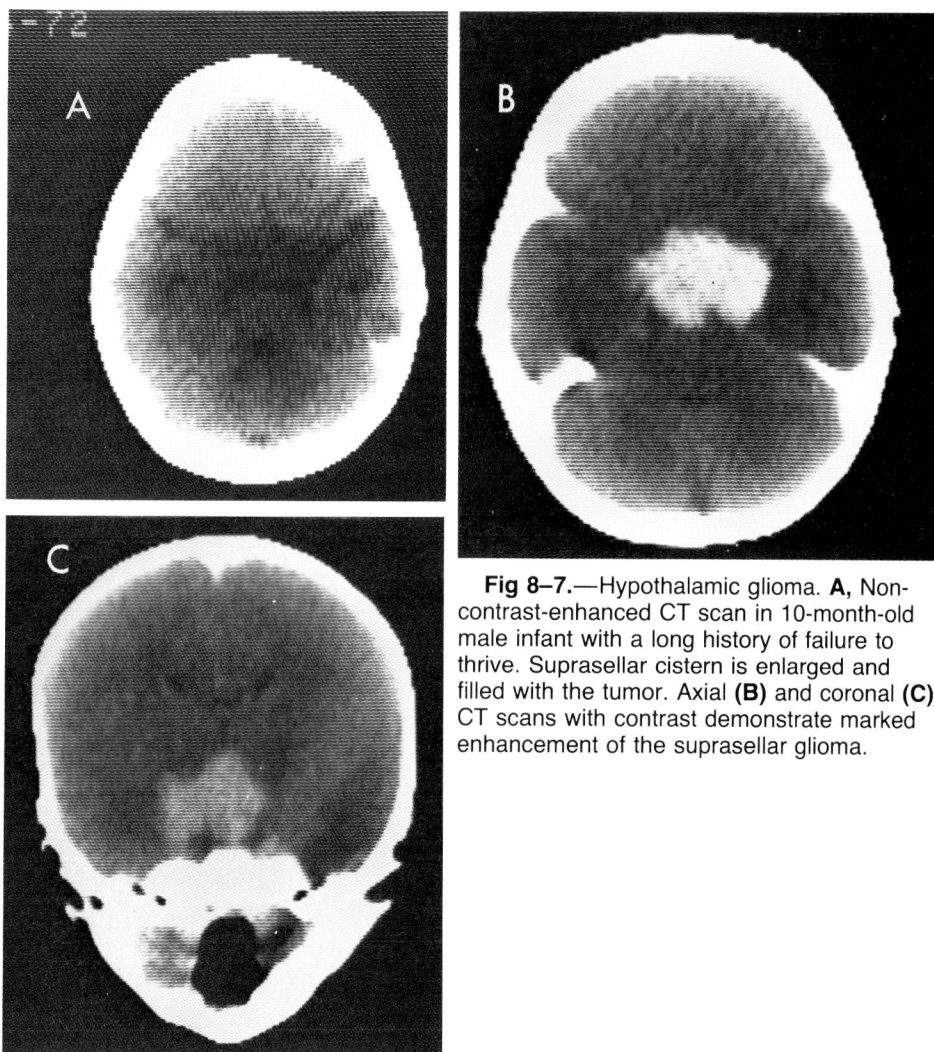

Fig 8–7.—Hypothalamic glioma. **A,** Non-contrast-enhanced CT scan in 10-month-old male infant with a long history of failure to thrive. Suprasellar cistern is enlarged and filled with the tumor. Axial **(B)** and coronal **(C)** CT scans with contrast demonstrate marked enhancement of the suprasellar glioma.

In another newborn, periventricular "hemorrhage" was demonstrated at age 10 days (Fig 8–6). At age 8 months, a scan was done for uncontrolled seizures. It demonstrated a mass encasing the ventricular system that had grown over the course of the scans to include most of one hemisphere. The pathologic diagnosis was *ganglioglioma*, which is one of the least common neonatal tumors. Infants with ganglioglioma have a life expectancy of less than 1 year.[1]

Failure to Thrive

In the suprasellar region, a *hypothalamic glioma* (Fig 8–7) usually presents as failure to thrive, since it causes the diencephalic syndrome, presenting clinically with a voracious appetite and weight loss. An intracranial neoplasm should be ex-

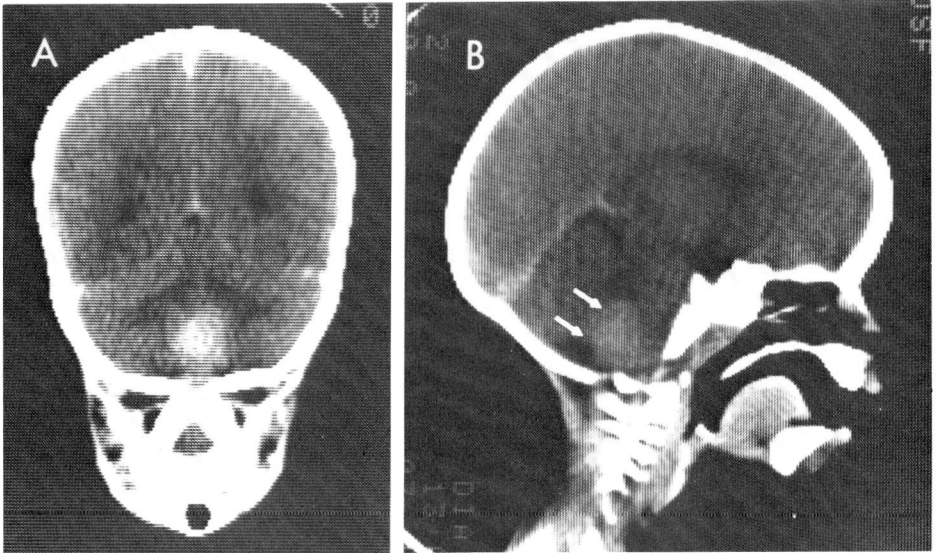

Fig 8–8.—Brain stem glioma. Coronal **(A)** and sagittal **(B)** CT scans demonstrate enhancing tumor *(arrows)* and surrounding edema compressing posterior fossa structures, including the fourth ventricle, but still normal ventricular size, in a 1-year-old girl with persistent vomiting.

cluded as a cause of failure to thrive in infants, especially if there is persistent, unexplained vomiting (Fig 8–8). The vomiting may be due to increased intracranial pressure from the tumor mass, but more likely it is the result of obstructive hydrocephalus caused by the tumor.

Intracranial tumors are best studied with contrast-enhanced computerized tomography. Even with CT, care should be taken to differentiate tumor and infection. The primary *neuroblastoma* shown in Figure 8–9 was found in a 3-month-old infant who was febrile and showed signs of increased intracranial pressure. The tumor extended bifrontally into the parenchyma and had cystic and solid components. The ring of enhancement made differentiation from abscess difficult, but the appearance was characteristic of a primary neuroblastoma.[25]

More computerized tomographic details can be obtained about these rare tumors and about those in older children in two excellent articles by Fitz and Rao[22] and Segall et al.[4] However, ultrasound characteristics of these tumors need to be defined because ultrasound is often the screening test in newborns. An unusual hemorrhage, hydrocephalus, or calcification on ultrasound, or unexplained neurologic finding should be evaluated with contrast-enhanced CT. Nuclear magnetic resonance imaging may add a great deal to our understanding of these tumors, especially brain stem gliomas.[26, 27]

INTRACRANIAL CYSTS

An intracranial cyst is a fluid-filled cavity within or adjacent to the brain which has mass effect.[1] Thus, a large cisterna magna is not a cyst. There are many cystic

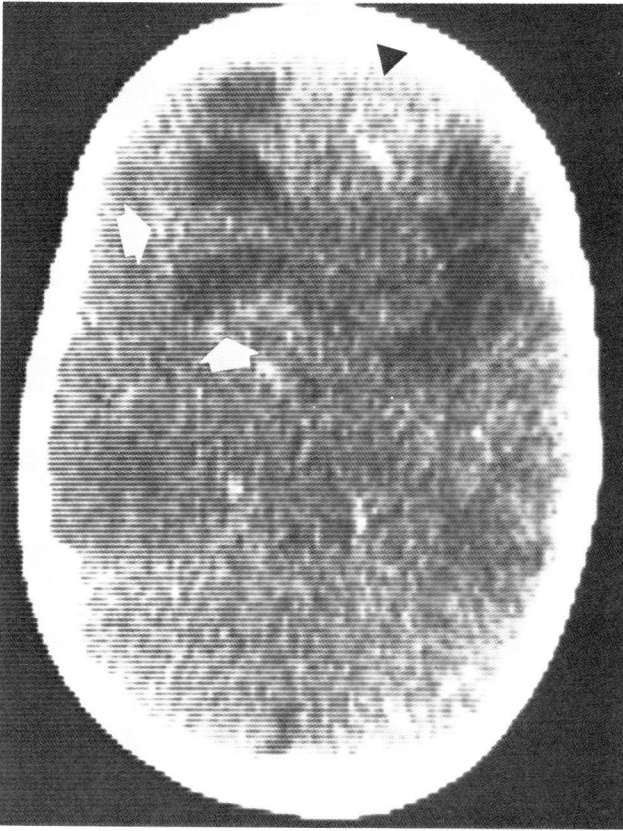

Fig 8–9.—Primary neuroblastoma. Axial CT scan demonstrates bifrontal mass with cystic and solid areas centrally. There is a faint enhancing rim *(white arrows)*, which in this febrile 3-month-old infant was initially thought to be an abscess. Biopsy of the brain was done for decompression and diagnosis, revealing the tumor. *Arrowhead* points to central enhancement.

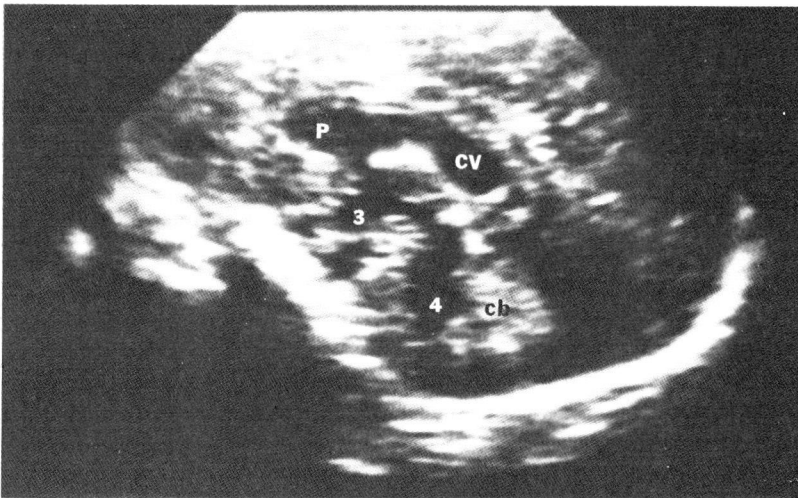

Fig 8–10.—Normal midline sagittal ultrasound scan shows cavum septi pellucidi *(P)*, cavum Vergae *(CV)*, third *(3)* and fourth *(4)* ventricles, and cerebellum *(cb)*.

TABLE 8–4.—INTRACRANIAL CYSTIC LESIONS

Congenital
  Large cisterna magna* (normal variant)
  Porencephalic cysts*
  Arachnoid cysts*
  Posterior fossa extra-axial cyst (Dandy-Walker variant)
  Dandy-Walker syndrome
  Hydranencephaly
  Holoprosencephaly
  Agenesis of the corpus callosum
Inflammatory
  Cerebral abscess
  Subdural empyema
Neoplastic*
  Cystic cerebellar astrocytoma
  Craniopharygioma
  Colloid cyst
  Teratoid tumors
Trauma
  Porencephalic cyst*
  Chronic subdural hygroma or hematoma
Vascular*
  Vein of Galen aneurysm
  Arteriovenous malformation

*Discussed in this chapter.
Table modified from Maravilla et al.[29]

lesions within the brain of infants (Table 8–4).[28, 29] Normal cystic areas include the ventricles, subarachnoid cisterns, and the cava septi pellucidi and Vergae in the septum pellucidum (Fig 8–10).[38] The cisterna magna can be quite large normally and should not be confused with an arachnoid cyst. Most of the acquired cystic lesions in infancy represent an area of brain necrosis following an intraparenchymal hemorrhage, infection, or infarction. If they communicate with the ventricular system or subarachnoid space, they should be termed porencephaly. *True* porencephalic cysts with mass effect are uncommon.

Congenital cysts are filled with clear fluid and localized within the arachnoid membrane. They may be classified according to their relationship to the neural axis or by their cyst wall, which may consist of arachnoid connective tissue or glioependymal tissue.[31] Cysts on the convexity of the cerebral hemispheres and in the spinal canal are nearly always arachnoid; both arachnoid and glioependymal cysts may be found in the supracollicular or retrocerebellar areas. Unless there is surgical or autopsy proof of the cell wall composition, it will be difficult to distinguish these lesions. They are discussed below under the general term of arachnoid cyst.

Porencephalic Cysts

Porencephalic cysts from hemorrhage, infarction, and infection are due to brain necrosis adjacent to a ventricle with the development of communication. In our experience, they most commonly result from a subependymal hemorrhage that has

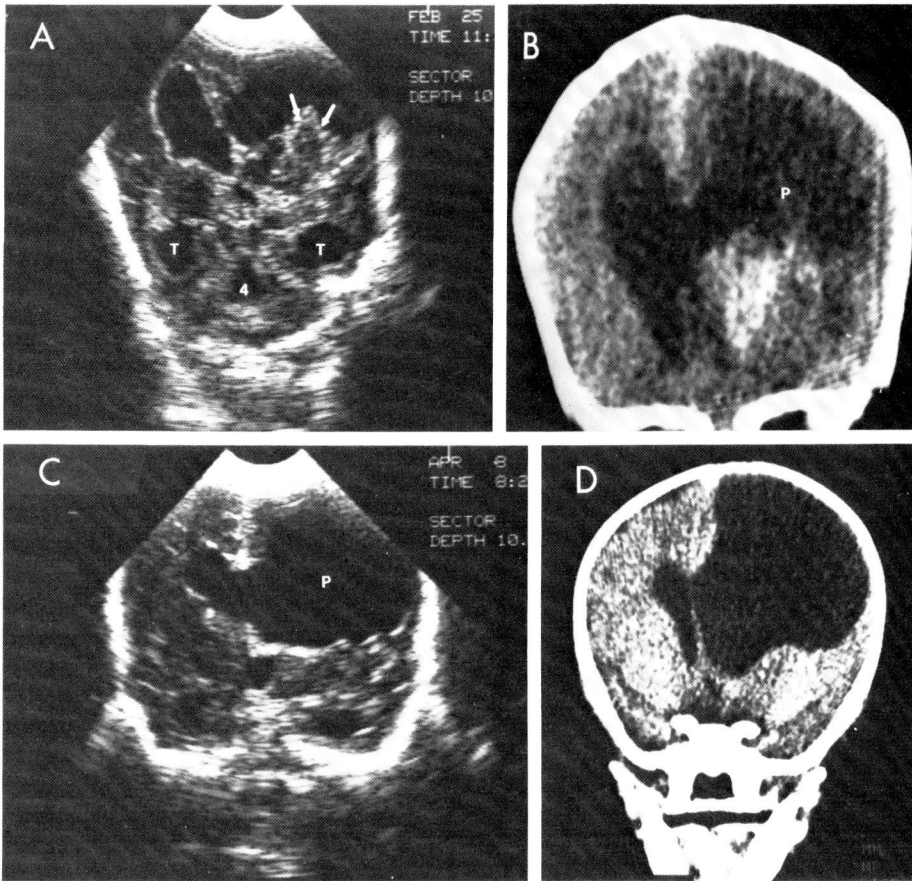

Fig 8–11.—Porencephalic cyst development. **A,** Coronal ultrasound scan in 2-week-old premature infant after intraventricular and intraparenchymal hemorrhage. On the left side the hematoma *(arrows)* has retracted and is settling onto the floor of the ventricle, which has been enlarged by necrosis of the hematoma. Note enlargement of the right lateral ventricle, temporal horns *(T)*, and fourth *(4)* ventricle, and slight left-to-right mass effect. **B,** Coronal CT scan demonstrates high-density hematoma and beginning of porencephaly *(P)*. **C,** Coronal ultrasound at age 2 months after complete resolution of the hematoma. The porencephalic area *(P)* resulting from the combined hematoma and ventricle has a slight left-to-right shift, presumably due to partial obstruction of CSF flow. **D,** Coronal CT scan comparable to **C.**

extended out from the ventricular system into a parenchymal lesion (Fig 8–11). Over 1–2 weeks, the hematoma demarcates from the surrounding brain and retracts from its edges. As the clot is resorbed over 1–3 months, a porencephalic cyst results. In these cases, the mass effect is most likely due to poor communication with the ventricular system. After shunting for hydrocephalus, the porencephalic area will become somewhat smaller.[32] It may be necessary to shunt or decompress the cyst if there is significant mass effect.[33]

Arachnoid Cysts

Arachnoid cysts lie between the brain and dura, contain clear CSF, and are lined by arachnoid.[31] A congenital arachnoid cyst does not communicate. It is usually found in the sylvian fissure and middle fossa or interhemispheric fissures.[34] Midline cystic lesions may be difficult to distinguish from portions of the ventricular system (Fig 8–12).[35]

Secondary or acquired arachnoid cysts are thought to be caused by arachnoid

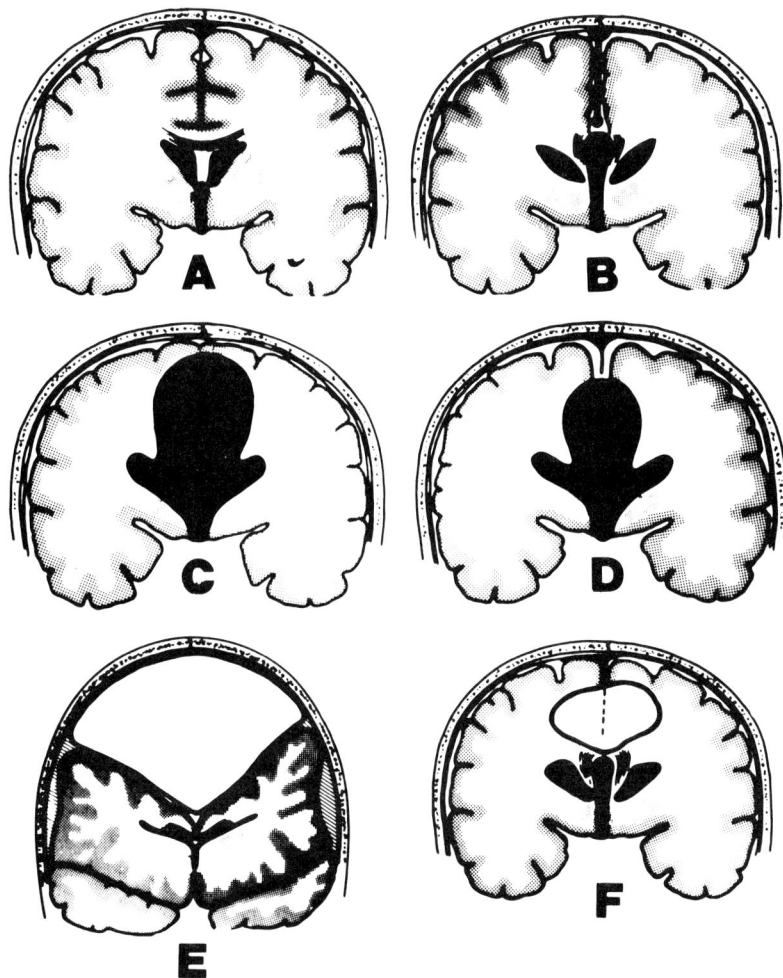

Fig 8–12.—Midline cystic lesions. **A,** Normal brain, coronal section. **B,** Agenesis of the corpus callosum. Third ventricle extends between lateral ventricles. **C,** Alobar holoprosencephaly. Note the single large central ventricle. **D,** Lobar holoprosencephaly. Note partial separation of a single central ventricle. **E,** Intradural interhemispheric cyst. **F,** Agenesis of the corpus callosum with interhemispheric cyst. (From Rao et al.[34] Reproduced with permission.)

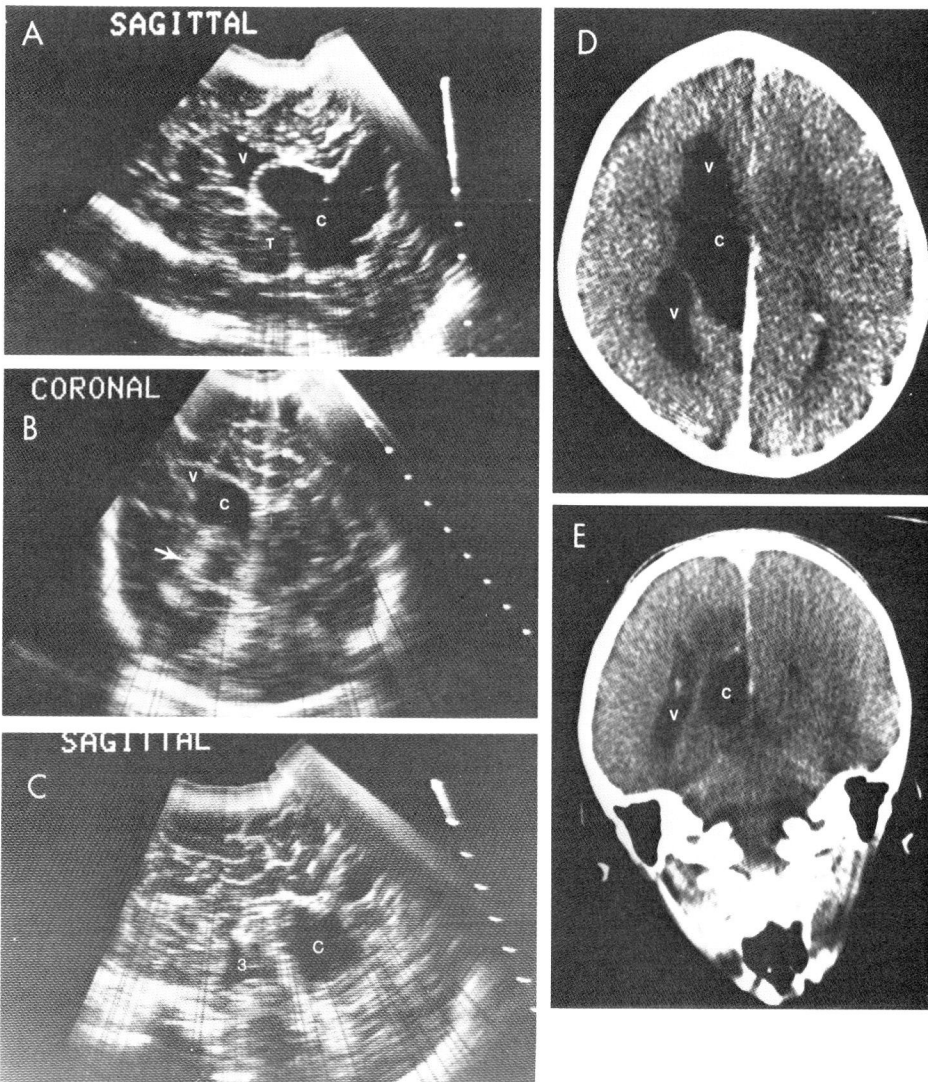

Fig 8–13.—Arachnoid cyst in region of quadrigeminal cistern. Sagittal **(A)** and coronal **(B)** ultrasound scans demonstrate multiseptated cyst *(C)* lying medial and inferior to the lateral ventricle *(V)* and superior and posterior to the thalamus *(T)*. The choroid plexus *(arrow, **B**)* at the trigone is inferior to the cyst. **C,** Midline sagittal ultrasound scan shows the cyst *(C)* behind the third *(3)* ventricle in the region of the quadrigeminal cistern. Axial **(D)** and coronal **(E)** CT scans demonstrate medial and inferior position of cyst.

lesions.[1] These occur in the cisterns, adjacent to the third ventricle and sella, and in the posterior fossa.

Most arachnoid cysts are diagnosed during evaluation for another anomaly. We have diagnosed several arachnoid cysts in utero during an ultrasound evaluation. One was a quadrigeminal cyst (Fig 8–13) and one was an interhemispheric cyst (Fig

8–14). The most common posterior fossa cystic lesion is actually an obstructed and dilated fourth ventricle in the Dandy-Walker syndrome (Fig 8–15).

Ventricular Cysts

Ventricular cysts occur when there is entrapment or obstruction of just one part of the ventricular system (see chapter on hydrocephalus). Sometimes after shunting for hydrocephalus, there may be entrapment of the fourth ventricle.[36, 37] As the sutures close in older children, the ventricles may form diverticula when they extend through the tentorial incisura with ventricular enlargement.[38]

Neoplastic Cysts

Cystic neoplasms such as a cystic astrocytoma usually have a thickened wall or distort the surrounding brain parenchyma. If an ultrasound scan reveals a cystic lesion with mass effect and the origin is uncertain, contrast-enhanced CT should be performed to exclude a brain tumor.

Vascular malformations may be confused with cystic lesions on ultrasound but Doppler evaluation with real-time examination permits immediate differentiation of

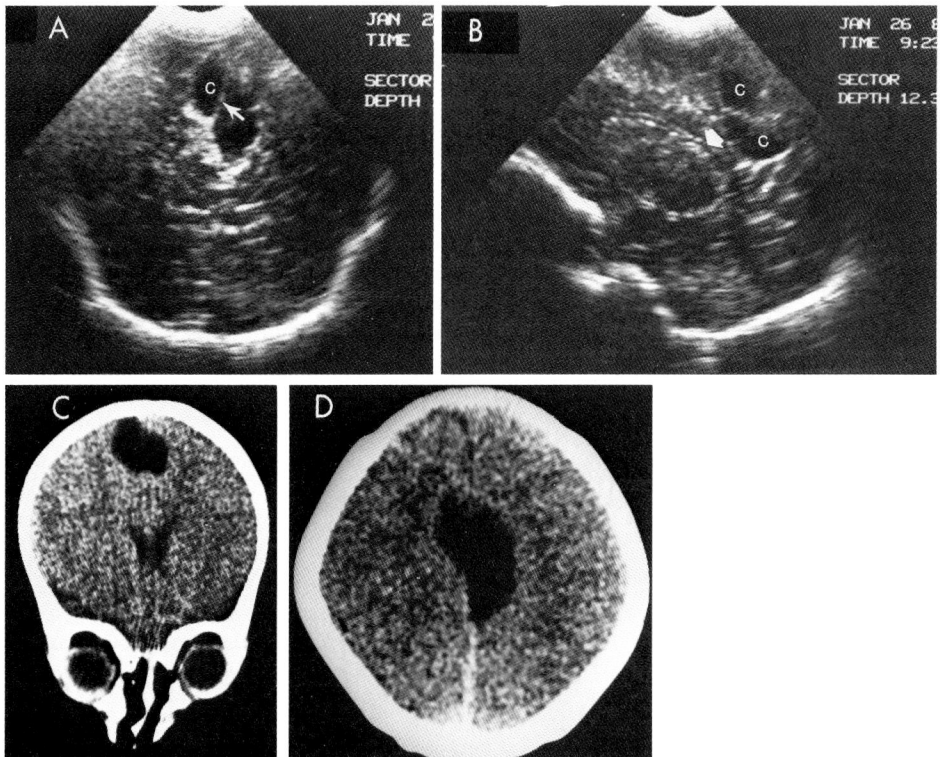

Fig 8–14.—Arachnoid cyst. Coronal **(A)** and sagittal **(B)** ultrasound scans in newborn infant demonstrate interhemispheric arachnoid cyst *(C)* with a central septum lying behind the lateral ventricle *(arrow)*. Coronal **(C)** and axial **(D)** CT scans demonstrate cyst but septum is not visualized.

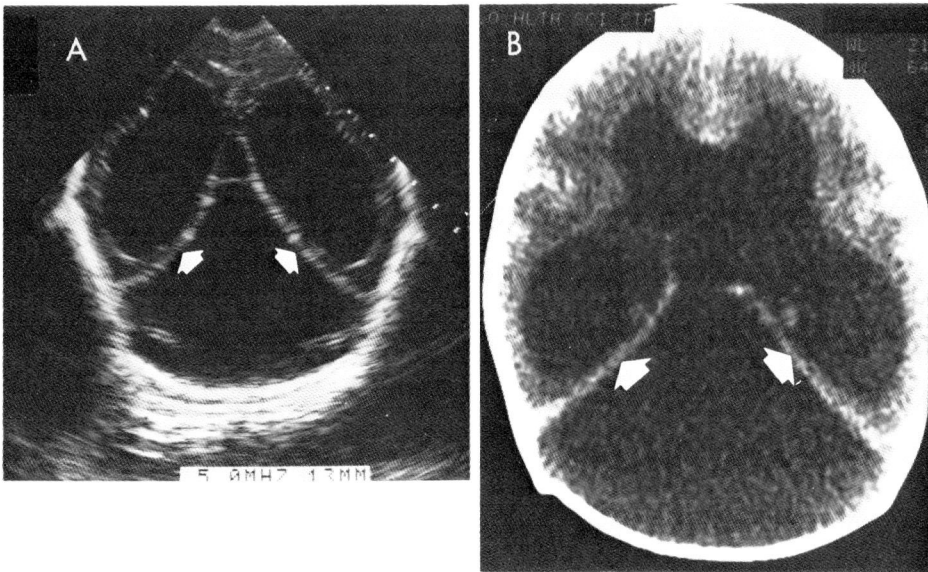

Fig 8–15.—Dandy-Walker cyst. Coronal ultrasound scan **(A)** and axial CT scan **(B)** demonstrate posterior fossa cyst which fills the entire posterior fossa and extends to the edges of the tentorium *(arrows)*.

a cyst from a vessel. Contrast-enhanced CT should be performed if Doppler imaging is not available, particularly if the cystic lesion has a rather linear appearance.

Intraoperative Shunts for Cysts

Cystoperitoneal shunts may be necessary when a cyst enlarges, causing mass effect and/or hydrocephalus. Intraoperative ultrasound may be extremely valuable for shunt placement in these difficult lesions.[39]

VASCULAR MALFORMATIONS

Arteriovenous malformations (AVMs) are the most common vascular lesions seen in children, and aneurysms are very rare.[1] An AVM is thought to be a congenital lesion that represents a failure of the embryonic arteriovenous shunts to be replaced by capillaries.[40] These malformations vary tremendously in size, but those that are present in infancy are typically larger. The most common is a vein of Galen aneurysm. Vascular malformations have been detected both on ultrasound and on CT.[41-44]

Vein of Galen Malformation

Multiple arterial feeders are frequently present that derive from the anterior and posterior cerebral circulations. These steal blood from the cerebral vasculature and result in a greatly dilated vein of Galen.

Cardiac failure due to the large arteriovenous shunt is the commonest presenting symptom. In fact, cardiac catheterization may be done for congestive failure but it will be normal except for rapid filling of the neck venous drainage due to the shunt. Congestive failure with an enlarged aorta on chest radiographs may be the first indication if the cranial bruit and head enlargement are not prominent features.

Hydrocephalus is the next most common presenting symptom. Dilatation of the vein of Galen can cause obstruction of the ventricular system at the level of the third ventricle and aqueduct. If there is subarachnoid or intraventricular bleeding, hydrocephalus may develop on this basis alone.

Symptoms and prognosis depend on the size of the arteriovenous shunt. A larger shunt presents earlier and has a poorer prognosis.

ULTRASOUND FINDINGS IN GALENIC MALFORMATION.—On coronal projections, a cystic structure is present posterior to the foramen of Monro and superior to the third ventricle; this structure is the dilated vein of Galen (Fig 8–16). Tortuosity of the vein of Galen is frequent and dilatation often extends posteriorly into the straight sinus.[41] It typically lies between the dilated lateral ventricles on the coronal scan and can be traced posteriorly into the straight sinus and torcular Herophili with sagittal real-time examination. Doppler studies will confirm the markedly increased flow in the vein of Galen and the carotid arteries. Hydrocephalus is frequently present.

Although the differential diagnosis includes many of the cystic lesions previously discussed, particularly quadrigeminal cyst, Doppler evaluation should exclude a cyst because there will be markedly increased flow unless there has been thrombus formation.

CT FINDINGS IN GALENIC MALFORMATION.—A cystic lesion lying between the dilated lateral ventricles will be evident on the non-contrast-enhanced axial CT scan. It may be partly or totally high density if there has been acute hemorrhage with clot formation. After contrast, the vein of Galen, straight sinus, and torcular Herophili will typically enhance in the areas of the freely flowing blood (Fig 8–17). The appearance of a vein of Galen aneurysm will vary with the amount of thrombosis.[42] After the neonatal period, thrombosis of the torcular and straight sinus is increasingly present so that only the vein of Galen may be apparent. Spontaneous thrombosis is apparently well tolerated and has been reported to have resulted in a cure in at least one patient.[44] The obstructed veins (distal to the vein of Galen) may cause aneurysmal enlargement leading to hydrocephalus or subarachnoid and intraparenchymal hemorrhage.

Brain damage may occur from a galenic malformation. It is apparently due to anoxia from ischemia when arterial feeders to the malformation steal blood from the posterior choroidal and thalamostriate arteries.[43] Calcification and edema of the basal ganglia may result. Enhancement of these areas indicates the presence of arterial feeders. Tiny arterial feeders are usually not evident on CT or ultrasound but may be visualized if they are large enough.

The differential diagnosis includes pineal tumors, which should present with different clinical symptoms. Pineal tumor enhancement will never extend into the

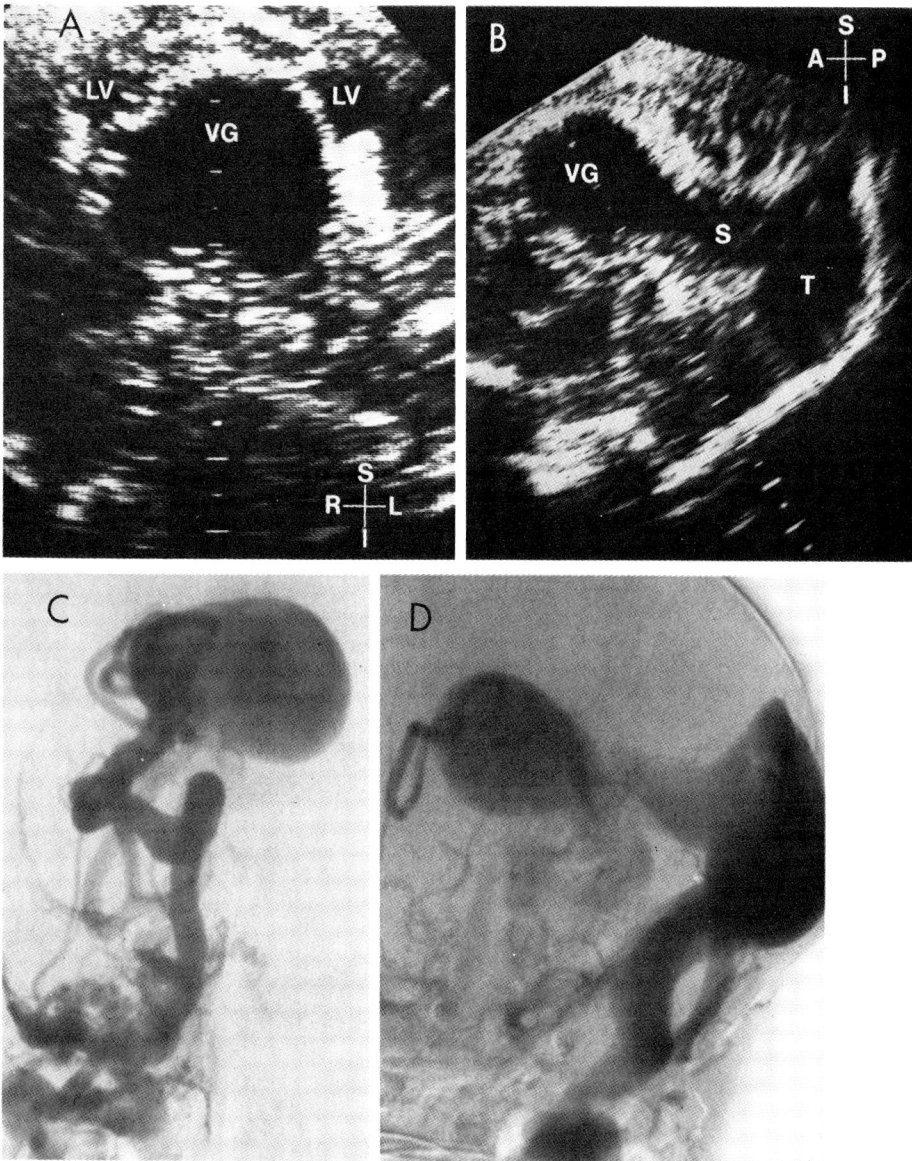

Fig 8–16.—Vein of Galen aneurysm, typical appearance. **A,** Coronal ultrasound scan showing dilated vein of Galen *(VG)* centrally placed between mildly dilated lateral ventricles *(LV)*. *S,* superior; *R,* right; *L,* left. **B,** Sagittal ultrasound scan showing dilated vein of Galen *(VG)* extending into the straight sinus *(S)* and torcular Herophili *(T)*. Small cystic structures inferior to vein of Galen pulsated on real-time scanning, representing arterial feeding arteries. *S,* superior; *I,* inferior. **C,** Anteroposterior projection of early arterial phase of right internal carotid injection. Note dilated feeding vessels and aneurysmal vein of Galen. **D,** Lateral projection of right internal carotid injection at midarterial phase. Vein of Galen aneurysm with rapid filling of straight sinus and torcular Herophili in projection like the sagittal scan. (From Cubberly et al.[41] Reproduced with permission.)

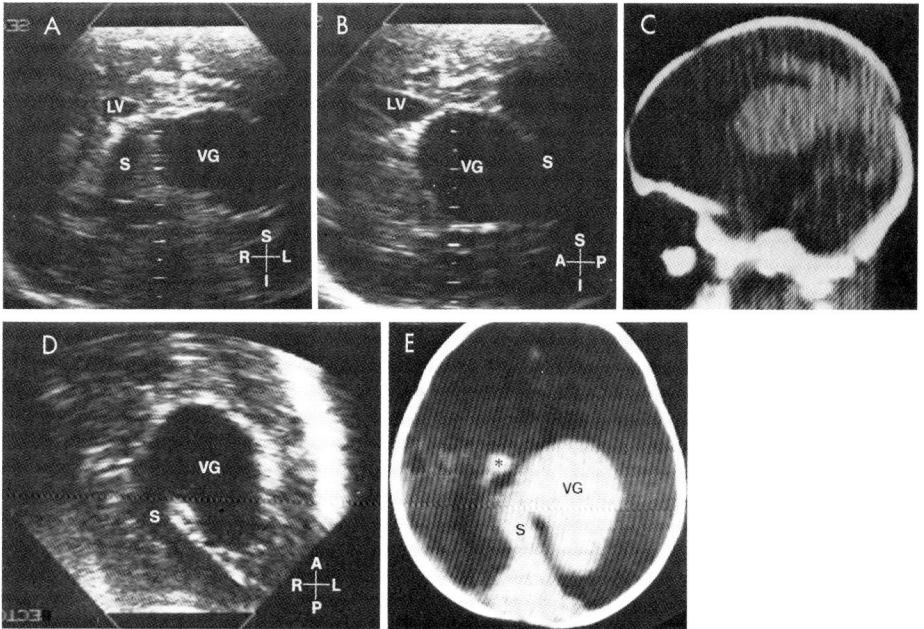

Fig 8–17.—Eccentric vein of Galen malformation with mild hydrocephalus. Coronal **(A)** and sagittal **(B)** ultrasound scans demonstrate eccentric dilated vein of Galen *(VG)*, part of straight sinus *(S)*, and lateral ventricle *(LV)*. S, superior; I, inferior; R, right; L, left. **C,** Sagittal CT scan comparable to **B** about 2 cm to left of midline. **D,** Axial ultrasound scan through posterior fontanelle, and **E,** axial CT scan demonstrate dilated, eccentric vein of Galen *(VG)* and straight sinus *(S). Asterisk* indicates part of arteriovenous malformation. (From Cubberly et al.[41] Reproduced with permission.)

straight sinus and torcular Herophili. Tumor calcifications will be quite different—frequently central.

Angiography will be necessary to guide possible embolic or operative intervention but ultrasound will be sufficient to make a firm initial diagnosis.

REFERENCES

1. Harwood-Nash D.C., Fitz C.R.: *Neuroradiology in Infants and Children.* St. Louis, C.V. Mosby Co., 1976.
2. Arnstein L.H., Baldrey E., Nafzinger H.C.: Case report and survey of brain tumors during neonatal period. *J. Neurosurg.* 6:315–319, 1951.
3. Koos W.T., Miller M.H.: *Intracranial Tumors in Infants and Children.* St. Louis, C.V. Mosby Co., 1971.
4. Segall H.D., Zee C.S., Naidich T.P., et al.: Computed tomography in neoplasms of the posterior fossa in children. *Radiol. Clin. North Am.* 20:237–253, 1982.
5. Grant E., Schellinger D., Richardson J.: Real time ultrasonography of the posterior fossa. *J.Ultrasound Med.* 2:73–87, 1983.
6. Shkolnik A.: B-mode scanning of the infant brain: A new approach case report. Craniopharyngioma. *JCU* 3:229–231, 1975.
7. Slovis T.L.: Real-time ultrasound of the intracranial contents. *Clin. Diagn. Ultrasound* 8:13–27, 1981.

8. Slovis T.L., Kuhns L.R.: Real-time sonography of the brain through the anterior fontanelle. *AJR* 136:277–286, 1981.

9. Han B.K., Babcock D.S., Oestrich A.E.: Sonography of brain tumors in infants. Presented at the Society of Pediatric Radiology, Atlanta, April 1983.

10. Shkolnik A., McLone D.G.: Intraoperative neurosonography of brain masses in the pediatric patient. Presented at the Society of Pediatric Radiology, Atlanta, April 1983.

11. Gooding G.A.W., Edwards M.S.B., Rabkin A.E., et al.: Intraoperative real-time ultrasound in the localization of intracranial neoplasms. *Radiology* 146:459–462, 1983.

12. Rubin J.M., Dohrmann G.J., Greenberg M., et al.: Intraoperative sonography of meningiomas. *AJNR* 3:305–308, 1982.

13. Knake J.E., Chandler W.F., McGillicuddy J.E., et al.: Intraoperative sonography for brain tumor localization and ventricular shunt placement. *AJR* 139:733–738, 1982.

14. Heimburger R.F., Eggleton R.C., Fry F.J.: Ultrasonic visualization in determination of tumor growth rate. *JAMA* 224:497–501, 1973.

15. Sugar O., Uematsu S.: The use of ultrasound in the diagnosis of intracranial lesions. *Surg. Clin. North Am.* 44:55–64, 1964.

16. French L.A., Wild J.J., Neal D.: The experimental application of ultrasonics to the localization of brain tumors. *J. Neurosurg.* 8:198–203, 1951.

17. Tsutsumi Y., Andoh Y., Inque N.: Ultrasound-guided biopsy for deep-seated brain tumors. *J. Neurosurg.* 57:164–167, 1982.

18. Masuzawa H., Kamitani H., Sato J., et al.: Intraoperative application of sector scanning electronic ultrasound in neurosurgery. *Neurol. Med. Chir. (Tokyo)* 21:277–285, 1981.

19. Taboada D., Froufe A., Alonso A., et al.: Congenital medulloblastoma: Report of two cases. *Pediatr. Radiol.* 9:5–10, 1980.

20. Zimmerman R.A., Bilaniuk L.T., Pahlajani H.: Spectrum of medulloblastomas demonstrated by computed tomography. *Radiology* 126:137–141, 1978.

21. Mahour G.H., Wooley M.M., Trivedi S.N., et al.: Teratomas in infancy and childhood. *Surgery* 76:309–318, 1974.

22. Fitz C.R., Rao K.C.V.G.: Primary tumors in children in *Cranial Computed Tomography*. New York, McGraw-Hill Book Co., 1983. (pp. 295–343).

23. Zimmerman R.A., Bilaniuk L.T., Bruno L., et al.: Computed tomography of cerebellar astrocytoma. *AJR* 130:929–933, 1978.

24. Lin S., Lee K.F., O'Hara A.E.: Congenital astrocytomas: The roentgenographic manifestations. *AJR* 115:78–85, 1972.

25. Zimmerman R.A., Bilaniuk L.T.: CT of primary and secondary craniocerebral neuroblastoma. *AJR* 135:1239–1242, 1980.

26. Gooding C.A., Brasch R.C., Brant-Zawadzki M.M., et al.: Nuclear magnetic resonance of the brain in children. Presented at the Society for Pediatric Radiology, Atlanta, May 1983.

27. Brant-Zawadzki M., Davis P.L., Crooks L.E., et al.: NMR demonstration of cerebral abnormalities: Comparison with CT. *AJNR* 4:117–124, 1983.

28. Mack L.A., Rumack C.M., Johnson M.L.: Ultrasound evaluation of cystic intracranial lesions in the neonate. *Radiology* 137:451–455, 1980.

29. Maravilla K.R., Kirks D.R., Maravilla A.: CT diagnosis of intracranial cystic abnormalities in children. *Computerized Tomography*, 2:221–235, 1978.

30. Shaw C.M., Alvord E.C.: Cava septi pellucidi et vergae: Their normal and pathological states. *Brain* 92:213–224, 1969.

31. Freide R.L.: *Developmental Neuropathology*. New York, Springer-Verlag, 1975.

32. Grant E.G., Kerner M., Schellinger D., et al.: Evolution of porencephalic cysts from intraparenchymal hemorrhage in neonates: Sonographic evidence. *AJR* 138:467–470, 1982.

33. Tardieu M., Evrard P., Lyon G.: Progressive expanding congenital porencephalies: A treatable cause of progressive encephalopathy. *Pediatrics* 68:198–202, 1981.

34. Rao K.C.V.G., Gunadi I.K., Diaconis J.N.: Interhemispheric intradural cyst. *J. Comput. Assist. Tomogr.* 6:1167–1171, 1982.

35. Armstrong E.A., Harwood-Nash D.C.F., Hoffman H., et al.: Benign suprasellar cysts: The CT approach *AJNR* 4:163–166, 1983.
36. Zimmerman R.A., Bilaniuk L.T., Gallo E.: Computed tomography of the trapped fourth ventricle. *AJR* 130:503–506, 1978.
37. Scotti G., Musgrave M.A., Fitz C.R., Harwood-Nash D.C.: The isolated fourth ventricle in children: CT and clinical review of 16 cases. *AJR* 135:1233–1238, 1980.
38. Naidich T.P., McLone D.G., Hahn Y.S., et al.: Atrial diverticula in severe hydrocephalus. *AJNR* 3:257–266, 1982.
39. Shkolnik A., McLone D.G.: Intraoperative real time ultrasonic guidance of ventricular shunt placement in infants. *Radiology* 144:573–576, 1982.
40. Kaplan H.A., Aronson S.M., Browder E.J.: Vascular malformations of the brain: An anatomical study. *J. Neurosurg.* 18:630, 1961.
41. Cubberly D.A., Jaffe R.B., Nixon G.W.: Sonographic demonstration of Galenic arteriovenous malformations in the neonate. *AJNR* 3:435–439, 1982.
42. Shirkhoda A., Whaley R.A., Boone S.C., et al.: Varied CT appearances of aneurysms of the Vein of Galen in infancy. *Neuroradiology* 21:265–270, 1981.
43. Martelli A., Scotti G., Harwood-Nash D.C., et al.: Aneurysms of the Vein of Galen in children: CT and angiographic correlations. *Neuroradiology* 20:123–133, 1980.
44. Spallone A.: Computed tomography in aneurysms of the vein of Galen. *J. Comput. Assist. Tomogr.* 3.779–782, 1979.

CHAPTER 9

Intracranial Infection

Carol M. Rumack, M.D.
Michael L. Johnson, M.D.

ANTENATALLY ACQUIRED INFECTIONS

THE MOST COMMON infections acquired in utero are cytomegalovirus,[1-3] herpes simplex virus,[4, 5] rubella, and toxoplasmosis.[6] Maternal symptoms are uncommon and the diagnosis is usually made in the neonatal period.[7]

Cytomegalovirus

Cytomegalovirus presents in the neonate with intracranial calcification, typically in the periventricular region. The active phase of encephalitis begins in utero and by birth may have caused calcification. One case of cytomegalovirus with intracranial calcification in utero has been reported.[3] Intrauterine infection in one of our patients caused moderately echogenic ventricular margins and distortion of the parenchymal architecture (see chapter on fetal intracranial diagnosis). One month later, periventricular calcification was seen on the first day of life (Fig 9–1).

ULTRASONIC FINDINGS.—Periventricular calcifications typically are scattered along the ventricular wall. They may be small enough that acoustic shadowing does not occur. However, the punctate, discrete nature of the calcifications are quite different from the appearance of subependymal hemorrhage, even when the calcifications occur in the typical region of the caudothalamic notch (see Fig 9–1). Atypical densities about the ventricles and in the brain parenchyma are due to calcification. Delayed studies in 1–2 weeks will show persistence of the highly echogenic density, whereas hemorrhage becomes less echogenic. Ventricular enlargement may be present due to diffuse brain damage with scattered areas of encephalomalacia.

COMPUTERIZED TOMOGRAPHIC (CT) FINDINGS.—Periventricular calcifications, ventricular enlargement, and scattered brain damage are common findings on CT after cytomegalovirus infection (see Fig 9–1).

197

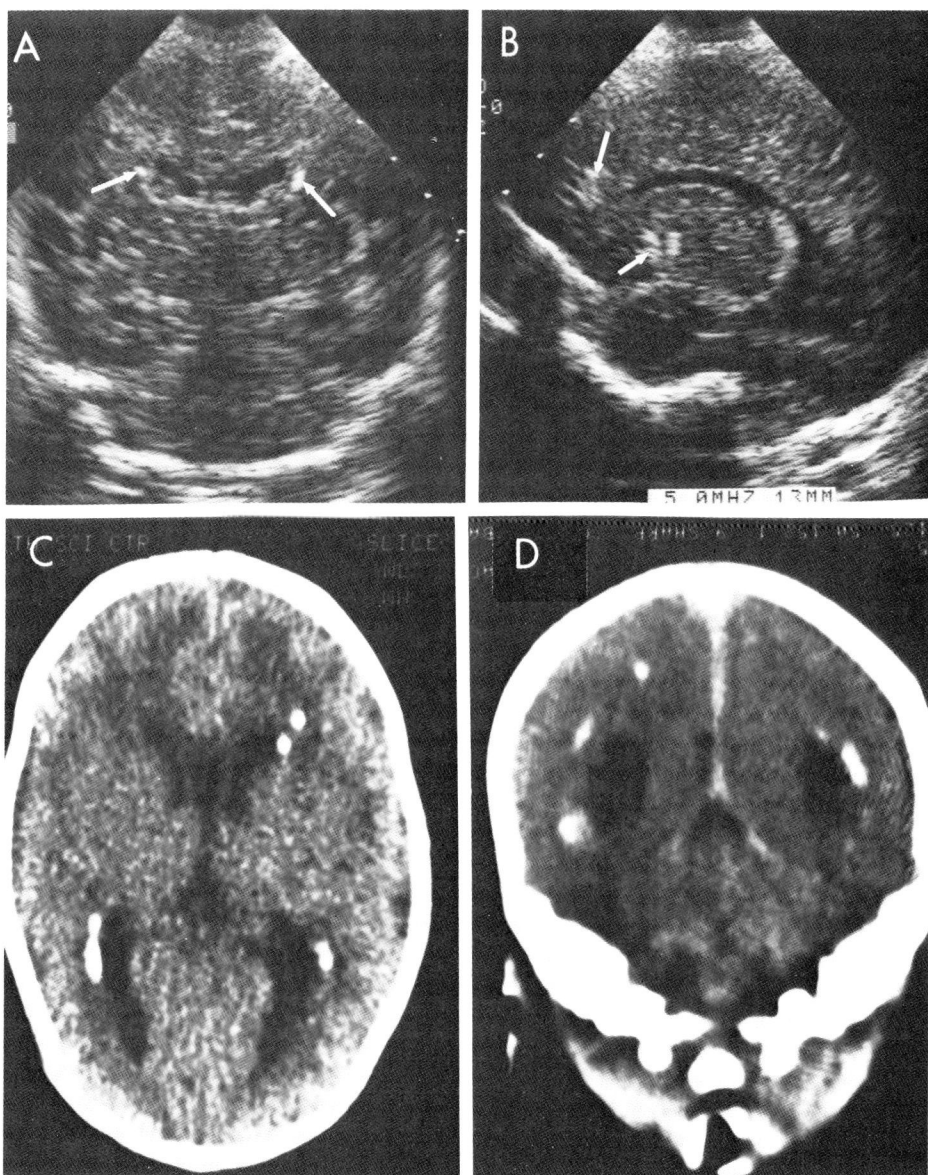

Fig 9–1.—Cytomegalovirus infection in 1-day-old infant. Coronal **(A)** and sagittal **(B)** ultrasound scans demonstrate discrete periventricular calcifications *(arrows)*. Axial **(C)** and coronal **(D)** CT scans at somewhat different levels show periventricular calcification. Tentorium and falx are enhanced with contrast on coronal CT scan. (From Rumack C.M., Johnson M.L., Role of CT and ultrasound in neonatal brain imaging. *J. Computed Tomog.* 7:17–29, 1983. Reproduced with permission.)

Toxoplasmosis

Congenital acquired toxoplasmosis is a parasitic infection that classically has been described as causing scattered parenchymal calcification and hydrocephalus. However, reports in the literature[6, 8] have demonstrated that toxoplasmosis may cause periventricular calcification. Clinically there is usually an associated chorioretinitis, hepatosplenomegaly, anemia, thrombocytopenia, skin rash, pneumonia, seizures, and microcephaly, although it may be a more subtle infection.

Herpes Simplex Virus (Type 2)

Infection in the neonate is caused by herpes type 2 virus[4, 5] acquired transplacentally or in the intrapartum period, rather than type 1 virus,[9–13] which causes sporadic encephalitis in older children and adults. Herpes type 2 virus causes a diffuse encephalitis and, if acquired in utero, has been associated with microcephaly, intracranial calcification, microphthalamia, and retinal dysplasia.[4] Herpes virus in utero causes marked neurotropic teratogenic changes. The severe brain damage may result in a compressed, thin brain mantle and periventricular calcifications.[5]

Rubella

Massive calcification of the brain has been reported by Harwood-Nash et al.[2] in what appeared to be a congenital rubella or cytomegalovirus infection. Usually rubella results in microcephaly without calcification. It has become extremely rare since the rubella vaccine became available in 1967.

NEONATALLY ACQUIRED INFECTIONS

The commonest intracranial infections in the neonate are bacterial and typically are caused by *group B Streptococci* or *Escherichia coli*. After the newborn period, *Hemophilus influenzae* becomes the most common cause of CNS infection in infants and children.[7]

Meningitis or meningoencephalitis are frequently associated, particularly in the infant less than 1 year old, who usually has difficulty containing an area of infection in the brain.[14] Areas of edema or cerebritis develop initially, although a scan in the first week may be normal. This disease process may develop into a widespread infection, heal completely, or localize in an abscess. Salmon reported that 8 of 12 neonates with meningitis developed ventriculitis.[15] Late complications include subdural effusions,[16, 17] ventricular enlargement, ventricular septations,[18, 19] encephalomalacia or atrophy, and, uncommonly, abscess. Occasionally hemorrhagic necrosis may complicate the course.[20]

ULTRASOUND FINDINGS.—Early changes in infection may cause distortion of the brain parenchymal architecture suggested only by the presence of cerebral edema and small, compressed ventricles.[18, 21] This can be a difficult diagnosis, and contrast CT may be more reliable in diagnosing early evidence of cerebritis (Fig 9–2).

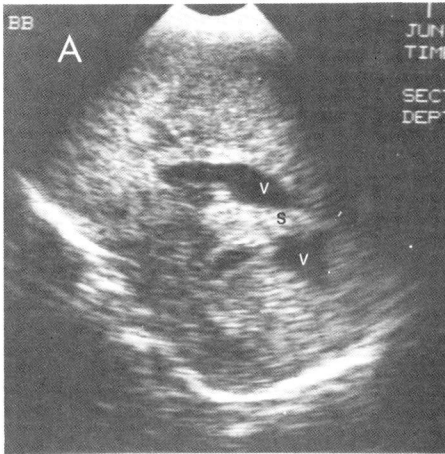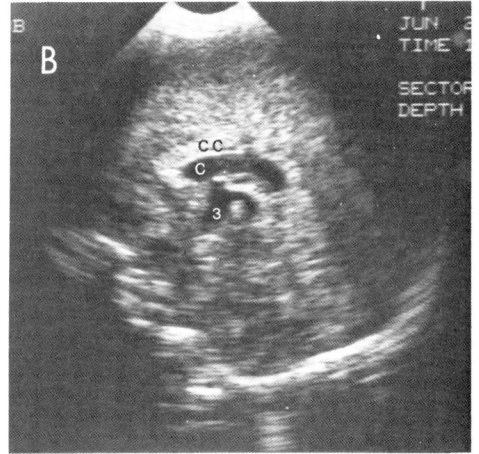

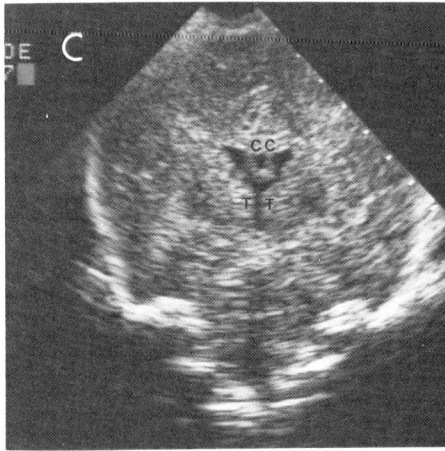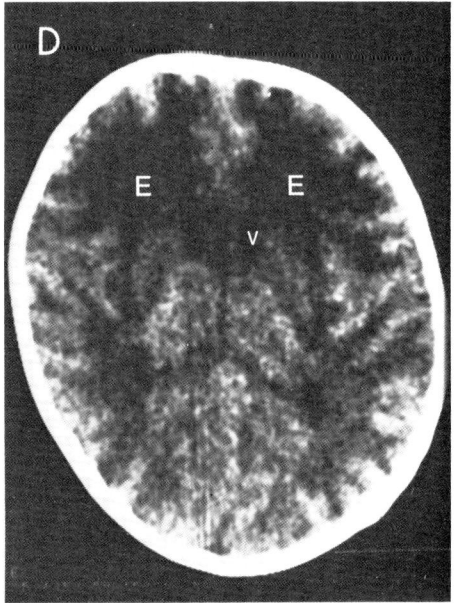

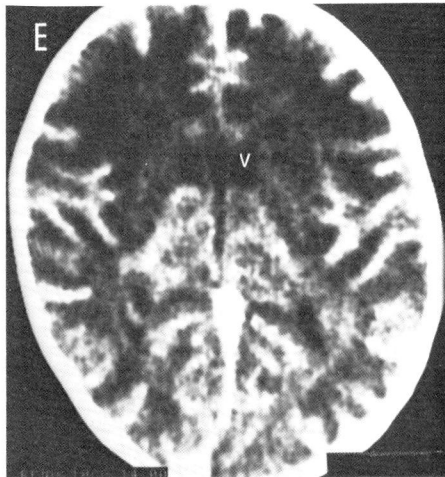

Complications can be followed by evaluating ventricular size and areas of encephalomalacia, which are typically cystic (Fig 9–3).[22] Subdural effusions can be diagnosed by ultrasound (Fig 9–4) but effusions less than 1 cm thick may be missed, owing to the initial transducer artifact. Ventriculitis may be recognized as debris diffusely within the ventricles (Fig 9–5) and may leave ventricular septations (see Fig 9–2).

CT FINDINGS.—The initial CT scan may be normal if done during the first week of the illness, but even then there may be evidence of focal areas of edema. Cerebritis presents with enhancement in these regions (Fig 9–6).[23] Ventricular compression may be present due to focal or diffuse edema. Hydrocephalus may develop from extraventricular obstruction, as the scarring causes an arachnoid block. Ventricular enlargement may also occur secondary to diffuse or focal brain damage (see Fig 9–5). Vasculitis is a well-known complication of meningitis that can result in an area of brain damage with a characteristic vascular distribution (see Fig 9–3).[16, 24]

Subdural effusions are a common complication of *H. influenzae* infection, and rapid head growth may be caused by subdural collection alone even without associated hydrocephalus (see Fig 9–4). Enhancement of the ventricular wall is evidence of ventriculitis (see Fig 9–5).

Cockrill et al. have reported a high incidence of frontal lobe involvement (72%) in leptomeningeal infections caused by *Hemophilus* organisms,[16] probably secondary to a vasculitis. Polycystic brain disease has been reported as a major complication of meningitis.[24]

Calcification has been reported following bacterial meningitis involving the cerebral cortex,[25] dura mater, and basal meninges.[26–29] Periventricular calcifications have been reported in two cases following meningitis (*E. coli* and group D *Salmonella*) complicated by ventriculitis and severe brain damage.[30] Marked hydrocephalus was present with calcification and a thin brain mantle.

Abscess formation is evident when an area of low density enhances in a ringlike fashion, but frequently the neonatal brain cannot contain an infection or abscess (Fig 9–7). CT has allowed the accurate diagnosis of brain abscess.[31–36] The therapy of choice now seems to be changing from surgery to medical treatment, with surgery indicated only if there is a strong possibility of tumor, failure to respond with

---

Fig 9–2.—*Salmonella* meningoencephalitis in 1-month-old infant. **A,** Sagittal ultrasound scan through the lateral ventricle *(V)* shows ventricular septum *(S)*, irregular ventricular walls, and diffusely increased echogenicity of the brain parenchyma. At this age there should be clearly demarcated sulci above the ventricle. **B,** Sagittal ultrasound scan through the midline. The corpus callosum *(CC)* is highly echogenic (usually echolucent) above the cavum septi pellucidi *(C)* and cavum vergae. The third ventricular margins *(3)* are irregular and the cerebellum is hard to differentiate from the abnormally highly echogenic cerebral cortex. **C,** Coronal ultrasound scan through the frontal horns of the lateral ventricles and cavum septi pellucidi. The thalamus *(T)* and corpus callosum *(CC)* are abnormally echogenic. Sulci are obscured. **D** and **E** are axial CT scans in same patient. **D,** Non-contrast-enhanced scan demonstrates diffusely edematous brain *(E)* with prominent gyri. **E,** Contrast-enhanced scan shows diffuse abnormal enhancement of the gyri from meningitis. Cerebritis is evident from deeper enhancement of the cerebral cortex and the thalami. Ventricles *(V)* are separated by the cavum septi pellucidi.

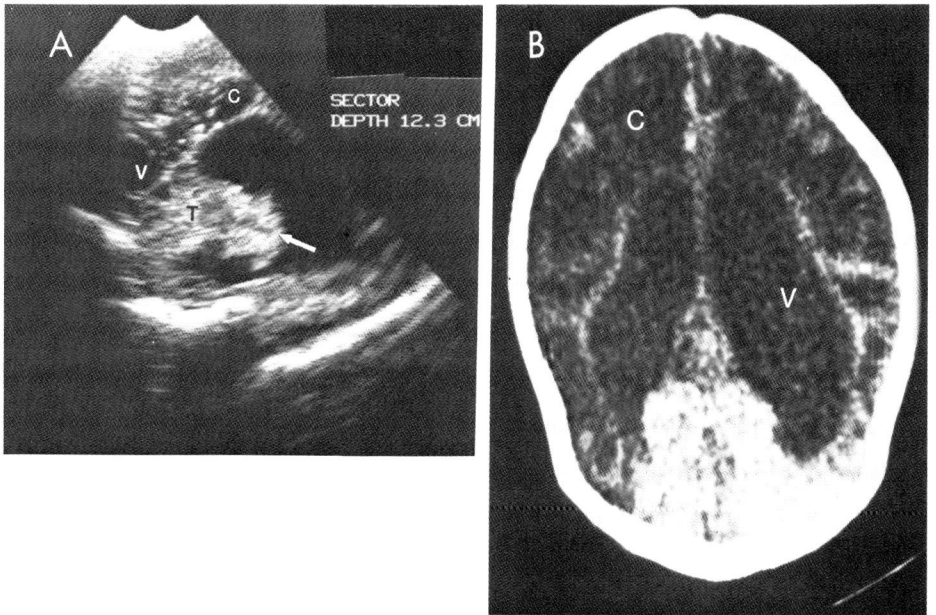

Fig 9–3.—Encephalitis of unknown etiology caused diffuse parenchymal brain damage of the entire cerebral cortex, leaving ventricular enlargement and cystic encephalomalacia with some sparing of the thalamus and posterior fossae. **A,** Sagittal ultrasound scan demonstrates enlarged lateral ventricle with cystic *(C)* areas above it. Note abnormally echogenic thalamus *(T)* and shaggy choroid plexus *(arrow)*. **B,** Axial CT scan demonstrates diffuse cystic *(C)* areas of brain sparing the subependymal lining of the ventricles *(V)*. Ventricular enlargement is most likely caused by brain destruction.

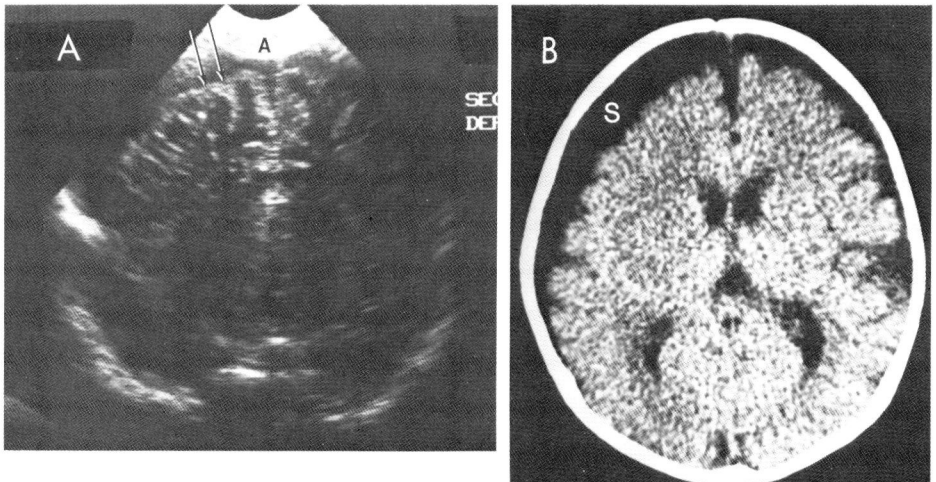

Fig 9–4.—Subdural effusions from *Hemophilus influenzae* meningitis. **A,** Coronal ultrasound scan with surface of gyri and sulci clearly evident *(arrows)*. Entire brain is displaced inferiorly by the effusion from the transducer artifact *(A)*. **B,** Axial CT scan with bilateral subdural *(S)* effusions (right greater than left). Ventricular size is normal, although hydrocephalus is a common complication of meningitis.

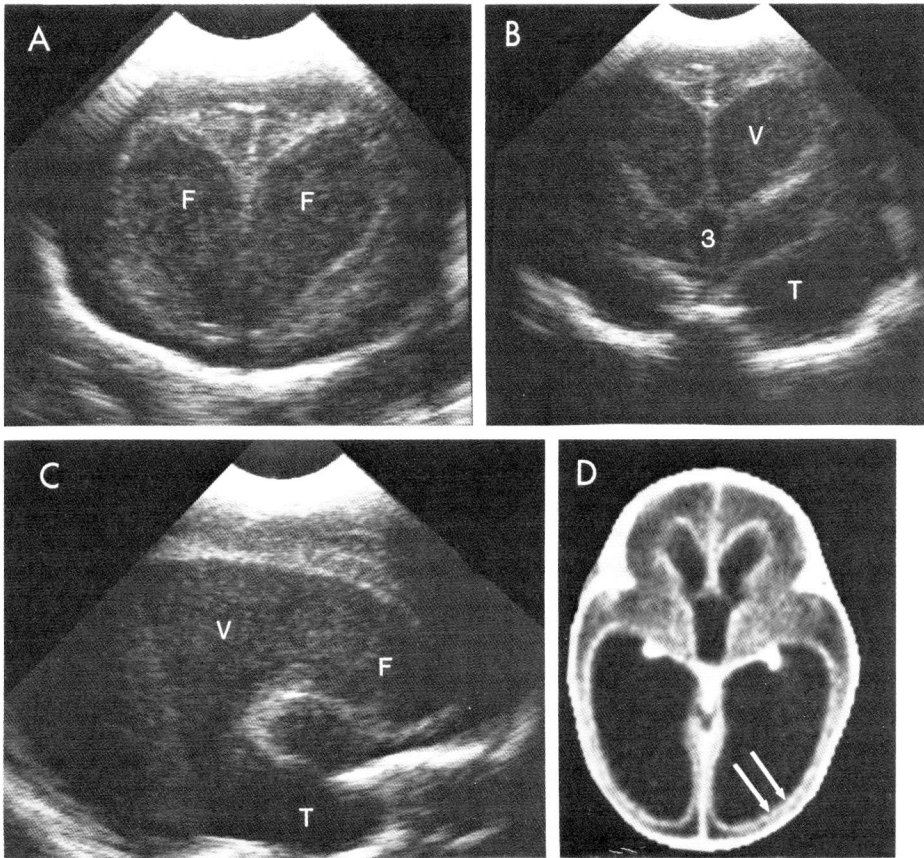

Fig 9–5.—*Bacteroides* ventriculitis. **A,** Coronal ultrasound scan quite far anterior shows massively enlarged frontal horns *(F)* filled with pus or white cells. **B,** Coronal ultrasound scan at the level of the foramen of Monro. The frontal horns *(F)*, temporal horns *(T)*, and third *(3)* ventricle are massively enlarged. **C,** Sagittal ultrasound scan shows massively enlarged lateral ventricle *(V)* with frontal *(F)* and temporal *(T)* horns extending quite far anteriorly. Sulcal pattern is obscured by encephalitis. **D,** Axial CT scan after contrast. Enhancement of ventricular walls *(arrows)* indicates ventriculitis. (Courtesy of Lawrence Mack, M.D.)

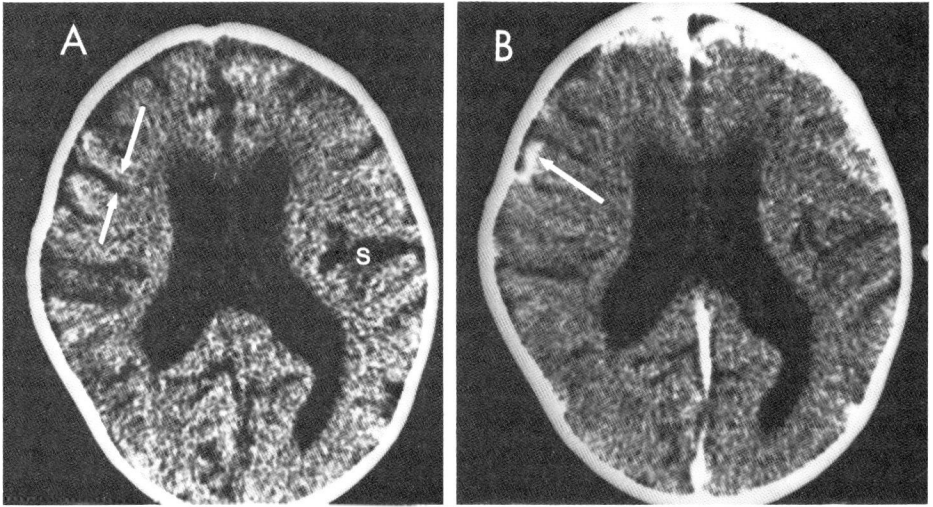

Fig 9–6.—*Hemophilus influenzae* meningitis on axial CT scans. **A,** Non-contrast-enhanced scan shows widened sulci, sylvian *(S)* fissures, and ventricular enlargement. **B,** Contrast-enhanced scan shows gyral enhancement *(arrow)* from meningitis but no ventricular enhancement.

development of an enlarging mass lesion, or clinical deterioration of the patient.[37, 38]

Parasites

Most of the parasitic infections are extremely uncommon in children in North America. However, immigration from Mexico, Central America, and Southeast Asia has introduced patients with cysticercosis. This disease is usually diagnosed in adults many years after their immigration (average, 8 years).[39–44] A 3-year-old Cambodian girl presented with focal seizures and on CT examination had ring enhancement in two vertex lesions, which represented an intense reaction of the brain to the dead larvae (Fig 9–8). This might be a difficult lesion for ultrasound to visualize, because it is in the area immediately under the transducer, but the other signs of cysticercosis may be amenable to ultrasound diagnosis. They include ventricular and subarachnoid cysts and hydrocephalus created by ventricular or periventricular compression.[42–44] In a patient with a history of immigration, a positive serologic test would be sufficient to make the diagnosis of cysticercosis. The differential diagnosis includes neoplasm, other granulomatous or bacterial infection, and, although less likely, infarction. In general, this ringlike lesion is most likely to be due to a pyogenic abscess.

Tuberculosis

Tuberculous meningitis is very rare. However, its peak presentation is before the fourth year of life, so it must be considered in children at risk for tuberculosis,[26–29] particularly if there is evidence of calcification or basilar arachnoiditis.

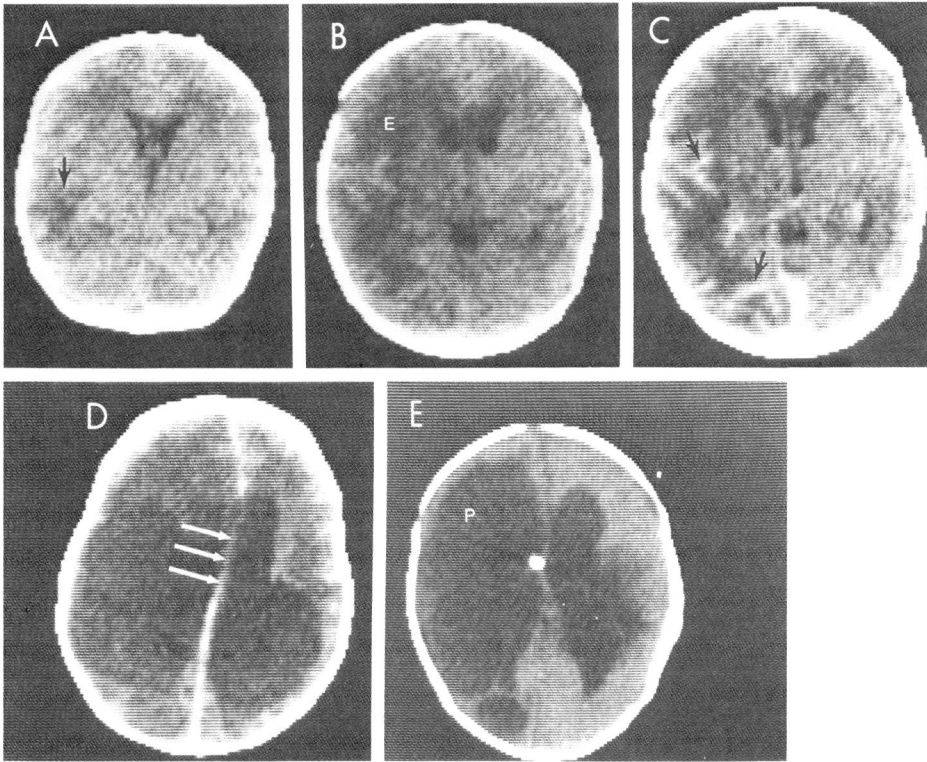

Fig 9–7.—Enterobacterencephalitis. Axial CT scans in premature infant with focal sei-zures. **A,** Initial scan at about 1 month of age demonstrates usual frontal low-density areas and slightly asymmetric low-density areas adjacent to the trigone, the right side slightly abnormal *(arrow).* **B,** Four days later there is extensive edema *(E)* in the right paren-chyma. **C,** Contrast-enhanced scan on same day as **B.** Note gyral enhancement from meningitis and some deep cortical enhancement in areas of cerebritis *(arrows).* **D,** Three weeks after the initial scan, marked hydrocephalus has developed. There is a midline shift from right to left *(arrows)* which is probably due either to extensive cerebritis or to obstruc-tion of the right lateral ventricle. Obstruction is more likely, since there is little cerebral enhancement on this contrast scan. **E,** After ventriculoperitoneal shunt was placed at age 3 months, left lateral ventricle decreased to moderate dilation. The right lateral ventricle is continuous with the area of porencephaly *(P),* which has developed secondary to brain necrosis.

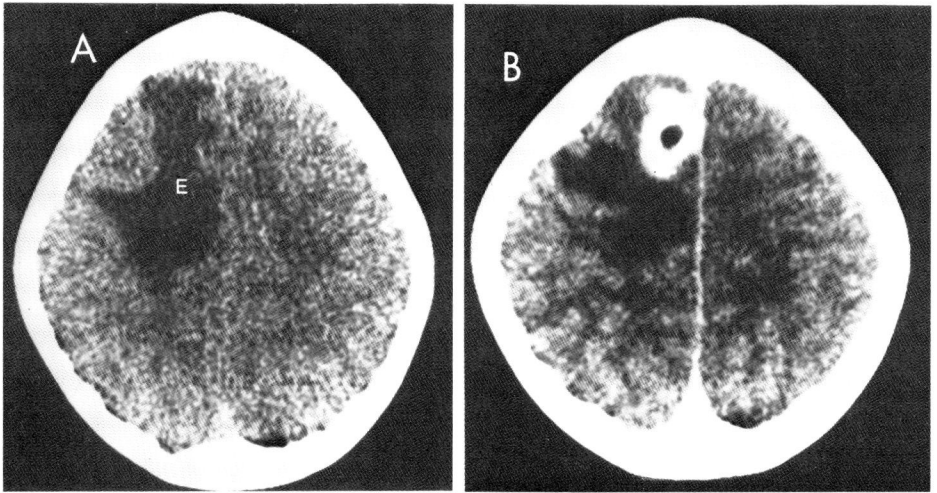

Fig 9–8.—Cysticercosis infection in a 3-year-old Cambodian girl. **A,** Non-contrast-enhanced axial CT scan shows extensive inflammatory edema *(E)* as extremely low-density white matter. **B,** Contrast-enhanced axial CT scan at slightly higher level shows intense ring enhancement of the encysted pork tapeworm *(Taenia solium)*.

SUMMARY

The diagnosis of intrauterine infections can be suggested by ultrasound evaluation and confirmed in the neonatal period with ultrasound and contrast CT, if necessary. Neonatally acquired infections should be studied with CT until ultrasound correlation is shown to be accurate enough to diagnose cerebritis consistently. Subdural fluid collections may require CT for diagnosis but the diagnosis will be made more frequently if high-frequency transducers are used, because they offer better near field detail.

REFERENCES

1. Marquis J.R., Lee J.K.: Extensive central nervous system calcification in a stillborn male due to cytomegalovirus infection. *AJR* 127:665–667, 1976.
2. Harwood-Nash D.C., Reilly B.J., Turnbull I.: Massive calcification of the brain in a newborn infant. *AJR* 108:528–532, 1970.
3. Graham D., Guidi S.M., Sanders R.C.: Sonographic features of in utero periventricular calcification due to cytomegalovirus infection. *Ultrasound Med. Biol.* 1:171–172, 1982.
4. South M.A., Tompkins W.A.F., Morris F., et al.: Congenital malformation of the central nervous system associated with genital (type 2) herpes virus. *J. Pediatr.* 75:13–18, 1969.
5. Dublin A.B., Merten D.F.: Computed tomography in the evaluation of herpes simplex encephalitis. *Radiology* 125:133–134, 1977.
6. Collins A.T., Cromwell L.D.: Computed tomography in the evaluation of congenital cerebral toxoplasmosis. *J. Comput. Assist. Tomogr.* 4:326–329, 1980.
7. Bell W.E., McCormick W.F.: *Neurologic Infections in Children*. Philadelphia, W.B. Saunders Co., 1981.
8. Malloy P.M., Leyman R.M.: The lack of specificity of neonatal paraventricular calcifications. *Radiology* 80:98–102, 1963.

9. Greenberg S.B., Taber L., Septimus E., et al.: CT in brain biopsy–proven herpes simplex encephalitis: Early normal results. *Arch. Neurol.* 38:58–59, 1981.
10. Leo J.S., Weiner R.L., Lin J.P., et al.: CT in herpes simplex encephalitis. *Surg. Neurol.* 9:313–317, 1978.
11. Davis J.M., Davis K.R., Kleinman G.M., et al.: CT of herpes simplex encephalitis with clinicopathological correlation. *Radiology* 129:409–417, 1978.
12. Enzmann D.R., Ranson B., Norman D., et al.: CT of herpes simplex encephalitis. *Radiology* 129:419–425, 1978.
13. Zimmerman R.D., Russell E.J., Leeds N.E., et al.: CT in the early diagnosis of herpes simplex encephalitis. *AJR* 134:61–66, 1980.
14. Lee S.H.: Infectious diseases in Lee S.H., Rao K.C.V.G. (eds.): *Cranial Computed Tomography.* New York, McGraw-Hill Book Co., 1981.
15. Salmon J.H.: Ventriculitis complicating meningitis. *Am. J. Dis. Child.* 124:35–40, 1972.
16. Cockrill H.H., Dreisbach J., Lowe B.: CT in leptomeningeal infections. *AJR* 130:511–515, 1978.
17. Jacobsen P.L., Farmer T.W.: Subdural empyema complicating meningitis in infants: Improved diagnosis. *Neurology* 31:190–193, 1982.
18. Edwards M.K., Brown D.L., Chua G.T.: Complicated infantile meningitis: Evaluation by real-time sonography. *AJNR* 3:431–434, 1982.
19. Schultz, P., Leeds N.E.: Intraventricular septations complicating neonatal meningitis. *J. Neurosurg.* 38:620–626, 1973.
20. Cussen L.J., Ryan G.B.: Hemorrhagic cerebral necrosis in neonatal infants with enterobacterial meningitis. *J. Pediatr.* 71:771–776, 1967.
21. Reeder J.D., Sanders R.C.: Ultrasonic diagnosis of ventriculitis in the neonate. Presented at the Second Special Conference on Perinatal Intracranial Hemorrhage, Washington, D.C., Dec. 1982.
22. Stannard M.W., Jiminez J.F.: Ultrasound diagnosis of multiple cystic encephalomalacia, unpublished manuscript.
23. Packer R.J., Bilaniuk L.T., Zimmerman R.A.: CT parenchymal abnormalities in bacterial meningitis: Clinical significance. *J. Comput. Assist. Tomogr.* 6:1064–1068, 1982.
24. Brown L.W., Zimmerman R.A.: Polycystic brain disease complicating neonatal meningitis: Documentation of evolution by CT. *J. Pediatr.* 94:757–759, 1979.
25. Yamanouchi Y., Soweda K., Tani S., et al.: Gyriform calcification after purulent meningitis. *Neuroradiology* 20:159–162, 1980.
26. Lorber J.: Intracranial calcifications following tuberculous meningitis in children. *Acta Radiol.* 50:204–210, 1958.
27. Price H.I., Danziger A.: CT in cranial tuberculosis. *AJR* 130:769–771, 1978.
28. Casselman E.S., Hasso A.N., Ashwal S., et al.: CT of tuberculous meningitis in infants and children. *J. Comput. Assist. Tomogr.* 4:211–216, 1980.
29. Enzmann D., Norman P., Mani J., et al.: Computed tomography of granulomatous basal arachnoiditis. *Radiology* 120:341–346, 1976.
30. Kotagal S., Tantanasirvongse S., Archer C.: Periventricular calcification following neonatal ventriculitis. *J. Comput. Assist. Tomogr.* 5:651–653, 1981.
31. Zimmerman R.A., Patel S., Bilaniuk L.A.: Demonstration of purulent bacterial intracranial infections by computed tomography. *AJR* 127:155–156, 1976.
32. New P.F.J., Davis K.R., Ballantine H.T.: Computed tomography in cerebral abscess. *Radiology* 121:641–646, 1976.
33. Stevens E.A., et al.: CT brain scanning in intraparenchymal pyogenic abscesses. *AJR* 130:111–114, 1978.
34. Kauffman D.M., Leeds N.E.: CT in the diagnosis of intracranial abscess. *Neurology* 27:1069–1073, 1977.
35. Enzmann D.R., Britt R.H., Yeager A.S.: Experimental brain abscess evolution: Computed tomographic and neuropathologic correlation. *Radiology* 133:113–122, 1979.
36. Whelan M.A., Hilal S.K.: Computed tomography as a guide in the diagnosis and follow-up of brain abscesses. *Radiology* 135:663–671, 1980.

37. Kamin M., Biddle D.: Conservative management of focal intracerebral infection. *Neurology* 31:103–106, 1981.
38. Rosenblum M.L., Hoff J.T., Norman D., et al.: Decreased mortality from brain abscesses since advent of computerized tomography. *J. Neurosurg.* 49:658, 1978.
39. Bentson J.R., Wilson G.H., Helmer E., et al.: Computed tomography in intracranial cysticercosis. *J. Comput. Assist. Tomogr.* 1:464–471, 1977.
40. Zee C.S., Segall H.D., Miller C., et al.: Unusual neuroradiological features of intracranial cysticercosis. *Radiology* 137:397–407, 1980.
41. Carbajal J.R., Palacios E., Azar-Ku B., et al.: Radiology of cysticercosis of the central nervous system including CT. *Radiology* 125:127–131, 1977.
42. Jankowski R., Zimmerman R.D., Leeds N.E.: Cysticercosis presenting as a mass lesion at the foramen of Monro. *J. Comput. Assist. Tomogr.* 3:694–696, 1979.
43. Loyo M., Klergia E., Estanol B.: Fourth ventricular cysticercosis. *Neurosurgery* 7:456–458, 1980.
44. Carbajak J.R., et al.: Radiology of cysticercosis of the central nervous system including CT. *Radiology* 125:127–131, 1977.

CHAPTER 10

Cerebral Ischemia and Infarction

Carol M. Rumack, M.D.
Michael L. Johnson, M.D.

NORMAL PREMATURE BRAIN—SULCI AND VASCULATURE

ON SERIAL ULTRASOUND examinations, the normal premature brain has increasing numbers of sulci that can be used as basic landmarks (Fig 10–1). These sulci may be obscured by several types of processes, including infection (Fig 10–2), infarction, and hemorrhage, so that their identification is quite important, particularly on the sagittal scan. The sulci are usually echogenic due to normal vascular structures.

On coronal real-time ultrasound scans, symmetric pulsations should be present in the sylvian fissures where the middle cerebral artery and its branches lie. Starting anteriorly, the anterior cerebral arteries can be evaluated in the midline in front of the frontal horns where the internal carotids bifurcate into middle and anterior cerebral arteries. As these vessels are traced posteriorly, the pulsations should remain symmetric (Fig 10–3).

On computerized tomography (CT) examination, the vasculature can be visualized with contrast material and should be symmetric. It is valuable always to use the same amount of contrast per kilogram (3 cc/kg), because an increased dose may cause confusion by enhancing more vessels. Sulci are usually not prominent except for the parieto-occipital sulcus, which normally may be cut on the high vertex slices (Fig 10–4).

Normal Premature Brain—Low-Density White Matter

With presently available ultrasound equipment, white and gray matter are not as readily distinguished as they are on CT. On CT, the white matter in the normal premature brain is usually of lower density than it is in adults. It may range from approximately 15 to 25 Hounsfield units (HU). Gray matter in infants is also not as dense (often a maximum of 30–35 HU) as in the fully developed adult brain. In newborns, there is typically such a striking difference between gray and white matter that the same appearance might be diagnosed as cerebral edema in children

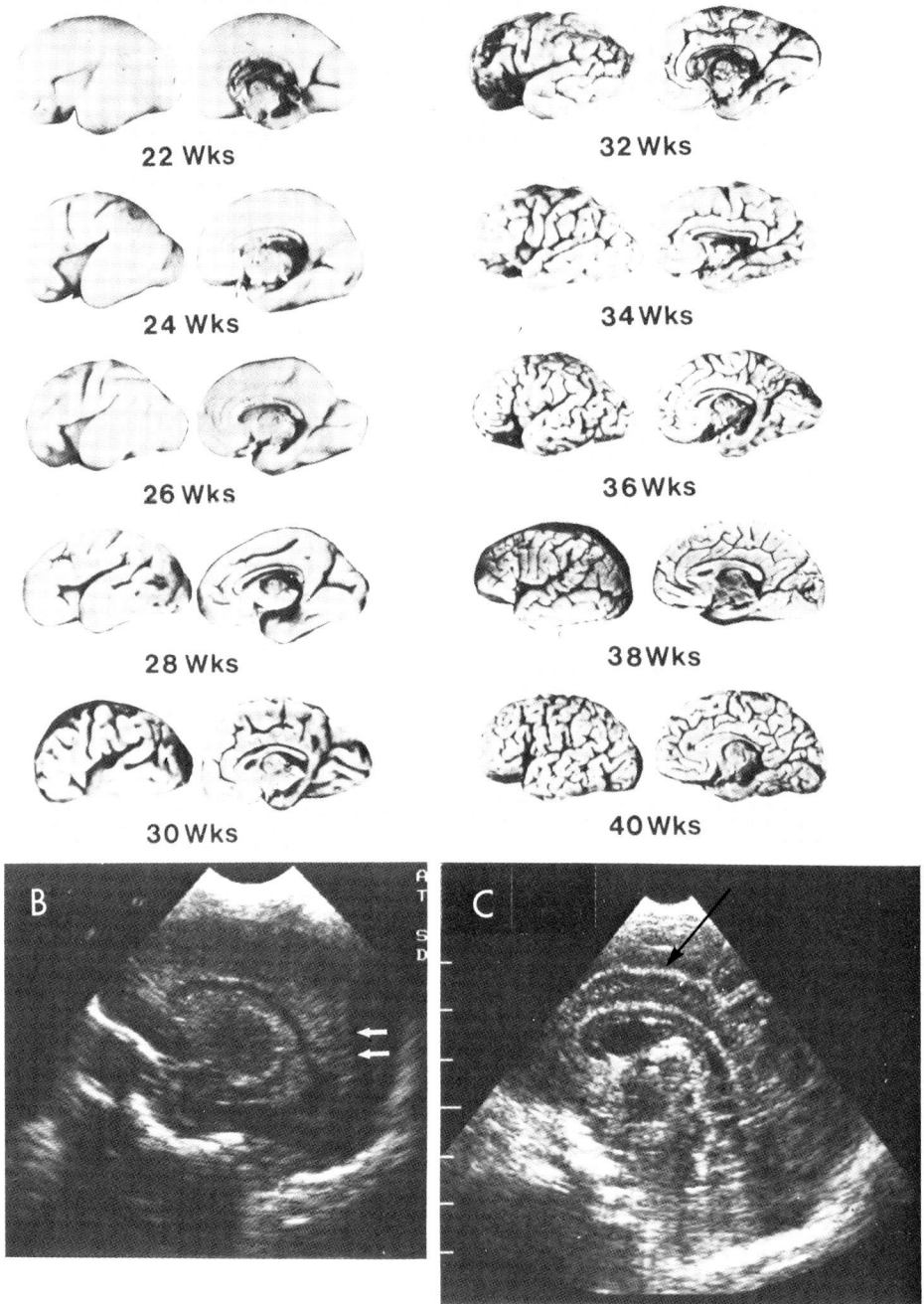

Fig 10–1.—Top, Normal sulcal pattern development, 22–40 weeks' gestation. (From Dolman et al.[26] Reproduced with permission.) **B,** Normal sagittal ultrasound scan at 26 weeks' gestation with only parieto-occipital sulcus developed *(arrows).* **C,** Cingulate sulcus *(arrow).* Many other sulci are also evident in this term infant.

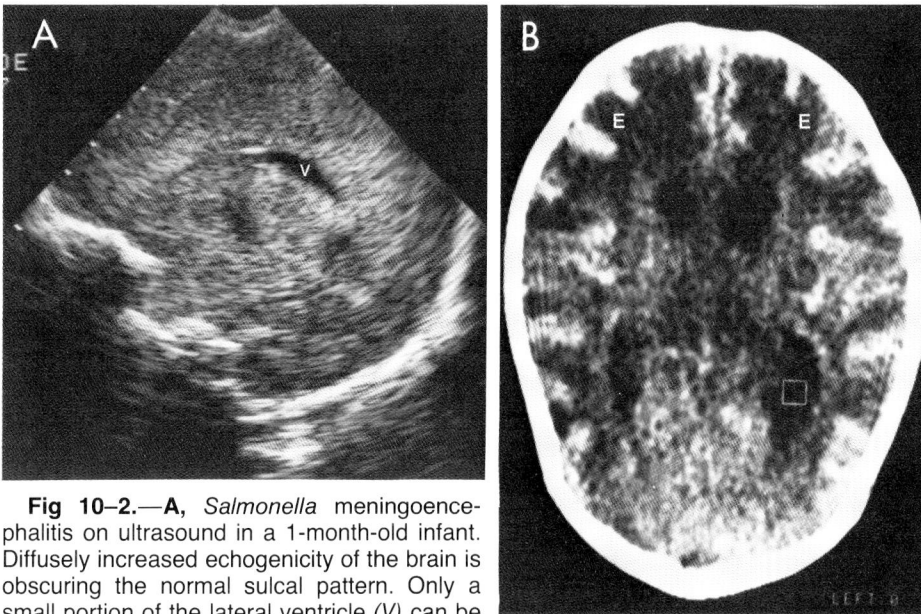

Fig 10–2.—A, *Salmonella* meningoence-phalitis on ultrasound in a 1-month-old infant. Diffusely increased echogenicity of the brain is obscuring the normal sulcal pattern. Only a small portion of the lateral ventricle *(V)* can be seen, due to the diffuse cerebral edema on this sagittal scan. B, *Salmonella* meningoen-cephalitis on axial CT scan on same infant at 1 month. There is diffuse cerebral edema *(E)* throughout the white matter and extending out into grey matter.

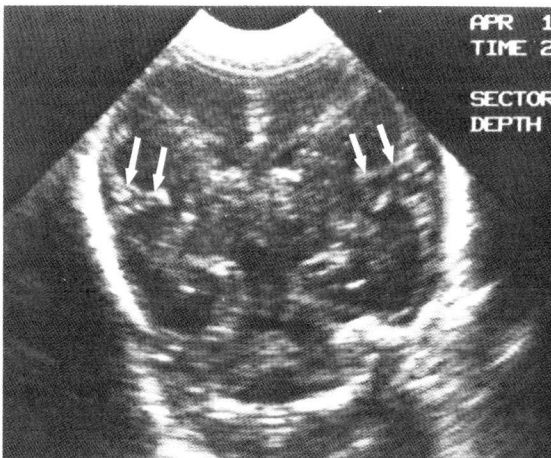

Fig 10–3.—Normal coronal ultrasound scan showing middle cerebral artery branches in the sylvian fissures *(arrows)*. On real-time scanning these should pulsate symmetrically.

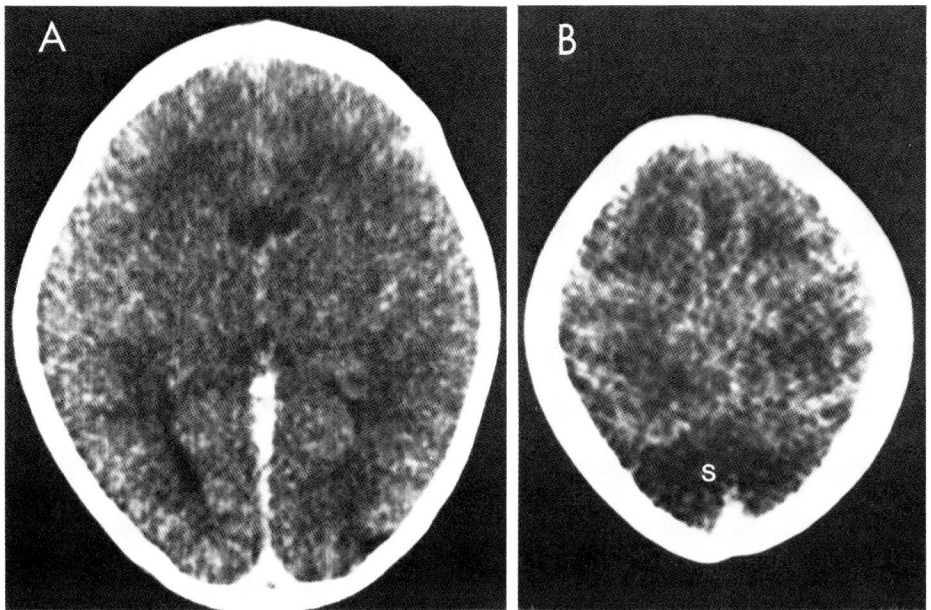

Fig 10–4.—A, Normal axial CT scan showing low-density white matter areas anterior to the frontal horns and adjacent to the trigone. **B,** Normal axial CT scan at the vertex cuts through the parieto-occipital sulcus, which is the CSF region seen posteriorly *(S)*.

and adults (see Fig 10–4). In premature infants, there are symmetric low-density areas in the frontal, parietal, and occipital white matter. These regions slowly become denser during development, leaving only bifrontal low-density areas in most term infants.[1]

Primate cerebral maturation studies confirm that in normal development, the frontal lobe white matter is less myelinated at birth and becomes increasingly denser over the first year of life. By age 1 year, the white matter is almost the same density as gray matter on CT.[2] Initial nuclear magnetic resonance studies demonstrate myelination so exquisitely that this method promises to define the normal stages of brain maturation in the living patient.[3]

CEREBRAL ISCHEMIA AND INFARCTION

Anoxic brain damage in infants may cause *diffuse cerebral edema*, which is difficult to diagnose on ultrasound or CT unless it is unilateral or very extensive (Fig 10–5). The normal low-density white matter may obscure small amounts of edema just presenting as lower-density white matter. As cerebral edema extends to involve all of the white matter and then out into the gray matter (see Fig 10–2,B) the diagnosis of diffuse cerebral edema can be made with good accuracy.[4] The best diagnostic criteria are decreased ventricular size and generally hypodense cerebral tissue. These findings of generalized ischemia have a poor prognosis.[4, 5] Typically the periventricular areas, basal ganglia, and cerebellum are spared.

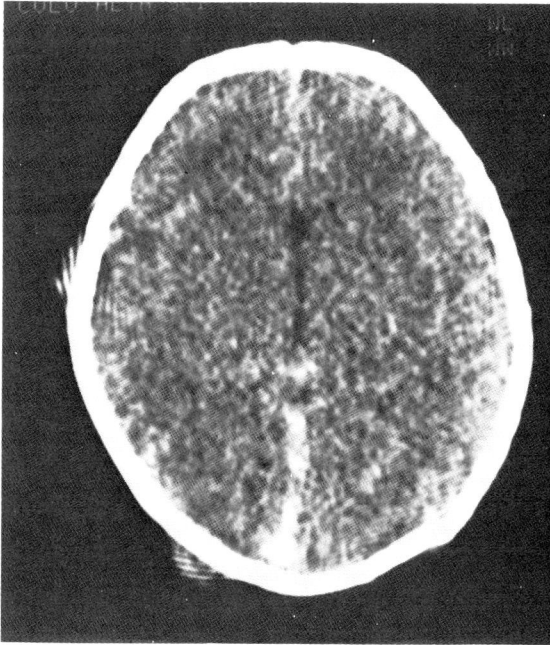

Fig 10–5.—Diffuse cerebral edema causing the entire cerebral cortex to be of low density, only sparing grey at the edges. The straight sinus is evident, although no contrast has been given. It stands out within the edematous brain. Scan was made in term infant with severe birth asphyxia who later died.

Generalized cerebral edema may lead to *multiple areas of infarction*, termed encephalomalacia, or porencephaly, if infarction extends to the ventricle (Fig 10–6). Large edematous areas in infancy may resolve to normal brain density with only cortical atrophy. In the most severe cases, anoxic damage leads to extensive areas of brain necrosis, leaving only strands of tissue extending from and about the ventricles (Fig 10–7).[6–8] Subarachnoid hemorrhage can be mistakenly diagnosed in the presence of diffuse edema on CT because the brain density drops so low that the normal vascular structures become prominent (see Fig 10–5). Measurement of specific areas of interest for density will demonstrate the edematous nature of the brain and prevent the examiner from diagnosing hemorrhage in areas of brain sparing and normal vessels.

On ultrasound, edema may present as a diffusely hazy density between sulci, but it can become so echogenic as to resemble hemorrhage.[9–11] A CT examination may be necessary to define whether an "unusual hemorrhage" on ultrasound is really a hematoma or an area of infarction (Fig 10–8). Cerebral edema has the opposite appearance on CT and ultrasound. It is echogenic on ultrasound and low density on CT. Once necrosis has occurred, the residual infarcted areas will appear cystic on both modalities.

Unilateral cerebral infarction may be more easily diagnosed with both modalities. In the first few days after an insult, changes in the brain may be quite subtle,

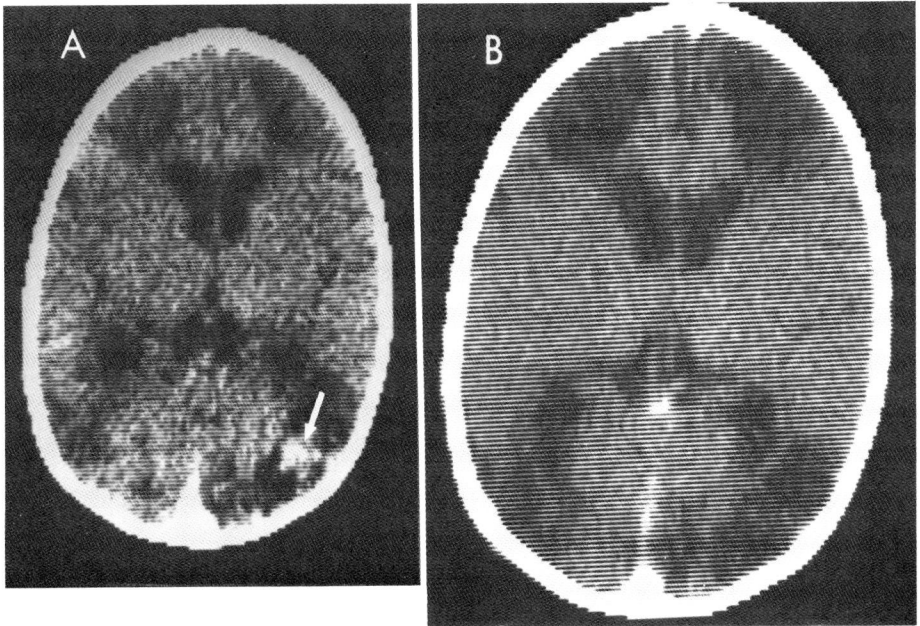

Fig 10–6.—Sudden infant death syndrome (aborted) in 6-month-old infant. Initial CT scans showed only faint low-density areas bifrontally. **A,** Axial contrast-enhanced CT scan at 6½ months shows enhancement of left occipital infarct *(arrow)* and slightly increased enhancement adjacent to the right trigone. Bifrontal infarcts do not enhance. **B,** Axial contrast-enhanced CT scan at 8½ months shows no residual enhancement of areas of infarction.

with only slightly obscured sulci on ultrasound. Asymmetric pulsations may be the first indications of ischemia and potential infarction.[12] After several days, the brain appears more echogenic (see Fig 10–8,A). This may be due to increased water content, causing more interfaces to be apparent, or it may actually represent luxury perfusion into an area of infarction, causing the vessels to become more apparent.[9–11]

On CT, unilateral edema will cause an asymmetry in low-density white matter (see Fig 10–8,B). This may be very subtle and of low density in the first few days after the insult but becomes more discretely demarcated over time if infarction occurs (Fig 10–9). There may be an isodense stage to an infarction at 2–3 weeks after the insult,[13] after which it demarcates from normal tissue.

Enhancement without mass effect is typically seen in infarction from about 3 days to 3 weeks.[14] Transient positive postictal enhancement has been reported within hours after seizures,[15] so CT scanning should be performed at a few days after the onset of symptoms, if possible.

PORENCEPHALY AND HYDRANENCEPHALY

Extensive areas of porencephaly may cause the most extreme form of porencephaly, termed hydranencephaly.[7, 8, 16] This occurs in utero and is thought to re-

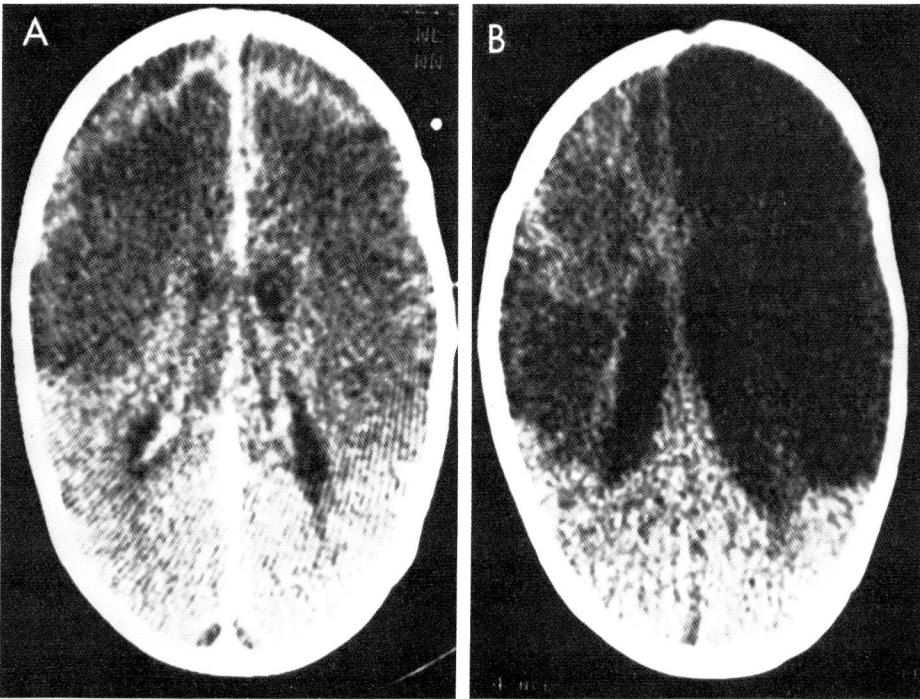

Fig 10–7.—Bilateral middle cerebral artery infarction with sparing of the posterior cerebral circulation and parts of the anterior cerebral in a patient with vasculitis from meningitis. **A,** Axial CT scan at about 2 weeks after onset of symptoms shows peripheral enhancement of the infarcting areas. **B,** Axial CT scan later showed more extensive loss of brain on the left than the right, with mild ventricular enlargement, probably due to atrophy.

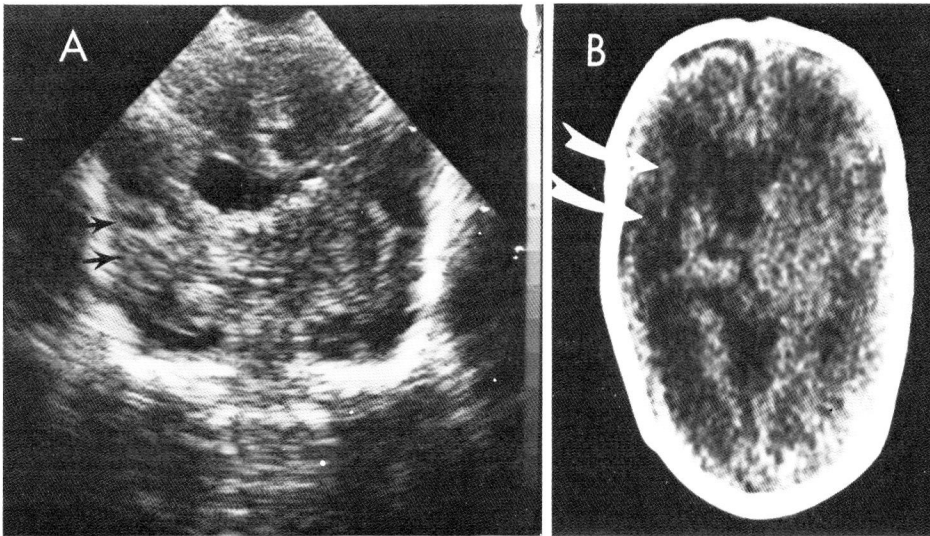

Fig 10–8.—Unilateral cerebral infarction following subependymal hemorrhage in 1-month-old infant. **A,** Coronal ultrasound scan through the bodies of the lateral ventricles shows increased echogenicity of right cerebral cortex *(arrows)* with unilateral ventricular enlargement toward that side. Entire hemisphere was extremely echogenic. **B,** Axial CT scan made on same day shows diffuse low-density brain from cerebral infarction *(arrows)* with unilateral ventricular dilatation.

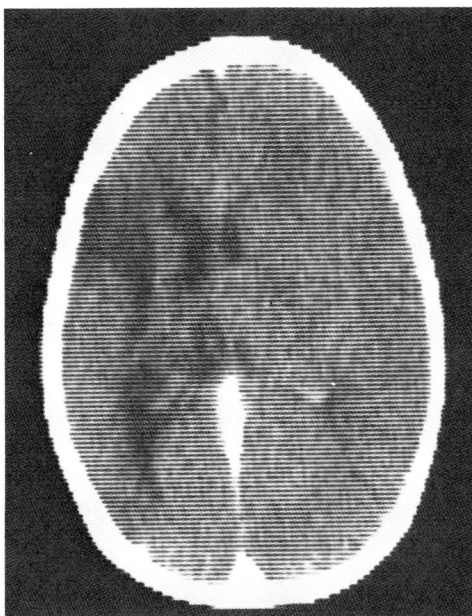

Fig 10–9.—Focal cerebral infarction following temporal artery catheterization. Axial CT scan at 6 months of age. Note the smaller hemisphere on the side of infarction due to diffuse loss of brain, in addition to the clearly demarcated focal infarction.

sult from bilateral internal carotid artery occlusion. The cerebral hemispheres are reduced to a thin layer of gliomatous tissue lined by leptomeninges.[17] The brain stem and cerebellum are relatively normal. The skull is filled with cerebrospinal fluid.

On ultrasound, there will be only a small amount of brain tissue at the level of the brain stem and cerebellum surrounded by fluid (Fig 10–10,A). Hydranencephaly may be diagnosed in utero on the basis of these findings.[18] On CT examination, normal-density brain will be at the base of the brain (Fig 10–10,B) with fluid around it.

Hydranencephaly can be differentiated from massive hydrocephalus or rare bilateral massive subdural hygromas only by angiography. All of these conditions cause major brain damage, and a shunt may be necessary for good nursing care even in a patient with hydranencephaly, who usually dies in the first year of life.

Positron emission tomographic scanning in a few newborns has demonstrated that ischemia or poor perfusion can be much more extensive than present studies suggest. In the region of a small subependymal hemorrhage, there may be unilateral hemispheric decreased perfusion.[19] The final brain damage may not include the entire hemisphere, but it may be more extensive than would be expected from the subependymal hemorrhage. The discrepancy between the amount of hemorrhage and the amount of real ischemic damage probably accounts for the lack of correlation between lower grades of hemorrhage and neurologic outcome. Since premature infants without hemorrhage often do as poorly as infants with small hematomas, there are undoubtedly ischemic changes occurring which are not well documented by present techniques.

Multiple causes of infarction have been described in the neonatal period, includ-

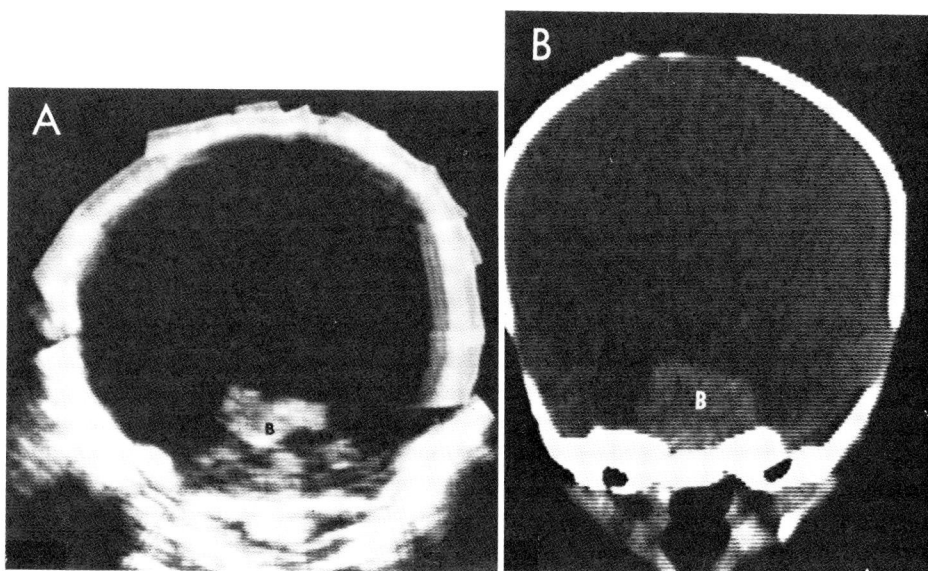

Fig 10–10.—Hydranencephaly. **A,** Coronal ultrasound scan at a few days of age shows only a small amount of brain *(B)* at the base of the skull. The skull is filled with cerebrospinal fluid. **B,** Coronal CT scan with comparable findings. (From Mack et al.[7] Reproduced with permission.)

ing anoxia, shock, dehydration, temporal arterial catheterization,[20–22] and hyperviscosity. The causes most commonly identified are those with a hemorrhagic component,[4] but neither ultrasound nor CT is good at defining neonatal infarction without hemorrhage. Nuclear magnetic resonance imaging and positron emission tomographic scanning will probably contribute to a more thorough understanding of neonatal ischemic damage.[23–25]

REFERENCES

1. Picard L., Claudon M., Roland J., et al.: Cerebral computed tomography in premature infants, with an attempt at staging developmental features. *J. Comput. Assist. Tomogr.* 4:435–444, 1980.
2. Quencer R.M., Parker J.C., Hinkle D.K.: Maturation of normal primate cerebral tissue: Preliminary results of a computed tomographic-anatomic correlation. *J. Comput. Assist. Tomogr.* 4:464–465, 1980.
3. Smith F.W.: NMR imaging in pediatric practice. *Pediatrics* 71:852–854, 1983.
4. Flodmark O., Becker L.E., Harwood-Nash D.C.: Correlation between computed tomography and autopsy in premature and full term infants that have suffered perinatal asphyxia. *Radiology* 137:93–103, 1980.
5. Magilner A.D., Wertheimer I.S.: Preliminary results of a CT study of neonatal brain hypoxia-ischemia. *J. Comput. Assist. Tomogr.* 4:457–463, 1980.
6. Brown L.W., Zimmerman R.A., Bilaniuk L.T.: Polycystic brain disease complicating neonatal meningitis: Documentation of evolution by computed tomography. *J. Pediatr.* 94:757–759, 1979.
7. Mack L.A., Rumack C.M., Johnson M.L., et al.: Ultrasound evaluation of cystic intracranial lesions in the neonate. *Radiology* 137:451–455, 1980.

8. Dublin A.B., French B.N.: Diagnostic image evaluation of hydranencephaly and pictorially similar entities with emphasis on CT. *Radiology* 137:81–91, 1980.
9. Hill A., Martin D.J., Daneman A., et al.: Focal ischemic cerebral injury in the newborn: Diagnosis by ultrasound and correlation with computed tomographic scan. *Pediatrics* 71:790–793, 1983.
10. Rumack C.M., Johnson M.L.: Role of CT and ultrasound in neonatal brain imaging. *J. Comput. Tomogr.* 7:17–29, 1983.
11. Babcock D.S., Ball W. Jr.: Ultrasound diagnosis and short-term prognosis of post-asphyxial encephalopathy in term infants. Presented at the Second Special Conference on Perinatal Intracranial Hemorrhage, Washington, D.C., 1982.
12. Williams J.L.: Intracranial vascular pulsations in pediatric neurosonology. Presented at the Second Special Conference on Perinatal Intracranial Hemorrhage, Washington, D.C., Dec. 1982.
13. Becker H., Desch J., et al.: CT fogging effect with ischemic cerebral infarcts. *Neuroradiology* 18:185–192, 1979.
14. Weisberg L.A.: Computerized tomographic enhancement patterns in cerebral infarction. *Arch. Neurol.* 37:21–24, 1980.
15. Rumack C.M., Guggenheim M.A., Fasules J.W., et al.: Transient positive post-ictal computed tomographic scan. *J. Pediatr.* 97:263–264, 1980.
16. Halsey J.H., Allen M., Chamberlain H.R.: The morphogenesis of hydranencephaly. *J. Neurol. Sci.* 12:187–217, 1971.
17. Harwood-Nash D.C., Fitz C.R.: *Neuroradiology in Infants and Children*. St. Louis, C.V. Mosby, Co., 1976.
18. Lee T.G., Warren B.H.: Antenatal diagnosis of hydranencephaly by ultrasound: Correlation with ventriculography and CT. *JCU* 5:271–273, 1977.
19. Volpe J., Perlman J., et al.: Positron emission tomography (PET) in the assessment of regional cerebral blood flow in the newborn. *Ann. Neurol.* 12:225(A), 1982.
20. Bull M.J., Schreiner R.L., Garg B.P., et al.: Neurologic complications following temporal artery catheterization. *J. Pediatr.* 96:1071–1073, 1978.
21. Prian G.W., Wright G.B., Rumack C.M., et al.: Apparent cerebral embolization after temporal artery catheterization. *J. Pediatr.* 93:115, 1978.
22. Lubchenco L.O., et al.: Warning: Serious sequelae of temporal artery catheterization. *J. Pediatr.* 92:284, 1978.
23. Deuel R.K: Pathophysiology, live. *Pediatrics* 70:312, 1982.
24. Volpe J.J., Perlman J.: Positron emission tomography (PET) in the assessment of regional cerebral blood flow in the newborn. *Ann. Neurol.* 12:225, 1982.
25. Delpy D.T., Gordon R.E., Hope P.L., et al.: Non-invasive investigation of cerebral ischemia by phosphorus. *Pediatrics* 70:310, 1982.
26. Dolman C.L., Dorovini-Zis: Gestational development of the brain. *Archives of Pathol. Lab. Med.* 101:192–193, 1977.

CHAPTER 11

Comparative Role of US and CT; Future Role of NMR Imaging

Carol M. Rumack, M.D.
Michael L. Johnson, M.D.

ROLE OF ULTRASOUND AND COMPUTERIZED TOMOGRAPHY

IMAGING OF THE FETAL BRAIN is best done with ultrasound. However, there are occasional indications for CT, and nuclear magnetic resonance imaging of the fetus may prove valuable in the future.

Neonatal brain imaging has become a very exciting field with the introduction of CT and ultrasound. Prior to CT, there were substantial risks in studying a newborn with angiography or ventriculography. Many major neurologic problems, including intracranial hemorrhage, were studied on the basis of autopsy data without a clear understanding of the incidence or outcome in the living patient. During this same period, improvements in neonatal care resulted in fewer respiratory deaths, and neurologic problems became the limiting factor in neonatal prognosis.[1] Intracranial hemorrhage and hypoxic encephalopathy are now common problems in the new-born nursery. The increased incidence of intracranial hemorrhage and the concern for the immediate survival and long-term outcome of affected infants have increased the need for neonatal brain evaluation. The understanding of these problems will lead to the ability to treat and, hopefully, to prevent these disorders. At this stage in the development of neonatal brain imaging it is valuable to consider the relative merits of these modalities in specific CNS lesions in order to give the best care.

Imaging of the neonatal brain should be done with the best available modality. Presently this includes primarily real-time sector ultrasound in the nursery and CT in the department. The choice of modality should be based on the characteristics of specific lesions and the probability of those lesions occurring in a particular age group. Understanding the relative merits of ultrasound and CT in detecting specific lesions, as we have outlined in the preceding chapters, should be an asset in the decision-making process.

Specific types of lesions are better evaluated with either ultrasound or CT, but

219

others can be evaluated equally well with either modality. When the expected abnormality can be evaluated with either modality, ultrasound is usually the method of choice because it is portable, less expensive, and requires no sedation. CT is the method of choice in the term infant, where posttraumatic peripheral lesions and ischemic lesions are more likely. Neonatal brain imaging is a rapidly developing field, and the introduction of techniques such as Doppler ultrasound within cross-sectional ultrasound and nuclear magnetic resonance (NMR) imaging will change the field tremendously, particularly in the evaluation of ischemic lesions.

BASIC TECHNICAL DIFFERENCES

With real-time ultrasound, sedation is not required because the transducer can be held on the fontanelle so that it moves with the head; thus, motion of the infant does not interfere with the formation of the image.

With CT, sedation is occasionally used in children aged 1–2 years. In the newborn and infant, restraints are usually sufficient. Details of sedation were discussed in chapter 2. Of course, evaluation of the brain in utero is the biggest challenge, since the only views possible may be upside down or at an oblique angle, but it is an interesting test of one's three-dimensional thinking process. It may be possible to hold the fetus in one position for short periods of time to obtain the best scanning projections.

Hypothermia and handling can be significant risks to a critically ill newborn and may require that evaluation be done with portable ultrasound equipment.

As medical costs become an increasing concern, the use of relatively less expensive procedures, including ultrasound, may be preferred, with CT reserved for specific indications. One must always consider the most likely diagnosis. If there was a traumatic term birth, then a CT scan should be done first, as the less expensive ultrasound technique cannot yet exclude small subdural or epidural hematomas.

The radiation effects of a CT scan are presently equivalent to a complete radiographic skull series and thus do not present a significant risk to the infant. Large numbers of CT scans in the newborn period should be avoided both because of cost and because of the minimal potential risk of radiation.

Equipment Limitations: Areas of Interest

With CT scanning, the entire neonatal brain and skull can be evaluated. With ultrasonic scanning, the skull cannot be studied as well due to attenuation of the sound beam, although the inner table of the skull can be seen and bony landmarks can be used to orient the scan.

Since ultrasound examination is best performed through the anterior fontanelle in coronal and sagittal planes, some areas of the brain are not usually visualized. Peripheral lesions, especially those directly below the transducer, may be missed due to an artifact in the ultrasound image, which obscures the first centimeter of tissue. A water path below the transducer or higher-frequency transducers (7.5-MHz) may be necessary to obtain better detail of superficial areas. Lesions adjacent to the skull may be subtle on CT, but with frequent changes of the window level between the bone and brain, even thin subdural and epidural hematomas can be diagnosed.

Real-time sector scans allow more lateral field of view than do linear array transducers which can only visualize the brain directly below the transducer. However, even a 90° sector scan cannot visualize the entire brain. It is possible to scan through the bone with ultrasound, but this results in degradation of the image quality. For this reason, most ultrasound examinations do not routinely include the superficial subdural space or superior lateral edges of the brain. The anterior fontanelle is the acoustic window and the bony fontanelle edges must be avoided for excellent scans.

Compared to the static scan of standard CT or ultrasound, real-time ultrasound allows a continuous sweep of brain so that contiguous structures can be studied and reviewed on videotape. Multiple views at different angles would be required on CT to obtain the same information. The basic difference is that ultrasound can reveal excellent brain detail in coronal and sagittal views, but it is limited by bone interference on axial scans. CT has the opposite problem—it yields excellent axial scans, and bony artifacts are more commonly seen in coronal and sagittal projections. Real-time CT should be able to yield the same images as real-time ultrasound but at an increased cost and radiation dose.

Vascular Detail: Contrast or Real-Time?

Vascular detail is not necessary in most neonatal lesions. On CT scans, vascular information can be obtained with 30% iodinated intravenous contrast medium at a dose of 3 cc/kg. Contrast medium should be used whenever there is an unusual hemorrhage to rule out an arteriovenous malformation or to exclude an abscess, ventriculitis, or neoplasm.

Real-time ultrasound can show normal vascular structures pulsating. Decreased pulsation has been reported with hydrocephalus in the anterior cerebral arteries.[2] Severe depression of vascular pulsations has been reported in association with edematous changes in the newborn brain.[3] Vascular lesions can be evaluated if the vessels are sufficiently large. Most of the lesions large enough to visualize with contrast-enhanced CT should be visible on real-time ultrasound. Vein of Galen aneurysms in particular have been diagnosed with real-time ultrasound.[4] Doppler flow study within vessels will be of great value in determining whether a cystic structure is vascular and in determining the type of flow, whether venous or arterial. The smallest possible Doppler ultrasound sample size should be developed in order to study cerebral blood flow and truly understand the causes of intracranial hemorrhage.

INTRACRANIAL HEMORRHAGE

Intracranial hemorrhage is a significant clinical problem in the newborn period. The incidence of hemorrhage is high, ranging from 40% to 60% in premature infants.[5] There are four major types of neonatal intracranial hemorrhage: subdural hemorrhage, subarachnoid hemorrhage, intracerebellar hemorrhage, and subependymal-intraventricular hemorrhage. With improvements in obstetric technique, the incidence of traumatic lesions such as subdural and subarachnoid hemorrhages has decreased tremendously. As neonatal intensive care has improved, however, subependymal germinal matrix hemorrhage has increased, due to the ability to support increasingly low birth weight infants and thus more premature infants.

SUBEPENDYMAL HEMORRHAGE.—Subependymal hemorrhage occurs in the germinal matrix, which lies above the caudate nucleus in the floor of the lateral ventricle, extending from the frontal horn posteriorly into the temporal horn. The germinal matrix is the most frequent site of hemorrhage in prematures.[6] The incidence of germinal matrix hemorrhage averages approximately 67% in infants born at 28–32 weeks' gestation and drops to less than 5% in infants born at 40 weeks' gestation, as the germinal matrix disappears in the normal maturation of the neonatal brain.[7] On CT, subependymal hemorrhage appears as a high-density (typically 60–80 HU) lesion bulging in the wall of the ventricle, usually adjacent to the frontal horn.[8] On ultrasound, subependymal hemorrhage is diagnosed by increased echogenicity in the region of the caudate nucleus.[9] Small subependymal hematomas may resolve or progress. As the hematoma enlarges, ventricular extension is most common, and with progressive enlargement the hematoma will extend into the brain parenchyma.[10]

ACUTE INTRAVENTRICULAR HEMORRHAGE.—Acute intraventricular hemorrhage will appear as high-density material within the ventricles. CSF-blood levels may be evident on both CT and ultrasound in the dependent portion of the ventricle, usually the occipital horns. Intraventricular hemorrhage often is difficult to diagnose after 2–3 weeks, as the ventricular clot resolves. However, the ventricular surfaces become very echogenic on ultrasound but are usually isodense on CT. The subependymal halo sometimes present on CT after an intraventricular hemorrhage may result from the same response of the ventricular wall.

PARENCHYMAL HEMORRHAGE.—Parenchymal hemorrhage may be diagnosed with either modality. In premature infants it typically lies adjacent to the ventricle, and thus ultrasound will always identify it. However, as hemorrhage presents more peripherally, especially near the edge of the skull, it may be missed on ultrasound. Intracerebellar hemorrhage can be a difficult diagnosis on CT because it may be confused with subarachnoid hemorrhage, which settles into the posterior fossa in the supine neonate. The cerebellum is normally quite echogenic on ultrasound, and care must be taken not to miss a posterior fossa hemorrhage. The pathogenesis of intracerebellar hemorrhage is not clear. It occurs much less commonly than other types of hemorrhage, and it has occurred after both trauma and asphyxia.

ISOLATED SUBARACHNOID HEMORRHAGE.—Isolated subarachnoid hemorrhage may occur secondary to trauma or asphyxia and is the most common intracranial hemorrhage in term infants. Subarachnoid hemorrhage has been a difficult diagnostic problem for both ultrasound and CT. When blood is clearly present in the suprasellar cistern and sylvian fissures on CT, it is easy to diagnose. However, in supine neonates, subarachnoid hemorrhage tends to collect in the posterior fossa and along the falx, where it may be confused with normal vascular structures on CT. Similarly, in ultrasound, subarachnoid hemorrhage may thicken the appearance of the vessels in the sylvian fissure, but a large amount of blood must be present to be certain structures are abnormally thick.

CHOROID PLEXUS HEMORRHAGE.—Choroid plexus hemorrhage is uncommon. It usually occurs in term infants. In premature infants, intraventricular hemorrhage frequently collects on the choroid but begins in the germinal matrix. The choroid is so echogenic that choroid hemorrhage may be difficult to diagnose unless a clot surrounds the choroid. On CT, the choroid is isodense with brain; thus, choroid hemorrhage is immediately evident.

PERIPHERAL HEMORRHAGE.—Peripheral hemorrhage, such as posttraumatic subdural hematoma or epidural hematoma, is best evaluated with CT as long as ultrasound scanners have an initial 1-cm artifact. When water paths are better developed, even thin hematomas immediately under the transducer should be diagnosable. If peripheral hematomas or effusions are large enough (at least 1 cm thick) they can be diagnosed with ultrasound.

Central lesions such as germinal matrix hemorrhage in premature infants can be diagnosed well by both ultrasound and CT, but hemorrhage will be detectable longer with ultrasound. Freshly clotted blood is highly echogenic but becomes less echogenic, although still detectable, for the following 30–45 days. An acute hemorrhagic clot is dense on CT initially but becomes isodense at 6–10 days.

Both CT and ultrasound have proved reliable and accurate in the detection of intracranial hemorrhage.[8, 11–13] The most likely site of hemorrhage should influence the choice of modality, with CT the method of choice in term infants and ultrasound in premature infants (Table 11–1). Ultrasound has the advantage of being able to follow the serial progression or resolution of intracranial hemorrhage.

CT examination should be done on any neonate with neurologic or hematologic signs suggesting hemorrhage but with a normal ultrasound scan.

TABLE 11–1.—HEMORRHAGIC LESIONS: DIAGNOSTIC CAPABILITIES*

LESION	CT	US	MODALITY LIMITATIONS
SEH	+	+	On late CT will be isodense and will be missed
IVH	+	+	Early, easier on CT, especially with normal ventricular size
IPH	+	+	Late, with US may visualize for several weeks
SAH	±	±	CT/US difficult to distinguish SAH from vessels; larger amounts of SAH easier to diagnose with CT
SDH	+	+ (big)	Peripheral, difficult with US unless > 1 cm
EDH	+	+	Peripheral, difficult with US unless > 1cm
Posterior fossa ICH	+	+	Highly echogenic cerebellum may make ICH diagnosis difficult
Choroid ICH	+	+	Highly echogenic choroid

*CT, computerized tomography; US, ultrasound; SEH, subependymal hemorrhage; IVH, interventricular hemorrhage; IPH, intraparenchymal hemorrhage; SAH, subarachnoid hemorrhage; SDH, subdural hematoma; EDH, epidural hematoma.

NONHEMORRHAGIC LESIONS (TABLE 11–2)

Hydrocephalus

Ventricular size and configuration are well shown on both ultrasound and CT, although CT may slightly underestimate the size of the ventricular system due to averaging at interfaces. Ultrasound sharply delineates the interfaces, particularly with fluid-filled spaces such as a ventricle, or other cystic spaces such as the subarachnoid cisterns. Correlation between ultrasound and CT in determining the severity of hydrocephalus has been excellent ($r = 0.99$ in our experience).[14] Ultrasound has become so reliable that screening of patients suspected of hydrocephalus can be performed with this modality. In neonates, hydrocephalus can be very subtle clinically, and significant cortical mantle may be lost before the condition is clinically suspected.[15, 16] Early ventricular evaluation by ultrasound is indicated for neonates at risk, particularly after intracranial hemorrhage, which would include at least all neonates of 1,500 gm or less birth weight or 32 weeks' or less gestational age at birth.

Once the fontanelle closes at 12–18 months, it is still possible to evaluate ventricular size through the squamosal portion of the temporal bone just above the external auditory meatus. In the presence of hydrocephalus, we have often been able to measure ventricular size on children up to age 4 years; however, parenchymal detail is very poor through the skull.

Contrast studies using metrizamide may be necessary to demonstrate continuity of structures. In most cases, however, ventricular shunting procedures can be done

TABLE 11–2.—NONHEMORRHAGIC LESIONS*

LESION	CT	US	MODALITY LIMITATIONS
Hydrocephalus	+	+	Easily diagnosed on all modalities
Subdural effusion	+	+ (big)	Peripheral, difficult to diagnose on ultrasound
Infarction			
Early	+ or −	+ or −	Difficult until texture changes
Subacute	+	+	US, increased density
			CT, decreased density, enhancing areas
Late	Cystic	Cystic	US delineates outlines and septa better
Fat	+	+	US, increased density
			CT, decreased density
Calcium	+	+	Small amounts may be difficult to diagnose on both US and CT; look for punctate appearance and increased density on both studies
Mass lesions			
Tumors	+	+ (some)	Texture changes may be difficult in either; contrast necessary
Infections	+	+ (some)	Texture changes may be difficult; contrast necessary
Cysts	+	+	Cystic nature more distinct with ultrasound; Doppler studies can exclude vascular malformations if cystic but no flow evident
Vascular lesions	+	+	Enhance with contrast CT; pulsate on real-time US; Doppler flow may be best method to evaluate large cystic areas

on the basis of the CT or ultrasound appearance alone.[17] In complicated lesions with unusual cysts, ventriculography may be required to delineate the anatomy.[18]

Many congenital forms of hydrocephalus can be well evaluated with either modality. The Chiari II malformation has been detected by both CT and ultrasound.[19–21] Ventricular enlargement typically involves the third and lateral ventricles, with displacement of the fourth ventricle into the spinal canal in the Chiari malformation. There is a batwing deformity with anterior pointing of the frontal horns, enlarged massa intermedia, and markedly enlarged occipital horns. The Dandy-Walker syndrome with obstruction of the foramina of Magendie and Luschka is characterized by a posterior fossa cyst that is actually a massively dilated fourth ventricle.[22] The entire spectrum of alobar holoprosencephaly can be studied with either modality, and as we gain more experience, CT may not be necessary, even in the most complex malformations.[23]

Cystic Lesions

Small cystic lesions, particularly when central, are much easier to diagnose with ultrasound than CT. The most common small lesion is a subependymal cyst that often forms as the last stage during the resolution of subependymal hematomas. Subependymal cysts are frequently found at autopsy in premature infants. Averaging of subependymal or small porencephalic areas in the ventricular margin frequently occurs on CT. Larger cystic lesions can easily be diagnosed with either modality. However, the thin septae within the ventricle or cyst can be more sharply defined on ultrasound.

Moderate-sized cysts that can be easily recognized are found in the posterior fossa in the Dandy-Walker syndrome, in the subarachnoid space as arachnoid cysts, and adjacent to the ventricles as porencephalic cysts.

Complicated cystic lesions, such as a subarachnoid cyst in the quadrigeminal plate cistern, can be localized by viewing the lesion from multiple angles. Visualization is particularly good with real-time ultrasound. Multiple reconstructions or at least two views may be required for CT localization.

CEREBRAL EDEMA AND INFARCTION

Normal Premature Brain

Cerebral edema is difficult to diagnose by CT at any age, but it is particularly difficult in premature infants, due to the usual low density of the white matter. The normal white matter in premature infants is of such low density that the same appearance in adults would allow a firm diagnosis of cerebral edema.[24] In fact, by age 6 months it would be abnormal to see these striking differences between gray and white matter. In premature infants and even some full-term infants, there are symmetric low-density areas in the frontal, parietal, and occipital regions. These areas slowly become denser during development, leaving only bifrontal low-density areas in most full-term infants.

Cerebral edema is more difficult to diagnose with ultrasound. Normal differentiation between gray and white matter is not as discrete.

Ischemia and Infarction

Anoxic brain damage in infants occasionally results in unilateral brain damage, but localized damage may be difficult to identify in the presence of the normal hypodense brain tissue of the neonate. In the most severe cases, anoxic damage leads to extensive areas of brain necrosis. Generalized cerebral edema and potential infarction—defined as decreased ventricular size and generally hypodense cerebral tissue—is a diagnosis that can be made with good accuracy on CT.[25] The diagnosis of cerebral edema requires very extensive white matter changes to low density extending into gray matter, sparing the periventricular areas and brain stem.

Ischemic lesions are frequently associated with intracranial hemorrhage. Generalized severe anoxic brain damage, when seen in premature infants, is usually a combination of subependymal and ventricular hemorrhage and ischemia. In term infants there may be generalized severe ischemia, sometimes associated with subarachnoid hemorrhage. A false-positive diagnosis of subarachnoid hemorrhage should be avoided when the brain is so edematous that vascular structures appear hyperdense in comparison. At any gestational age, cerebellar hemorrhage is uncommon and cerebellar ischemia even less common.

Unlike edematous lesions in the adult brain, large edematous areas on CT in the neonatal brain may resolve to normal brain density, or progress to infarction. Only cortical atrophy may be visible later. This differs from hemorrhage in that large parenchymal hemorrhages almost always result in porencephaly rather than returning to normal brain density. The challenge is to predict how extensive the final brain damage will be. Recent reports by Flodmark et al. and Magilner and Werth have shown that generalized ischemia in the newborn is associated with a poor prognosis.[25, 26]

On ultrasound examination, the diagnosis of edema is more difficult. In a few cases CT has shown unilateral edema that was of low density on CT and was highly echogenic on ultrasound. As the edema resolved into brain infarction and necrosis, the density of the brain dropped to that of water on a CT scan and became cystic or poorly echogenic on ultrasound. Edema has been described in a few patients as a diffuse haziness with lack of normal intracranial detail and increased echogenicity.[3] Thus, cerebral edema seems to have the opposite appearance on CT and ultrasound: echogenic on ultrasound and hypodense on CT. Clearly demarcated necrosis will appear cystic on both examinations.

NEONATAL CNS INFECTION

CNS infections are very serious in the first year of life, particularly in the neonate.[27] Many of the infections begin in utero, and the pregnant mother may not be symptomatic even though the infant has extensive disease. Unfortunately, most of the organisms causing CNS infection in the neonate are either difficult or impossible to treat. In addition, the neonate appears to have more difficulty in limiting the extension of meningitis, so that it frequently becomes encephalitis. The differential diagnosis of antenatally acquired infection includes rubella, cytomegalovirus, syphilis, toxoplasmosis, and herpes encephalitis.

Cytomegalovirus and toxoplasmosis usually will present in the newborn period with microcephaly and calcifications in the brain. Because the active infection occurs in utero, brain texture changes may indicate active encephalitis during a phase of maternal and fetal infection. Toxoplasmosis may appear in a similar fashion, but in the classic presentation it causes scattered calcifications, whereas cytomegalovirus causes periventricular calcifications.

Acute neonatal CNS infection is a difficult diagnosis in its early stages. Both CT and ultrasound may appear completely normal, and symptoms may be minor, suggesting only sepsis. Encephalitis may be visualized as a change in the brain texture, but it may be quite subtle on both modalities. When an abscess develops, this may be more easily diagnosed. Localized brain abscesses are uncommon in neonates. Infection may extend to include large areas of the brain. Herpes is usually acquired neonatally during vaginal delivery of an infected mother. It may result in very extensive encephalitis and has been reported to cause calcification.

The most common neonatal bacterial meningitis infections are caused by group B *Streptococcus* and *Escherichia coli*. Neither of these organisms causes specific lesions early, but cerebritis may be diagnosed on CT after contrast medium enhancement of specific areas of the brain. Extension of the infection frequently leads to extraventricular obstruction and hydrocephalus. Direct extension into the ventricles may cause contrast medium enhancement of the ventricular walls with ventriculitis or fluid levels due to white cells in the CSF.

INTRACRANIAL CALCIFICATIONS

Discrete, high-density lesions on CT usually represent calcification, but small flecks of calcium may not appear to be any denser than freshly clotted blood. These areas of calcification appear as discrete, highly echogenic lesions on an ultrasound scan. To be certain of calcification on ultrasound, an acoustic shadow should be present deep to the lesion, because calcium totally reflects sound and causes a loss of deeper information. Small flecks of calcium do not always cause acoustic shadowing. With either CT or ultrasound, serial examinations may be necessary to differentiate intracranial calcification from hemorrhage, which becomes less dense with time.

NEONATAL TUMORS

Neonatal tumors are extremely rare, and a CT scan should be obtained if this is a likely diagnosis. Solid lesions are more difficult to detect with either modality and will be significantly easier to diagnose if they enhance with contrast media on CT. Disturbance of the architecture and texture of the brain may be apparent on ultrasound but it is usually more subtle; no ultrasound contrast medium is yet available for use in the brain. Since all tumors studied with neonatal or intraoperative ultrasound have been more echogenic than brain, further investigation may suggest ultrasound as a good method of tumor evaluation.

SUMMARY AND RECOMMENDATIONS

The roles of CT and ultrasound in the examination of the neonatal brain depend on the characteristics of specific lesions and the most likely lesion in different age groups (Table 11–3).

In premature infants, central hemorrhagic lesions are most likely—such as subependymal germinal matrix hemorrhage, which often extends into the ventricles and, less commonly, into brain parenchyma. Therefore, a screening ultrasound scan at 5–7 days is recommended for all premature infants born at less than 32 weeks' gestation or weighing less than 1,500 gm at birth. If there are positive neurologic or hemorrhagic signs and the ultrasound scan is negative, a CT examination should be performed.

In term infants, trauma and asphyxia are more common, causing peripheral hematomas and subtle white matter ischemia. For this reason, CT is recommended when clinical indications occur. Ultrasound can effectively exclude hydrocephalus, and it should be considered to study major CNS malformations, intracranial calcifications, and vascular lesions.

CT and ultrasound have had a tremendous impact on the understanding of neonatal brain pathology. Prevention and treatment of these major neurologic problems should become possible as we study etiologic factors in relation to specific incidents of brain damage.

TABLE 11–3.—ROLE OF COMPUTERIZED TOMOGRAPHY AND ULTRASOUND IN NEONATES: RECOMMENDATIONS

PREMATURE INFANTS
  Ultrasound at 5–7 days to screen for intracranial hemorrhage all infants of ≤32 weeks' gestation or ≤1,500 gm birth weight
  Ultrasound at any time in a premature infant with neurologic or hematologic signs of intracranial hemorrhage or hydrocephalus
  CT if ultrasound scan appears normal, or if peripheral lesion is most likely and neurologic signs are present.
TERM INFANTS
  CT as indicated for:
    Birth asphyxia
    Birth trauma (peripheral lesions common)
    Neonatal seizures
    Neonatal infection or tumor (contrast required)
  Ultrasound as indicated for:
    Enlarging head size
    Clinical signs of major CNS malformations
    Infection (in utero)—for intracranial calcification
    Possible arteriovenous malformation: real-time diagnosis

FUTURE ROLE OF NMR: THE NEW FRONTIER

The future role of NMR imaging holds great potential in the fetal and neonatal brain. NMR has an advantage over both CT and ultrasound in the evaluation of the CNS because of the increased contrast sensitivity. Both ultrasound and NMR have the ability to make sagittal and coronal sections easily. However, the detail in the brain stem and peripheral areas of interest is an additional advantage to NMR.

The lack of bone artifact with NMR is an advantage over both CT and ultrasound. This is a unique property and allows good imaging of structures adjacent to bone.

NMR presently has the disadvantage of being slower than both CT and ultrasound. Use of NMR in newborns may require a return to previous sedation methods until the scan time can be shortened.

Semiquantitative data can be obtained in regard to the movement of blood and the deposition of paramagnetic elements. No other modality can evaluate the paramagnetic elements. The movement of blood can be studied with Doppler ultrasound, but the present sample size is not small enough to study accurately small intracranial vessels. Real-time ultrasound with evaluation of vascular pulsations is the only presently available direct imaging technique other than angiography.

NMR contrast agents are being developed that may alleviate some of the present limitations of NMR. Some lesions may be evaluated better with NMR than with any other modality. Present studies suggest a role for NMR in a number of specific disease processes, as described below.

Cerebral Infarction

The loss of gray-white matter contrast may be seen earlier on inversion recovery scans than on CT, so the diagnosis of infarction can be made earlier with more confidence. Anoxic brain damage in infants unaccompanied by hemorrhage may be recognizable much earlier and should not be confused with the normal gray-white matter pattern, as it is on CT. Present systems of grading intracranial hemorrhage do not include ischemic changes which occur in association with ICH because ischemic damage is difficult to recognize on both CT and ultrasound. Ischemic damage that is much more extensive than hemorrhage can be recognized with NMR, and a more exact prognosis can be made. NMR may indicate ischemic damage early enough to allow intervention to help prevent further brain damage.

Hydrocephalus

With a better delineation of periventricular edema it may be possible to determine whether there is a significant increase in intracranial pressure, rather than brain destruction, resulting in large ventricles. Periventricular edema is one of the first signs of hydrocephalus under pressure causing transudation of fluid into the periventricular white matter. This occurs when the normal absorption of CSF can no longer keep up with the increased pressure.

The cerebral aqueduct and fourth ventricle are better delineated than with any other modality but this probably will not affect the diagnosis of the level of obstruction in hydrocephalus. However, and particularly with sagittal scans of the brain stem, a small tumor may be delineated as the cause of hydrocephalus.

Extracerebral Fluid Collections

Subarachnoid hemorrhage may be more easily recognized due to the lack of bone artifact since it tends to layer against the bone (Fig 11–1). Subdural hemorrhage appears to be less likely to have an isointense phase with either gray or white matter. Changes within the hematoma not apparent on CT may be recognizable earlier with NMR as liquefaction occurs and presumably will correlate with the stages of hemorrhage resolution as studied with ultrasound.

Calcification

It should be possible to differentiate between calcium deposits and hemorrhage because acute hemorrhage appears white, owing to its short T_1 and long T_2, but calcium is black, owing to its low proton density. Very small amounts of calcium may be difficult to evaluate with any modality.

Arteriovenous Malformations

High blood flow can be seen by decreasing the sequence cycle in saturation recovery images, which will highlight the blood flowing into an area.

Venous Sinus Occlusion

The lack of flowing blood in a sinus can be demonstrated with saturation recovery sequences. The absence of bone artifact should make evaluation of the venous sinuses much easier.

Brain Stem Injuries

Detailed evaluation of brain stem hemorrhage, infarction, or shearing injuries have been very difficult with present techniques. NMR may be very valuable in infants, particularly in the search for cervical cord and brain stem injuries after breech deliveries.

Encephalitis

Edematous changes, which may be the first very subtle changes seen in CT and ultrasound, should be much easier to diagnose with NMR. However, it may require contrast agents for NMR to differentiate between breakdown of the blood-brain barrier and separate edema from areas of actual infection.

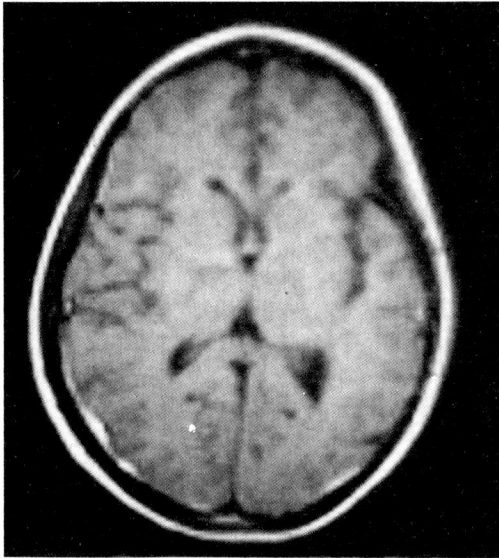

Fig 11–1.—NMR, axial scan in neonate with subarachnoid hemorrhage. CT missed the diagnosis. (From Gooding et al.[29] Reproduced with permission.)

Brain Maturation

The normal development of the brain may be studied potentially in utero and neonatally. Defective maturation from brain damage or genetic diseases may be identifiable with NMR. A more exact understanding of normal brain maturation should allow us to recognize abnormalities not previously described (Fig 11–2).

Atrophic Changes

Due to the lack of bone artifact, the cortical surfaces and cerebellum will be easier to evaluate with NMR than with CT and ultrasound.

Leukodystrophies

White matter diseases may be diagnosed more easily with NMR than with any previous method. Such diseases are uncommon in the first 2 years of life but become more important later.

Brain Tumors

Although brain tumors are quite rare in the first 2 years of life or in utero, displacement of normal structures may be recognizable earlier with NMR due to the detailed gray and white matter differences (Fig 11–3). This may not be quite as useful in infants, however, because of the lack of myelinization.

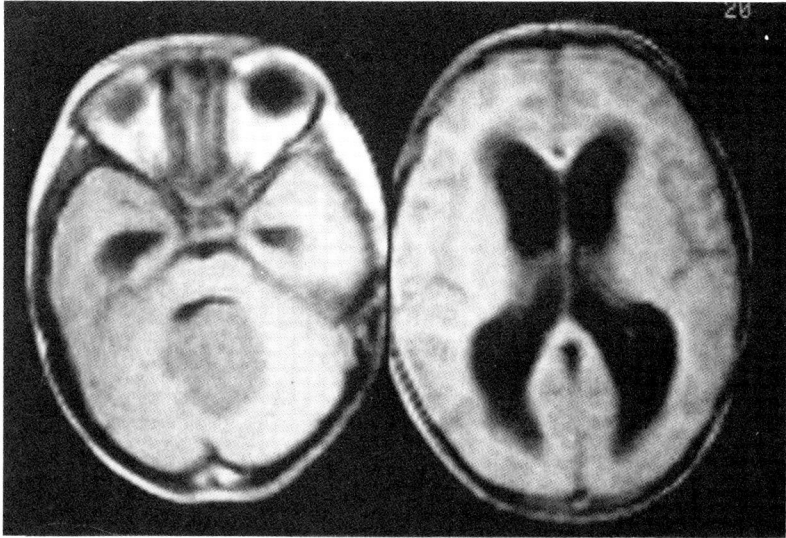

Fig 11–2.—NMR, axial scans of child with medulloblastoma compressing the fourth ventricle, causing hydrocephalus. Note that tumor behind fourth ventricle is easy to diagnose even without contrast. (From Gooding et al.[29] Reproduced with permission.)

Evaluation of the Course of Disease

Changes in hemorrhage over time and changes in tumors with therapy have been recognized more readily with NMR than with CT. Improved recognition may contribute substantially to understanding the effects of therapy. Correlation with ultra-

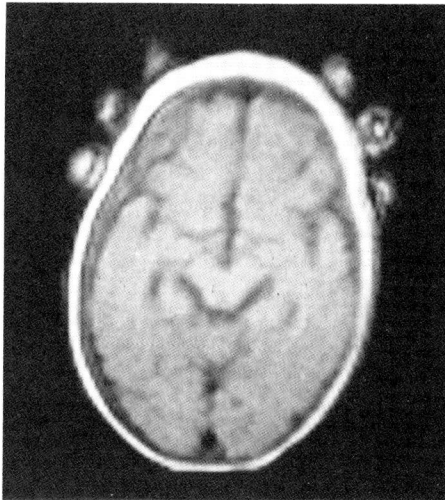

Fig 11–3.—NMR, axial scan, premature neonate. Normal unmyelinated brain. (From Gooding et al.[29] Reproduced with permission.)

sound scans will also be valuable to determine the comparable stages of change in each modality.

NMR imaging has been available clinically for only a relatively short time. It is already comparable to many of the latest CT images in the highest resolution scanners. The present problems of longer scan time, need for contrast media, and interference by magnetic materials should be overcome in the future development of NMR and should stimulate the development of a new frontier in radiology.

REFERENCES

Ultrasound and Computerized Tomography

1. Volpe J.J.: Neurology of the newborn, in Shaffer A. (ed.): *Major Problems in Clinical Pediatrics.* Philadelphia, W.B. Saunders Co., 1981.
2. Hill A., Volpe J.J.: Decrease in pulsatile flow in the anterior cerebral arteries in infantile hydrocephalus. *Pediatrics* 69:4–7, 1982.
3. Williams J.: Intracranial vascular pulsations in pediatric neurosonology. Presented at the 25th annual meeting of the Society for Pediatric Radiology, New Orleans, 1982.
4. Cubberly D.A., Jaffe R.B., Nixon G.W.: Sonographic demonstration of galenic arteriovenous malformations in the neonate. *AJNR* 3:435–439, 1982.
5. Papile L., Burstein R., Kaffler H.: Incidence and evolution of subependymal and intraventricular hemorrhage: A study of infants with birthweights less than 1500 grams. *J. Pediatr.* 92:529–534, April 1978.
6. Pape K.E., Wigglesworth J.: *Hemorrhage, Ischemia and the Perinatal Brain.* London, Lavenham Press, 1979.
7. Guggenheim M.A., Rumack C.M., Langendorfer S., et al.: Clinical factors associated with neonatal germinal matrix hemorrhage. *Biol. Neonate,* to be published.
8. Rumack C.M., McDonald M.M., O'Meara O.P., et al.: CT detection and course of intracerebral hemorrhage in premature infants. *AJR* 131:493–497, 1978.
9. Rumack C.M., Johnson M.L.: Ultrasonic evaluation of intracranial hemorrhage. *Semin. Ultrasound* 3:209–215, 1982.
10. Rumack C.M., Johnson M.L., Johnson J.A., et al.: Patterns of intracranial hemorrhage resolution: Correlation with clinical prognosis, unpublished manuscript.
11. Johnson M.L., Rumack C.M., Mannes E.J., et al.: Detection of neonatal intracranial hemorrhage utilizing real time and static ultrasound. *JCU* 9:427–433, 1981.
12. Babcock D.S., Han B.K.: Accuracy of high resolution real time ultrasonography of the head in infancy. *Radiology* 139:665–676, 1981.
13. Mack L.A., Wright K., Hirsch J.H., et al.: Intracranial hemorrhage in premature infants: Accuracy of sonographic evaluation. *AJR* 137:245–250, 1981.
14. Johnson M.L., Mack L.A., Rumack C.M., et al.: B-mode echoencephalography in the normal and high risk infant. *AJR* 133:375–381, 1979.
15. Korobkin R.: The relationship between head circumference and the development of communicating hydrocephalus in infants following intraventricular hemorrhage. *Pediatrics* 56:74–77, 1975.
16. Volpe J.J., Pasternak J.F., Allan W.C.: Ventricular dilation preceding rapid head growth following neonatal intracranial hemorrhage. *Am. J. Dis. Child.* 131:1212–1215, 1977.
17. Kirks D.R., Harwood-Nash D.C.: Computed tomography in pediatric radiology. *Pediatr. Ann.* 9:53–69, 1980.
18. Naidich T.P., Schott L.H., Baron R. L.: Computed tomography in evaluation of hydrocephalus. *Radiol. Clin. North Am.* 20:143–167, 1982.
19. Naidich T.P., Pudlowski R.M., Naidich J.B., et al.: Computed tomographic signs of the Chiari II malformation: I. Skull and dural partitions. *Radiology* 134:391–398, 1980.

20. Naidich T.P., Pudlowski R.M., Naidich J.B., et al.: Computed tomographic signs of the Chiari II malformation: II. Ventricles and cisterns. *Radiology* 134:657–663, 1980.
21. Babcock D.S., Han B.K.: Cranial ultrasound findings in patient with myelomeningocoele. *AJNR* 1:493–499, 1980.
22. Mack L.A., Rumack C.M., Johnson M.L.: Ultrasound evaluation of cystic intracranial lesions in the neonate. *Radiology* 137:451–455, 1980.
23. Rumack C.M., Johnson M.L., Mack L.A., et al.: The spectrum of alobar holoprosencephaly, unpublished manuscript.
24. Picard L., Claudon M., Roland J., et al.: Cerebral computed tomography in premature infants, with an attempt at staging developmental features. *J. Comput. Assist. Tomogr.* 4:435–444, 1980.
25. Flodmark O., Becker L.E., Harwood-Nash D.C., et al.: Correlation between computed tomography and autopsy in premature and full term infants that have suffered perinatal asphyxia. *Radiology* 137:93–103, 1980.
26. Magilner A.D., Werth I.S.: Preliminary results of a computed tomographic study of neonatal brain hypoxia-ischemia. *J. Comput. Assist. Tomogr.* 4:457–463, 1980.
27. Bell W.E., McCormick W.F.: Neurologic infections in children, in Shaffer A. (ed.): *Major Problems in Clinical Pediatrics.* Philadelphia, W.B. Saunders Co., 1981.
28. Gooding C.A., Brasch R.C., Brant-Zawadski M.M., et al., in Margulis A.R., Crooks L.E., Kaufman L., et al. (eds.): *Clinical Magnetic Resonance in Medicine.* University of California Printing Services, 1983.

Nuclear Magnetic Resonance

Brant-Zawadzki M., Davis P.L., Crooks L.E., et al.: NMR demonstration of cerebral abnormalities: Comparison with CT. *AJNR* 4:117–124, 1983.

Brasch R.C.: NMR imaging in children: Initial experiences. Submitted for publication to J. Pediatr.

Brasch R.C., Gooding C.A., Wesbey G.E., et al. NMR imaging in children: Initial experience with extracranial lesions. Presented at the 26th Annual Meeting of the Society for Pediatric Radiology, Atlanta, May 1983.

Budinger T.F.: Nuclear magnetic resonance (NMR) in vivo studies. Known thresholds for health effects. *J. Comput. Assist. Tomogr.* 5:800–811, 1981.

Bydder G.M.: Nuclear magnetic resonance of the brain. *Applied Radiology* Jan/Feb: 27–33, 1983.

Bydder G.M., Steiner R.E., Young I.R., et al.: Clinical NMR imaging of the brain: 140 cases. *AJR* 129:215–236, 1982.

Crooks L.E., Mills C.M., Davis P.L., et al.: NMR: Visualization of cerebral and vascular abnormalities by NMR imaging. The effects of imaging parameters on contrast. *Radiology* 144:843–852, 1982.

Crooks L.E., Ortendahl D.A., Kaufman L., et al.: Clinical efficiency of nuclear magnetic resonance imaging. *Radiology* 146:123–128, 1983.

Currie C.M., Partain C.L., Price R.R., James A.E.: The clinical potential of NMR-CT imaging. *Diagnostic Imaging* Nov.:46–50, 1981.

Delpy D.T., Gordon R.E., Hope P.L., et al.: Noninvasive investigation of cerebral ischemia by phosphorus nuclear magnetic resonance. *Pediatrics* 70(2):310–313, 1982.

Deuel R.K.: Pathyphysiology, Live. *Pediatrics* 70(4):650-652, 1982

Gooding C.A., Brasch R.C., In: Margulis A.R., Higgins C.B., Crooks L.E., et al. (eds.): *Clinical Magnetic Resonance Imaging.* San Francisco, University of California Printing Services, 1983.

Gooding C.A., Brasch R.C., Brant-Zawadzki M.M., et al.: Nuclear magnetic resonance of the brain in children. Presented at the 26th Annual Meeting of the Society for Pediatric Radiology, Atlanta, May 1983.

Hawkes R.C., Holland G.N., Moore W.S., Worthington B.S.: Nuclear magnetic resonance

(NMR) tomography of the brain: A preliminary clinical assessment with demonstration of pathology. *J. Comput. Assist. Tomogr.* 5:577–586, 1980.

Hinshaw W.S., Bottomley P.A., Holland G.N.: Radiographic thin-section imaging of the human wrist by nuclear magnetic resonance. *Nature* 270:722–723, 1977.

Holland G.N., Hawkes R.C., Moore W.S.: Nuclear magnetic resonance (NMR) tomography of the brain: coronal and sagittal sections. *J. Comput. Assist. Tomogr.* 4:429–433, 1980.

Holland G.N., Moore W.S., Hawkes R.C.: Nuclear magnetic resonance tomography of the brain. *J. Comput Assist. Tomogr.* 4:1–3, 1980.

Kaufman L., Crooks L.E., Margulis A.R. (eds): *Nuclear Magnetic Resonance Imaging in Medicine.* Igaku-Shoin, New York, 1981.

Morgan C.J., Hendee W.R.: Nuclear magnetic resonance imaging: An overview. Submitted for publication.

Oldendorf W.H.: NMR imaging: Its potential clinical impact. *Hospital Practice* Sept.: 114–128, 1982.

Sauders R.D.: The biological hazards of NMR, in Witcofski R.L., Karstaedt N., Partain C.L. (eds): *NMR Imaging.* Winston-Salem: Bowman Gray School of Medicine, 1982, pp. 65–71.

Shulman R.G.: NMR spectroscopy of living cells. *Scientific American,* January: 86–93, 1983.

Singer J.R., Crooks L.E.: Some magnetic studies of normal and leukemic blood, *J. Clin. Engin.* 3:237, 1978.

Smith F.W.: NMR imaging in pediatric practice (Commentary) *Pediatrics* 71(5): 852–853, 1983.

Taveras J.M.: Nuclear magnetic resonance imaging. *AJR* 139:406, 1982.

Young I.R., Burl M., Clarke G.J., et al.: Magnetic resonance properties of hydrogen: Imaging the posterior fossa. *AJR* 137:895–901, 1981.

Young I.R., Hall A.S., Pallis C.A., et al. Nuclear magnetic resonance of the brain in multiple sclerosis. *Lancet* 2:1063–1066, 1981.

Index